Genetic Justice

GENETIC JUSTICE

DNA Data Banks, Criminal Investigations, and Civil Liberties

Sheldon Krimsky
and Tania Simoncelli

Columbia University Press *New York*

Columbia University Press
Publishers Since 1893
New York Chichester, West Sussex

Library of Congress Cataloging-in-Publication Data

Krimsky, Sheldon.
Genetic justice : DNA data banks, criminal investigations, and civil liberties / Sheldon Krimsky and Tania Simoncelli.
p. cm.
Includes bibliographical references and index.
ISBN 978-0-231-14520-6 (cloth : alk. paper)—ISBN 978-0-231-51780-5 (electronic)
1. Criminal investigation—Cross-cultural studies. 2. DNA data banks—Cross-cultural studies. 3. Evidence, Criminal—Cross-cultural studies. I. Simoncelli, Tania. II. Title.

HV8073.K668 2011
363.25'62—dc22 2010007236

Columbia University Press books are printed on permanent and durable acid-free paper.

This book was printed on paper with recycled content.

Printed in the United States of America

c 10 9 8 7 6 5 4 3 2 1

CONTENTS

PART III: Critical Perspectives: Balancing Personal Liberty, Social Equity, and Security

FOREWORD

On a Saturday morning in March 2009, a young woman named Lily Haskell attended a peace rally in San Francisco to express her opposition to the Iraq war. During the course of the demonstration Haskell was arrested on suspicion of trying to help another protester, who was being held by police. Within hours Haskell was released. Although she was not charged with any crime, she was required to submit a DNA sample simply on the basis of her arrest.

Haskell is one of thousands of innocent individuals who have had their DNA forcibly collected, analyzed, and stored by police. She is the lead plaintiff in a class-action lawsuit filed by the ACLU of Northern California that seeks to put an end to a policy that requires anyone arrested for a felony to turn over their DNA. In recounting her experience she said, "Now my genetic information is stored indefinitely in a government database, simply because I was exercising my right to speak out."

Rapid developments in technology—the Internet, cell phones, digital photography, and computer networks, among others—unquestionably have brought significant benefits to our daily lives. Yet these same technologies, applied in the form of warrantless wiretaps, e-mail monitoring, video surveillance, and data mining, threaten to trump our most basic freedoms of privacy, autonomy, and self-expression. Too often—and perhaps nowhere more in history than in this post-9/11 era—we have seen the government wield the power of technology to gain information about its citizens, placing the delicate balance of safety and freedom at risk.

A similar pattern is true for forensic DNA technology. Twenty years ago the FBI initiated a pilot project with a handful of states and local crime laboratories to create a shared database system of the DNA profiles of violent sex offenders. By the new millennium every state law-enforcement agency in the United States had DNA collection systems up and running and connected through the FBI's system. Today more than 8 million people have had their DNA collected and permanently retained in this system, making the U.S. forensic data bank the largest in the world.

No one would deny that forensic DNA technology has brought significant benefits to law enforcement. Forensic DNA analysis has irrefutably contributed to the resolution of a number of crimes as well as to the exoneration of wrongfully convicted persons. But today's national DNA data bank is a far cry from the one that was envisioned 20 years ago. Aggressive promotion of the DNA technology under the rubric of public safety, along with a gradual erosion of basic privacy and due process protections, has led us to a system where increasing numbers of innocent people are being identified, tracked, or monitored by way of their DNA. Forensic DNA technology has moved well beyond its initial role as an important tool for use in criminal investigation and has emerged instead as an instrument for surveillance that reaches broadly across the population to individuals who have never been charged or convicted of a crime.

How have we gotten to this point? What does it mean for civil liberties and the pursuit of justice? And where do we go from here? *Genetic Justice: DNA Data Banks, Criminal Investigations, and Civil Liberties* seeks to answer these questions. Part I describes the history and expansion of DNA data banking in the United States and provides detailed analyses of some of the most controversial law-enforcement practices involving DNA that are used today, such as DNA dragnets, surreptitious DNA sampling, and arrestee testing. Part II focuses on comparative aspects with special emphasis on the United Kingdom, which has proceeded perhaps even more aggressively than the United States in building and maintaining its DNA data-banking system. Part III provides a thoughtful critique of society's efforts to balance personal liberty, privacy, social equity, and security in the development and use of forensic DNA technology. After analyzing in detail the implications of forensic DNA data banking for privacy and racial justice, the underacknowledged issue of fallibility in DNA identification, and the uncertainties associated with DNA data-banking efficacy, the authors prescribe a series of principled recommendations for ensuring that decisions about the uses of DNA technology reflect a vision of justice where privacy, autonomy, and fairness are placed in the context of responsible criminal investigation.

Genetic Justice is highly accessible to the general reader while it also provides a wealth of legal analyses and policy discussions that will be highly informative to lawyers, policy makers, judges, and law-enforcement officers who want a comprehensive understanding of both the power of the technology as well as its technical and constitutional limitations. From this book a broad spectrum of readers will gain a deeper and more balanced appreciation of the issues involving forensic DNA databases.

This book emerged out of the collaborative efforts of two eminently qualified and distinguished experts, and I am especially proud that the ACLU played a significant role in its fruition. Tania Simoncelli, the ACLU's first ever Science Advisor, brings to the book a mastery of scientific and technological aspects of genetic technologies together with their implications for civil liberties and social justice, as well as tremendous insights gained from her extensive advocacy work in this area. Sheldon Krimsky, who has published many important books in the areas of science and policy, was invited to the ACLU as a visiting scholar during 2007–2008, during which time the initial concept and research for the book took shape.

Genetic Justice should be read by everyone who cares about constitutional rights and criminal justice. We all want to solve crime. But the fact that crime occurs does not justify the erosion of principles that are at the core of our constitutional identity. One of the most fundamental principles of this country is that a balance must be retained between safety and freedom. Where technological developments challenge or disrupt that balance, a strong call for the protection of civil liberties and a clear, rational proposal for restoring the balance of personal liberty, social equity, and security are urgently needed. *Genetic Justice* will help you grapple with and perhaps make up your mind about where that balance should be.

Anthony D. Romero
Executive Director
American Civil Liberties Union

ACKNOWLEDGMENTS

We are grateful to a number of friends and colleagues who have contributed to this project. We wish to acknowledge the following people for their thoughtful help, support, and advice.

We are most appreciative of the feedback we received from the following U.S. reviewers who commented on early chapters: Bruce Budowle, Michael Risher, and Professors Joseph Dumit, Troy Duster, Dan Krane, and Patricia Williams. A number of people outside the United States also made important contributions to this book. For chapter 9 (the United Kingdom), Dr. Helen Wallace and Professor Carole McCartney. For chapter 10 (Japan), we are especially indebted to Emi Omura for researching and translating many documents, initiating contacts, and setting up and translating interviews with Japanese officials, lawyers, and professors. Others in Japan who provided useful knowledge include Professors Hikaru Tokunaga, Kenji Sasaki, Katsunori Kai, and Tatsuhiko Yamamoto, and Hiroshi Sato. For chapter 11 (Australia), Professors Jeremy Gans and Charles Lawson. For chapter 12 (Germany), Professors Michael Jasch and Peter Schneider for providing detailed background on Germany's forensic DNA database system. For chapter 13 (Italy), Gianna Milano contributed primary research, interpreted Italian reports, and wrote multiple drafts. Others in Italy—Magistrate Amedeo Santosuosso, Professor Giuseppe Novelli, Luca dello Iacovo, and Dr. Luciano Garofano—also read and commented on the chapter.

Our thanks to Dara Jospe and Claire Giammaria for their editorial work and research on early chapters.

Peter Neufeld and Barry Scheck, directors of the Innocence Project in New York, and their staff, especially Madeline DeLone, Stephen Saloon, Rebecca Brown, Emily West, and Vanessa Potkin, provided background material and data that contributed to several chapters. Chapter 16 benefited greatly from the work of and discussions with Professor William C. Thompson.

Sheldon Krimsky wishes to express his gratitude to the Department of Sociomedical Sciences at Columbia University, and especially Professor David Rosner, for hosting him as a visiting scholar and providing him with an office during 2007–2008; to the Benjamin Cardozo Law School for hosting him as a visiting scholar and for the use of its law library during 2007–2009;

and to the American Civil Liberties Union (ACLU) for hosting him as a visiting scholar and providing him with an office during the period this book was in its early gestation. Professor Margaret M. Wallace at John Jay College of Criminal Justice welcomed Sheldon to her laboratory course on forensic DNA analysis to watch her students go through all the steps in preparing a biological sample for a DNA profile and interpreting the output from a DNA analyzer.

Tania Simoncelli would like to thank the ACLU for providing her with the opportunity to engage in national policy debates and initiatives that are central to the contents of this book. She is grateful in particular to the following people in the organization who supported this effort: Anthony Romero, Barry Steinhardt, Terrence Dougherty, Robert Perry, Michael Risher, Jay Stanley, Claire Giammaria, Noam Biale, Mary Bonventre, Chris Hansen, and Sandra Park. She also would like to thank Troy Duster for hosting her as a visiting scholar at the Institute for the History of the Production of Knowledge at New York University during 2008.

Finally, we are greatly appreciative of our editorial team at Columbia University Press, Patrick Fitzgerald and Bridget Flannery-McCoy, for their support, patience, and guidance, and of the superb copyediting of Charles Eberline and John Donohue of Westchester Book Services.

INTRODUCTION

> We have 30 crime scenes where we have found traces of her DNA, but we have no face. . . . It's a huge mystery and it's incredible that the suspect has managed to hide herself for so long.
>
> — Rainer Koeller, German police spokesman[1]

On May 23, 1993, a 62-year-old woman was strangled to death in her home in Idar-Oberstein, Germany. The only clue that surfaced from the police investigation was a DNA sample obtained from a brightly colored teacup that was found on the woman's kitchen table. An analysis of the DNA revealed that it had come from a woman.

Eight years later a 61-year-old antiques dealer was strangled with garden twine in the southwestern city of Freiburg. DNA collected from items in the shop and the shop's door handle was found to be identical with that of the 1993 teacup DNA. A few months later the woman's DNA was found again, this time on a discarded heroin syringe.

In 2007 a 22-year-old German policewoman was killed in the city of Heilbronn while she and her colleague were on a lunch break in their patrol car. Two people climbed into the back seat and shot the officers from behind, killing the woman and seriously injuring her partner. Once again, DNA traces taken from the vehicle matched that of the DNA collected from the 1993 murder scene.

The murder of the young officer sparked one of the most extensive criminal investigations in German history. The more the police looked, the more DNA they found to match the "Phantom of Heilbronn," as the woman was quickly dubbed. Her DNA was found on an abandoned biscuit at the site of a burglary, a gas cap of a stolen car, a bullet used in a "gypsy feud," and beer bottles and a toy pistol found at the scenes of two robberies. In 2008 traces of her DNA appeared in an old car that belonged to an Iraqi suspect accused of killing three Georgian car dealers.

By 2009 a person known only by her DNA and described by police as "Germany's most dangerous woman"[2] was wanted for at least 40 crimes in three countries (Germany, Austria, and France) that included six murders, several muggings, and dozens of break-ins and petty robberies. The German government deployed more than 100 police officers and prosecutors in three separate teams across Germany, backed by DNA analysis from the Bundeskriminalamt (BKA), Germany's equivalent of the U.S. Federal Bureau of Investigation (FBI), and spent more than $18 million in tracking down the "woman without a face." The chief inspector of one of the German teams claimed that he alone had traveled more than 37,000 miles across Europe in search of the cold-blooded murderer.[3] Heilbronn police alone racked up 16,000 hours of overtime pursuing the culprit. A reward of 300,000 euros (approximately $400,000) was offered publicly for any help leading to her arrest. With seemingly nothing else to go on, police rounded up and collected DNA samples from more than 3,000 women in Germany, France, Belgium, and Italy—especially drug users and homeless women—in a "DNA dragnet." The more the woman's DNA surfaced at the scenes of crimes, the more the police assured the public that they were "closing in on her."[4]

Then a curious thing happened. The woman's DNA surfaced again, but this time even the most creative scenario could not explain its appearance. In this case investigators were attempting to establish the identity of a burned corpse that had been found in 2002. They reexamined fingerprints from a male asylum seeker's application to see whether a DNA sample could be lifted off the fingerprints and compared with that of the corpse. Thoroughly expecting any DNA found at all to be from a male source, they were shocked to find instead the DNA of the phantom woman.

The explanation was simple enough, but surprising to police and many DNA analysts. The woman, assumed to be a cold-blooded serial murderer, junkie, and burglar, was none other than a perfectly innocent woman who happened to work in the factory in Austria where the swabs used for collecting DNA at crime scenes were fabricated. She was never physically present at a single one of the crimes for which she had become a suspect; her skin particles, sweat, saliva, or combinations thereof had merely contaminated a large number of swabs during production that were then used in DNA collection. Her DNA thus "appeared" in an endless number of DNA tests, sometimes alongside the DNA of other individuals. Those swabs were "sterilized" before use in collecting DNA samples, but sterilization does not destroy DNA (it only destroys the viability of bacteria, viruses, and fungi). The entire law-enforcement community in Germany had spent 15 years on a wild-goose chase, and meanwhile the actual criminals had, in most cases, gone scot free.

On its face, this story seems perhaps nothing more than an extraordinary blunder, one we might dismiss with all its absurdity as an outlier that would never occur in most sophisticated pursuits of justice. But another way to look at it is to ask: What assumptions must have been operating in order for such an oversight to have occurred? How is it that an investigation as extensive as this one did not raise seriously the possibility of contamination as the source of the woman's DNA? Why would investigators go to such lengths—in euros, man-hours, and DNA dragnets—to search for a suspect in the face of significant contradictory evidence? What does it take to believe that one woman could have been involved in such an absurdly disparate array of criminal activity, especially considering that she never once in 40 crimes was seen by a witness, or that she could have worked with different accomplices in those cases where others were involved (all of whom claimed when questioned that she did not exist)?

Under this line of questioning, this case becomes a lesson of caution rather than a mere bungle, a warning that investigations driven exclusively by DNA evidence may lead us astray. It provides evidence of a kind of tunnel vision that is both extreme and systemic and demonstrates that—at least under some circumstances—DNA has come to trump other evidence and simple common sense and to serve as a teller of truth, a replacement for eyewitness identification, and a stand-in for an individual person or suspect.

More broadly, this case touches on a central question that this book seeks to answer: *Has an increasing reliance on DNA in our criminal justice system furthered our pursuit of justice?*

When DNA identification was first introduced into the criminal justice system in the 1980s, it was used in specific cases where DNA evidence had been obtained from a scene of a crime. An individual suspected of a crime would knowingly (and under a court-issued warrant) provide a DNA sample for comparison with DNA collected from the crime scene, and the results could be used as evidence to support his or her guilt or innocence.

The fact that no two people, with the possible exception of identical twins, share the exact same DNA sequence in their cells made DNA analysis an extraordinarily enticing tool for identification and surveillance. In the 1990s law enforcement in the United Kingdom and the United States began to permanently warehouse in data banks DNA taken from people who had been convicted of crimes. By generating a profile of the DNA that can be stored electronically, law enforcement can routinely compare a growing bank of DNA samples taken from offenders with DNA picked up at any crime scene. Anyone who had done wrong and had his or her DNA banked and went on to commit a second crime could be identified if his or her DNA was left behind. With the birth of DNA data banks, crime scenes have become a reservoir of biological trace materials like bloodstains, hair, semen, and skin cells from which DNA can be obtained and potentially matched with that of on-file suspects. Rapid developments in molecular genetics have made DNA analysis the fastest-growing forensic tool in criminal investigation since fingerprints were first introduced more than a century ago.

The last 15 years have witnessed an extraordinary expansion in the collection and storage of DNA by law enforcement. In the United States, state DNA data banks, initially limited to sexual offenders, have expanded dramatically so that today almost all states collect DNA from all felons, more than half from juveniles, and two-thirds from misdemeanants. States and federal agencies have adopted different criteria for human DNA data banking, resulting in a patchwork of standards for collection, retention, expungement, and access to and use of the information. Enthusiasm for DNA collection has prompted some state legislatures to expand their data banks beyond convicted offenders to include arrestees and other innocent individuals. By March 2010 over 8 million Americans have had their DNA forcibly collected and retained in a law-enforcement DNA data bank. Such collections have not been limited to the United States; many industrialized countries have established DNA databases of varying degrees of inclusion. In England children as young as 10 years old who are picked up by police for misdemeanors and petty theft have their DNA placed in the national data bank. Similarly, some U.S. states have collected DNA from teenagers. A few policy leaders, including former British prime minister Tony Blair, have called for a nationwide, universal DNA data bank in which everyone's DNA would be collected and stored from birth.

As collections of DNA have expanded, so too have the ways in which DNA is used in criminal investigation. Initially DNA analysis was used by law enforcement to compare DNA from a suspect with DNA found at a particular crime scene. Today law enforcement is increasingly relying on DNA dragnets, where hundreds, if not thousands, of individuals living in the vicinity of a crime are pulled aside and asked to provide a DNA sample for purposes of exclusion. We have also witnessed an increasing use of "DNA profiling," where genetic testing is used in an attempt to predict physical characteristics (eye color or hair color, for example) of a person whose DNA was left at the scene of a crime. In addition, law enforcement routinely picks up so-called abandoned DNA off coffee cups or cigarette butts left by individuals who are under suspicion without their knowledge or consent. Most recently the United Kingdom and some states in the United States have started to engage in "familial searching," where police investigate family members of individuals in the database whose profiles are similar but not identical to that of DNA collected from a crime scene. These major changes in DNA databanking laws and police practices have occurred with little, if any, public involvement.

The pursuit of justice involves more than simply the resolution and reduction of crime. Fairness, equality, and protection of basic civil liberties are as much a part of justice as are conviction and punishment of the guilty. If technology is to be used in the pursuit of justice, it should be used in ways that reflect a society's commitment to maintaining privacy and autonomy, minimizing racial discrimination and injustices, and contributing to overarching fairness in the criminal justice system. The limitations, as well as the promises, of the technology should be appropriately defined and considered in decisions about the application of the technology to criminal investigation. Decision making should occur openly and with considerable public input. And resources dedicated to the technology should be consistent with other social priorities and human rights within and beyond the criminal justice system.

This book seeks to identify and define appropriate uses of DNA by law enforcement that will result in the furtherance of justice. Part I is largely descriptive and documents the history, applications, and expansion of the use of DNA in law enforcement. After a basic overview of the technology of forensic DNA typing (chapter 1), we trace the history and development of DNA data banking in the United States (chapter 2). We examine the current policy landscape, including the recent expansion of DNA data banks beyond individuals convicted of felony offenses to include arrestees, juveniles, misdemeanants, and individuals who voluntarily submit a DNA sample to exclude themselves as suspects in a crime. Part I also addresses the major ethical, sociopolitical, and legal concerns associated with increasing uses of DNA dragnets (chapter 3), familial searching (chapter 4), phenotyping of crime-scene DNA evidence (chapter 5), and the use of so-called abandoned DNA (chapter 6). We highlight sensationalized success stories that have served to foster dramatic expansions of DNA data banks and forensic DNA techniques. Part I also examines the use of DNA evidence in the exoneration of falsely convicted persons and, in particular, the impediments to using DNA evidence to prove innocence (chapter 7). We complete this part of the book with a discussion of the debate about the issue of universal DNA data banks, where the DNA of every citizen is taken and stored at birth (chapter 8).

Part II provides a transnational perspective on DNA data banking. We chronicle key trends in five other countries—the United Kingdom (chapter 9), Japan (chapter 10), Australia (chapter 11), Germany (chapter 12), and Italy (chapter 13)—and compare in particular the ways in which these governments have sought to balance privacy and other social and ethical considerations in creating, maintaining, and using DNA in solving crimes. We selected these particular systems because each case highlights some key differences in the ways in which industrialized countries have considered the dignity and rights of individuals in the establishment of a DNA-collection program. A comparison of the main features of each of these forensic DNA data-bank systems, as well as that of the United States, is provided in the appendix.

After tracing the expanding uses of DNA by law enforcement both in the United States and abroad, Part III provides the authors' critical perspectives on society's effort to balance personal liberty, privacy, social equity, and security in the development and use of forensic DNA technology. We examine the privacy (chapter 14) and racial justice (chapter 15) implications of DNA database expansions. We uncover the myth of infallibility, exploring places for contamination and error not only in extreme cases of the sort described earlier, but in those far less likely to be righted (chapter 16). We analyze the efficacy of DNA data banking and question the prevailing assumption that "the more DNA, the better" (chapter 17). Finally, in the concluding chapter we propose that decisions about forensic uses of DNA be driven by a vision of justice that extends beyond crime and punishment and looks comprehensively at notions of privacy, autonomy, and fairness in the context of responsible criminal investigation (chapter 18).

Part I

DNA in Law Enforcement: History, Applications, and Expansion

Chapter 1

Forensic DNA Analysis

DNA analysis is one of the greatest technical achievements for criminal investigation since the discovery of fingerprints. Methods of DNA profiling are firmly grounded in molecular technology.

—Committee on DNA Forensic Science,
National Academy of Sciences[1]

For those who can benefit from a primer on genetics and DNA profiles, this chapter reviews the nomenclature and genetic technology that form the basis of forensic DNA analysis. After a brief discussion of the basics of the genetic code, we explain such topics as DNA typing methods, short tandem repeats (STRs), and random-match probabilities. This chapter is designed for people who have very little background in molecular biology and forensic DNA analysis. Those who already possess this knowledge can proceed directly to chapter 2.

What Is DNA?

Deoxyribonucleic acid, or DNA for short, is the chemical in cells that specifies the composition of proteins and, along with other cellular components, contributes to their synthesis. DNA is also largely responsible for the inherited characteristics of organisms. The structure of DNA, as first postulated by James Watson and Francis Crick in 1953, is often compared with a spiral staircase or double helix with rungs or steps (see figure 1.1). The spine or backbone of the helix (analogous to the banister of the spiral staircase) consists of

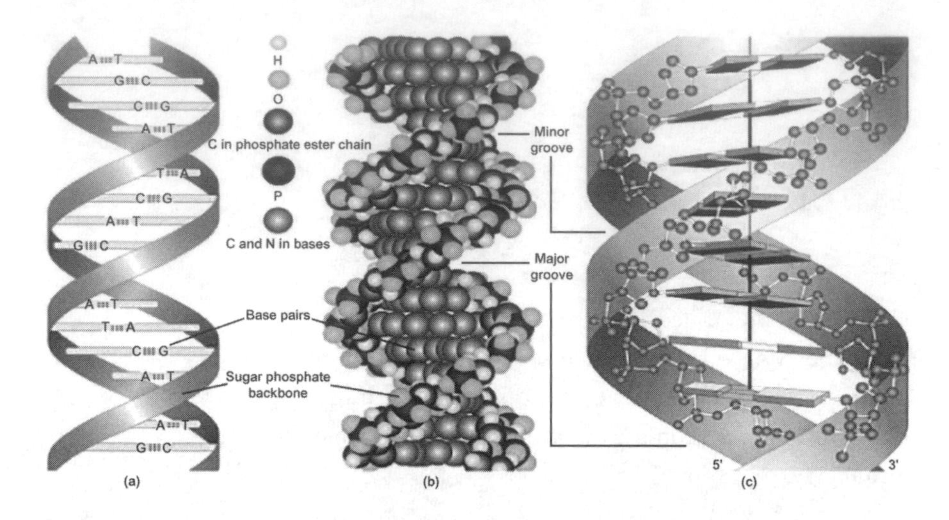

FIGURE 1.1. The double helix DNA structure. *Left*: A helixlike structure with two ribbons representing the sugar-phosphate groups and the horizontal steps or bases of the DNA molecule. *Center*: The hydrogen bond connecting complementary bases. *Right*: A model of a DNA molecule represented by a twisted lattice of spherical components. *Source*: From *Modern Genetic Analysis* by A. J. F. Griffiths, W. M. Gelbart, J. H. Miller, and R. C. Lewontin, ©1999, by W.H. Freeman and Company. Used with permission.

sugar-phosphate groups that link the steps of the spiral staircase and thus are constant throughout the length of the DNA strand for all individuals. The steps of the spiral staircase are composed of chemicals that are called bases or nucleotides. There are only four possible bases, adenine, guanine, thymine, and cytosine, denoted by the letters A, G, T, and C, respectively. The pattern or arrangement of these letters determines a person's genotype, or genetic identity, as opposed to a person's physical identity or phenotype (defined as the physical appearance or biochemical characteristics of an organism). No two people, with the possible exception of identical twins, have exactly the same series of letters that make up their DNA.

When a multicellular organism reproduces, it is the DNA within the reproductive cells (gametes or sperm and eggs) of the organism that serves as the template for the development of the fertilized egg. This fertilized egg develops into an organism of the same species, with species-similar but not necessarily identical physical (phenotypic) properties.

The complete set of human DNA is found in virtually every one of our cells (except the sperm and egg cells, which contain one-half of the DNA, and red blood cells, which have no nucleus), in every organ, and in our blood and immune system. DNA is found both in the nucleus of our cells and in the cell's mitochondria (a component of the cell outside the nucleus that resides in the cytoplasm). The long, continuous nuclear DNA molecules are distributed on chromosomes, which also contain ribonucleic acid (RNA) and proteins. Humans have 23 pairs of chromosomes that are packaged in the nucleus of each cell. The DNA in the chromosomes is packed tightly, wound up and coiled into the nucleus of the cell. Proteins called histones help stabilize the tightly packed DNA within each chromosome.

Although DNA is thought to provide essential instructions for the functioning of our cells, it is not self-effectuating—it does not act by itself. DNA responds to prompts from the cell's proteins, the body's enzymes and hormones, RNA molecules, and sometimes external environmental factors. As Barry Commoner notes in "Unraveling the DNA Myth," "Genetic information arises not from DNA alone but through its essential collaboration with protein enzymes."[2]

Outside an organism, DNA can persist for many years under optimal conditions (see box 1.1). However, prolonged exposure to sunlight, warm temperatures, high humidity, and bacterial and fungal activities can result in DNA degradation. Some of the chemical enzymes released upon cell death may initiate the degradation of DNA. It is also more likely to degrade when it is on soiled rather than clean materials.

BOX 1.1 Brown's Chicken Massacre and the Persistence of DNA Evidence

In 1993 seven people were ruthlessly killed in a robbery and left in two walk-in refrigerators at Brown's Chicken and Pasta, a suburban Chicago restaurant. Collected from a trash can at the scene of the crime was a partly eaten dinner (two half-eaten chicken pieces). DNA testing techniques at that time were not sophisticated enough to produce any DNA profiles from traces of human saliva, so the chicken was frozen in hopes

(*continued*)

that developments in DNA testing would allow for future testing. Seven years later, that testing occurred, producing two DNA profiles. The profiles did not match any of the crime victims or suspects that police had at that time. Two years later, in 2002, a woman came forward with important details of the crime to police, including names of people who, she claimed, spoke about their involvement in the robbery. Police obtained a DNA sample from one suspect and matched it to a sample of saliva from the chicken dinner. A month later, Juan Luna and James Degorski were arrested and charged with the murders. Luna's defense argued unsuccessfully that the DNA evidence against him should not be allowed because it had been mishandled over the years, including that it had been retested on multiple occasions and handled by scientists who acknowledged that they had not worn gloves. Both Luna and Degorski were found guilty of all seven counts of murder and were sentenced to life in prison.

Source: Authors.

What Is the Size of DNA?

If you uncoiled the entire nuclear DNA of a single human somatic cell (any cell other than sperm, egg, or red blood cells), holding the strands of the double helix end to end, its length would be about 2 meters (around 6 feet). The DNA is so thin and so tightly coiled that it can be this long but reside within the nucleus of the cell, which is about 5 millionths of a meter in diameter.

The length of a thread of DNA is usually measured in units called kilobases (kb). One kb is the molecular length equal to 1,000 base pairs of double-stranded DNA (there are two strands in the double helix), or 1,000 pairs of the bases (A, G, C, or T).

How many kilobases are there in the entire human genome? A human genome is made up of approximately 6 million kb (or 6 billion base pairs; see figure 1.1), 3 billion base pairs for each set of 23 chromosomes. To give some sense of this length, typing out the letters in a complete strand of our nuclear DNA would take up 57 million lines (where each line contained approximately 53 letters) and would fill about 1.2 million single-spaced pages.

Our DNA is distributed unequally among our 23 pairs of chromosomes. For example, chromosome 1, the largest human chromosome, has a length of 245,000 kb and would require 4.6 million lines of print or approximately 100,000 pages, while chromosome 22 of about 49,000 kb would require 925,000 lines of print or approximately 20,000 pages.

What Is a Gene?

A gene is usually defined as a segment of DNA that can be used by the machinery of the cell to synthesize a protein. In humans most of the genes are in the nucleus of the cell, dispersed across the 23 pairs of chromosomes (see figure 1.2). Genes can range in size from 100 bases (.1 kb) to as large as 2 million bases (2,000 kb). An average gene is about 5,000 bases long (5 kb).

A gene located on one of a pair of chromosomes (a particular form of a gene or a noncoding DNA sequence is called an allele) may be the same but is not necessarily identical to that of its "copy" located on the other chromosome of the pair. One of the genes residing on a chromosome was contributed by the egg and the other by the sperm. A person can often be perfectly healthy with one "good" and one "defective" copy of a gene.

The totality of the DNA in an organism or cell is called its genome. The human genome can be thought of as a set of encyclopedias with 23 volumes, where each chromosome represents one volume. The DNA code comprises the text of those volumes, and the genes make up discrete chapters or paragraphs inside each volume.

The stretch of DNA on which the gene resides has more DNA than is required to encode a protein. The region of the DNA that is used to synthesize the protein is called the coding region (also called exon). The extraneous DNA on the segment, which is excised during the process of transcription (when the protein is being synthesized), is called the noncoding region (or the intron). To use the analogy of the page of text, imagine a sequence of letters (each representing a single nucleotide) such as AAGTA**CAT**ATGAA**CAT**. Suppose that the letters CAT represent the noncoding text (intron). When the gene is read and copied into a usable message for synthesizing a protein, the noncoding regions are removed, and the remaining segment in our example is AAGTA-ATGAA. This segment (representing the functional gene) is used to make a protein product.

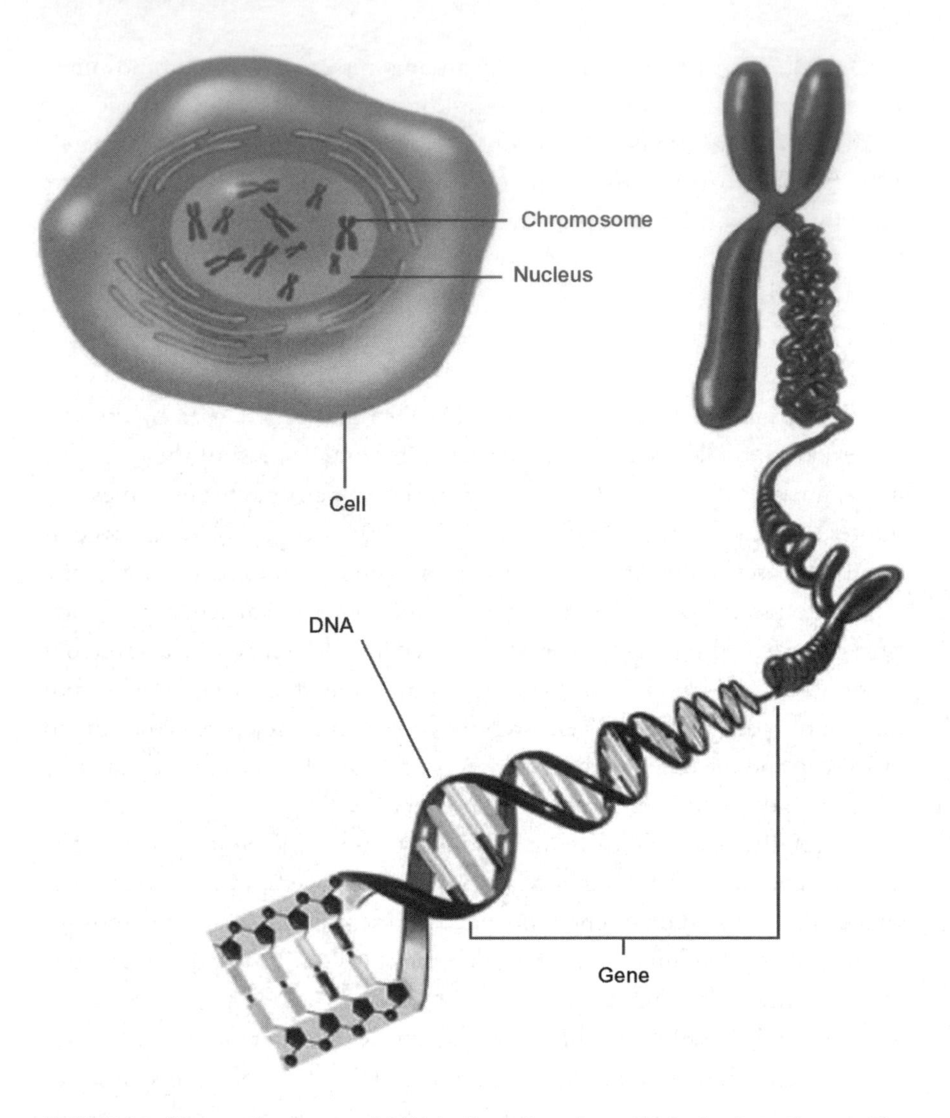

FIGURE 1.2. Human chromosomal DNA in the cell nucleus. *Clockwise from the top*: The cell and its nucleus, one of the 23 chromosomes, and the DNA molecule. *Source*: National Institute of General Medical Sciences, National Institutes of Health.

In human cells, where there are 3 billion base pairs of DNA in the nuclei, the number of genes, defined as functional segments of DNA that are used to encode proteins, is still uncertain. The International Human Genome Sequencing Consortium placed the number of protein-coding genes at 20,000 to 25,000. Other groups predict larger numbers. It is estimated that 97 to 98 percent of our DNA consists of noncoding regions.[3] Some of these sequences may have other functions. For example, promoters reside near the genes and initiate gene expression starting with transcription (the process of preparing a readable RNA message of the DNA so a protein can be synthesized by a process known as translation). Other DNA sequences, called enhancers, can raise the amount of product produced. There is also noncoding DNA, sometimes referred to as "junk DNA" or "evolutionary debris," that allegedly has no function, or at least not one that is known.

More recently, through a project called ENCODE (meaning the encyclopedia of DNA elements), scientists have discovered that a large amount of the DNA previously considered useless is essential to regulatory processes in the so-called functional part of the genome where gene transcription takes place.[4] In the words of geneticist Francis Collins, formerly director of the U.S. National Human Genome Research Institute, "Transcription appears to be far more interconnected across the genome than anyone had thought."[5] The so-called noncoding regions probably play a role in coding small RNA molecules.

The Typing of DNA for Forensic Identification

Sequencing DNA means reading its code, or the series of four letters (A, G, C, and T) that make up the bases of the DNA molecule. As discussed earlier, humans have about 3 billion nucleotides in each of the 23 pairs of chromosomes. The sequence of nucleotides (reading of the human DNA) is 99.9 percent identical for all people. In other words, human genetic variation is accounted for by only 0.1 percent of DNA, or about 3 million bases.[6] It is this variation in a segment of DNA that allows forensic scientists to determine whether two DNA samples could have come from the same individual.

The process of DNA analysis always begins with a sample of biological material: it could be a strand of hair, blood, sperm, tissue, skin cells, or saliva. To be useful for the analysis, the sample must have intact cells or DNA that has been removed from the cell. Unless the samples are very carefully handled, they can become contaminated by elements from the environment, such as DNA from plants, animals, insects, bacteria, or other human beings.[7]

If the biological sample contains cells, the first task is to break open the cells (lysing the cell membrane) or to dissolve the matrix surrounding the DNA (as in a hair shaft) to retrieve the DNA. The separation of the DNA from the biological sample is called extraction, while the separation of sperm-cell and non-sperm-cell DNA is called differential extraction. Breaking the membrane of the cell to release its DNA can be done by exposing the cells to detergents and other chemicals; different chemical recipes are effective for different cell types. Once the DNA is removed from the cell, it has to be broken down into manageable pieces so that it can be identified by its unique sequence of bases. One method to accomplish this involves the use of proteins isolated from bacteria called restriction enzymes (*Eco*RI in figure 1.3) that cut DNA at specified sites, leaving fragments of DNA varying in length and defined by the presence of a restriction site.

To identify specific fragments with the genetic variation of interest, scientists use probes—short, single-stranded fragments of DNA that are synthesized in a laboratory and labeled radioactively or with some other detectable

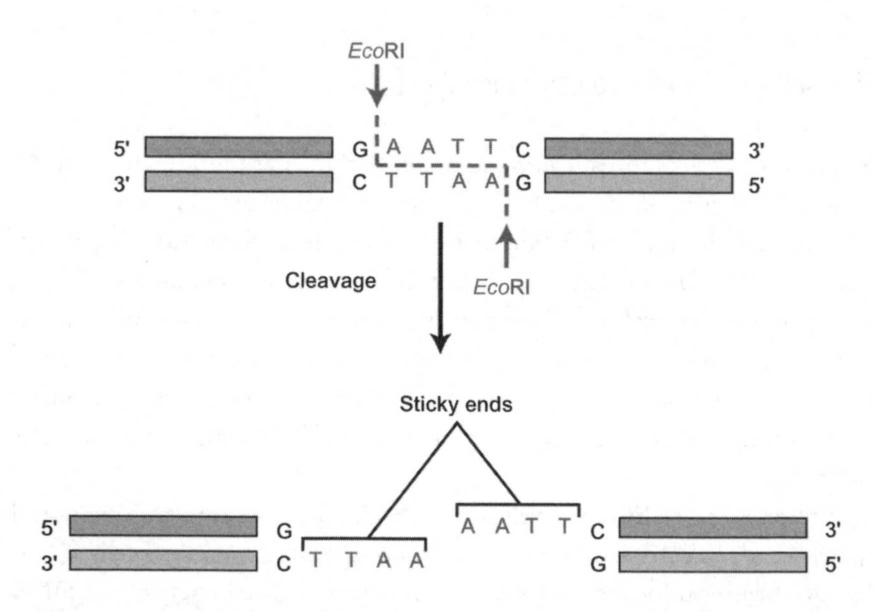

FIGURE 1.3. A restriction enzyme *Eco*RI is used to cut double-stranded DNA at a specific site, leaving the ends available for reattachment to complementary pairs. *Source*: From *Molecular Cell Biology*, 5th ed. by Harvey Lodish, Arnold Berk, Paul Matsudaira, Chris A. Kaiser, Monty Krieger, Mathew P. Scott, S. Lawrence Zipursky, and James Darnell. © 2004 by W.H. Freeman and Company. Used with permission.

molecule. A probe will seek out and attach to its complementary sequence (if it is present) to form a double-stranded sequence of DNA. Within the DNA code, base A is always complementary to base T, and G is always complementary to C. Thus a probe consisting of the nucleotide AGTTAGC is the complementary strand to TCAATCG.

Early Method of DNA Analysis

The first widely used method of forensic DNA analysis was based on restriction enzyme digestion and was called restriction fragment length polymorphism (RFLP). Introduced in 1988, RFLP involves cutting relatively large segments of the DNA molecule into smaller fragments and sorting those fragments by their size using a process called gel electrophoresis (see figure 1.4). RFLP analysis requires that a minimum amount of DNA from a sample be within a particular range from 50 to 500 nanograms (ng; 1 nanogram = 1 billionth of a gram or 10^{-9} gm). This corresponds to a bloodstain about the size of a dime to a quarter.

The restriction enzymes break the long DNA strand at specific sites. Because the DNA from two different individuals may have fragments that differ in size when they are excised at a specific place or locus (the term "locus" is used to designate the position or site of a DNA sequence on the genome), measuring the size of the fragments can distinguish DNA samples from different individuals. The difference in size of a fragment at a locus between individuals is due to the presence or absence of a restriction site and/or the number of short repeating noncoding sequences contained within a fragment. Once the DNA strand taken from a sample is broken up into discrete segments by restriction enzymes, it is then put through a process that separates the segments by weight. The DNA is placed in small wells on one end of a flat gelatin surface and then exposed to an electric field. The separation of the DNA segments by size is based on the fact that each DNA strand is negatively charged. When placed in an electric field, the negatively charged DNA will move toward the positive side of the electric field. Smaller DNA fragments move more quickly than larger ones because the latter experience more resistance migrating through the sieving gel medium, thus allowing the fragments to be separated according to size. After the fragments are put through gel electrophoresis, placed on a membrane, and exposed to x-ray

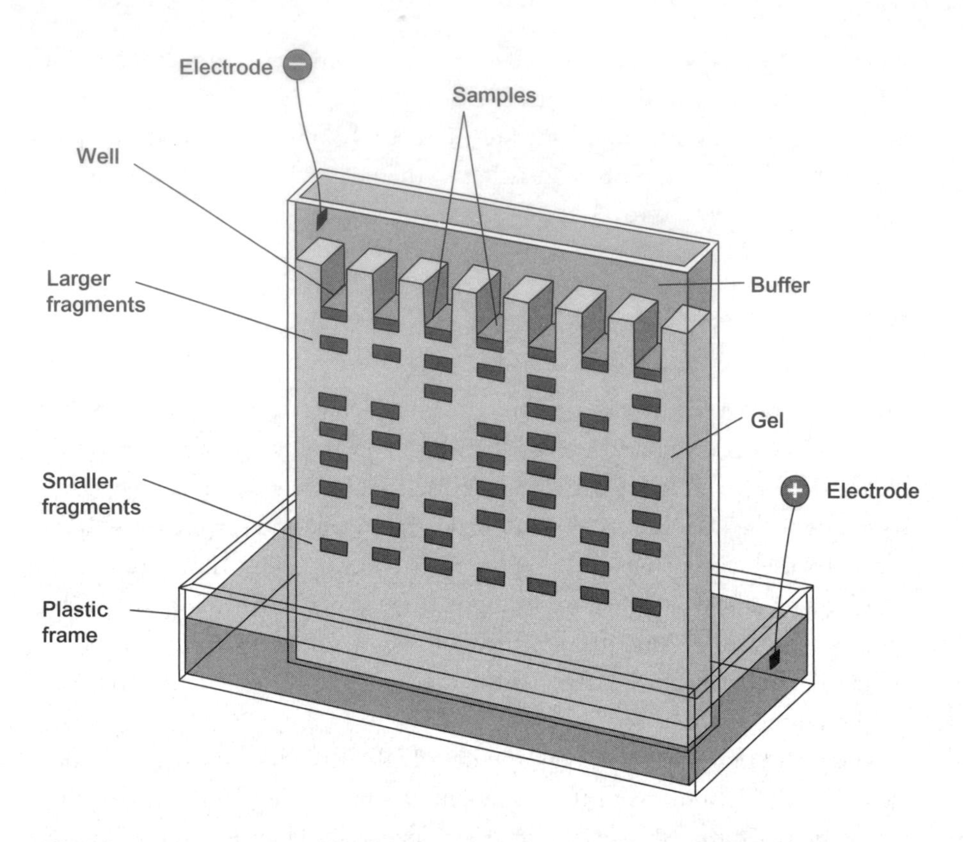

FIGURE 1.4. A rendering of a gel electrophoresis device in which gene fragments exposed to an electric field over a gelatin surface are separated by their size. *Source*: From *Genetics: Principles and Analysis*, 4th ed. by Daniel T. Hartl and Elizabeth Jones, 1998, Jones & Bartlett Publishers.

film, the result is a series of black bands like the bar code in consumer products. When the two profiles with dark bands (each representing the size of a DNA segment) match, this suggests that the DNA samples may have come from the same individual. If the two profiles are different, then the two samples could not have originated from the same individual. The x-ray photograph showing the position of the DNA segments is called an autoradiograph or autorad (figure 1.5).

The RFLP analysis given in figure 1.5 is of DNA from suspects in a sexual assault case. The columns show the DNA segments of the victim (column 4), suspect 1 (column 5), suspect 2 (column 6), and the crime-scene sperm DNA

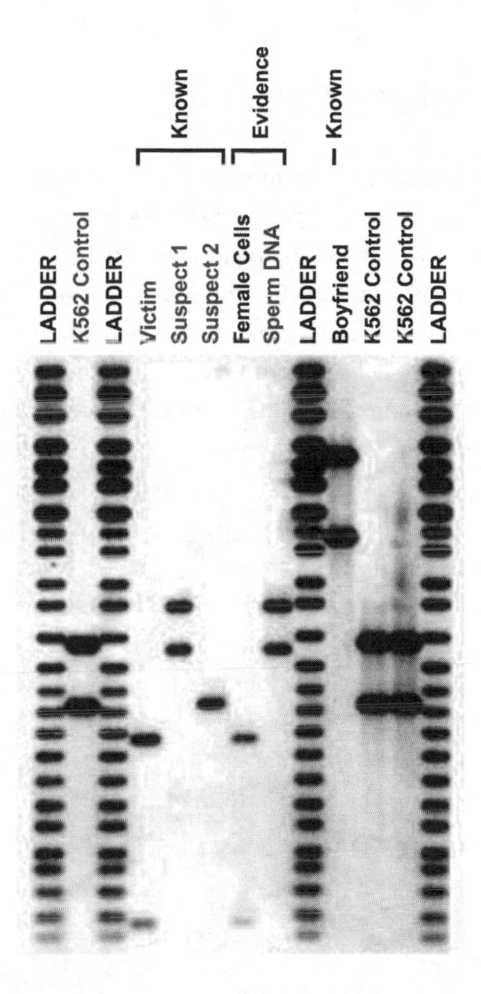

FIGURE 1.5. An example of an autoradiograph of radio-labeled DNA segments in a sexual assault case, developed from RFLP analysis. It includes the DNA typing of two suspects and a victim with results for control samples and ladders that are used to estimate the size of the DNA fragments. *Source*: Genelex.Com, http://www.healthanddna.com/genelex/about.html (accessed May 23, 2010).

(column 8). There is a matching profile at this locus between the crime-scene sperm DNA and suspect 1. The other columns are controls with known DNA patterns to validate the process. The columns marked ladders contain known sizes of the band sequence fragments that are used to estimate the size of the resulting profile bands.

PCR Method for Copying and Amplifying DNA

RFLP has been replaced by a far more sensitive method of DNA analysis, based on a technique called the polymerase chain reaction (PCR). PCR involves replicating, by a form of chemical copying, tiny defined segments of an individual's DNA (see figure 1.6). The PCR technique has revolutionized forensic DNA analysis because it is far more efficient and can be applied to the analysis of very small DNA samples. While the RFLP method could take upwards of six to eight weeks, PCR can complete a DNA analysis in one to two days. More important, RFLP requires a relatively large DNA sample; however, typical sample sizes for PCR analysis range from .5 to 2.0 ng. Thus PCR is equipped to analyze DNA samples 500 times smaller in quantity than RFLP requires. Because it is so much more sensitive, PCR is more useful than RFLP for analyzing degraded samples of DNA from blood, saliva, hair, semen, and other sources. On the other hand, because of its high degree of sensitivity, PCR is also more vulnerable to contamination (see the introduction and chapter 16).

The basic idea behind PCR is that copies of a small segment of DNA are made, and then copies of the copies are made through a cyclic process until

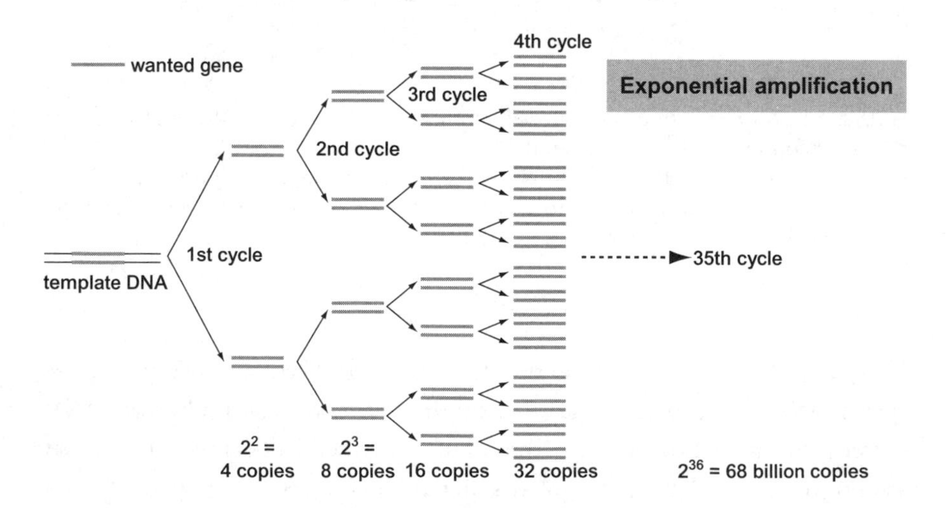

FIGURE 1.6. A representation of the DNA amplification process using the polymerase chain reaction (PCR). *Source*: Andy Vierstraete.

a sufficient quantity of the DNA is obtained for analysis. A simple analogy to the PCR amplification process is given as follows. Imagine an all-school dance, where the number of boys and girls is equal. The event begins with one pair of dancers (one boy and one girl) in the center of the room. Surrounding the dancers on the perimeter of the room are girls and boys in no special order. At some point in the dance, prompted by a change in the music, the dance pair divides (analogous to the split of a double-stranded DNA molecule into two single strands). The boy meanders to the perimeter to select another partner, his new complement—while the female does the same. The result is that from one dance pair come two, four, eight, and so on. The process can be repeated many times until the dance floor is filled to capacity and no one is left in the periphery (all complementary pairings are actualized).

In the PCR process the DNA extracted for analysis is mixed with a group of chemicals including primers (chemicals that will identify the specific DNA targets to copy and initiate DNA replication) and enzymes. The primers are single-stranded nucleotides synthesized in the laboratory and attached to specific target sites of the DNA sample of interest. The extracted DNA and these chemicals are then placed in a machine called a thermal cycler. This machine runs the sample through a series of heating and cooling cycles. The thermal cycling consists of three steps. In step 1 the sample is heated to 95°C, which is the temperature at which double-stranded DNA unzips to form two single strands. This process is sometimes called denaturing the DNA. In step 2 the sample is cooled to 55–60°C, which is the temperature range best suited for the primers to attach to the single strands. In step 3 the temperature is raised to 72°C, and an enzyme called Taq DNA polymerase acts as a catalyst to synthesize a complementary target to the single-stranded template DNA. A complete double-stranded DNA molecule is created. This is called the synthesis step. This process duplicates segments of DNA. A cycle of the PCR takes less than a few minutes. The three-step process is usually repeated two to three dozen times (using commercially available kits), and with each process the number of copies of the target DNA is doubled. By the 28th cycle the PCR process can make billions of copies of a particular DNA molecule that may have been present only a few times in the original sample. Once the DNA segments from two or more samples are replicated, these segments have to be measured and compared, as described in the following section.

Polymorphisms in DNA Sequences

Polymorphisms are defined as variations in DNA sequences at a particular position (locus) on the human genome, when the locus varies in at least 1 percent of the population. The variation in the DNA sequence can take multiple forms. It can be a change in one base of a sequence. For example, the difference between the DNA sequences AGACCTAG and AGACCTAC is that the last base G in the first sequence is replaced by the base C in the second sequence. Because the difference in the sequences is one base, this is called a point (or site) mutation or a single-nucleotide polymorphism (SNP, pronounced "snip"). A mutation at the site of a DNA sequence recognized by a restriction enzyme may prevent the enzyme from cutting the DNA at that site. Thus, if an enzyme cuts a DNA sequence every time it sees the base sequence TAG, it will cut the first sequence but not the second site if that individual carries a SNP that changed the sequence.

A second type of polymorphism is represented by repeated short sequences that lie adjacent to one another on the DNA thread (tandemly repeated). For example, in the two DNA sequences listed below, one has three repeats of AGTCA, and the other has five repeats of the same base sequence at a particular locus of the genome. This is called a "length polymorphism." A polymorphism may or may not affect the physical characteristics of an individual. Those used in forensic identification do not seem to have an effect on the physical characteristics of an individual. Human DNA loci that are highly polymorphic (variable) are especially useful for identifying individuals when the length polymorphisms are made up of STRs because they are more likely to display differences between two randomly selected individuals. Thus the more variable the locus, the better chance there is of excluding a person wrongly associated with a forensic evidentiary item.

AGTCAAGTCAAGTCA (three repeats of AGTCA)
AGTCAAGTCAAGTCAAGTCAAGTCA (five repeats of AGTCA)

Repeats of short contiguous segments of DNA (i.e., STRs) are used to ascertain the source (identity of the contributor) of a biological sample in most forensic DNA laboratories. In the U.S. forensic DNA database system, 13 STR loci are chosen that have high variability. The names and locations of these 13 core STR loci on the chromosomes, as well as the loci on the X and Y chromosomes that are used to determine the gender of the DNA donor, are shown in figure 1.7.[8]

These loci are named by scientists within the genetics community by alphanumeric terms like D3S1358 or FGA. The nomenclature does not follow a logical form, but the letters and numbers do have significance for the chromosome in which the sequence is found, whether the sequence is part of an intron (noncoding region of DNA) or resides outside a functional gene, and when the sequence was discovered. The automated DNA analyzer determines the number of STRs for each of the two alleles (one for each chromosome) in a designated locus. In table 1.1 the evidence sample shows 13 and 15 repeats for the STR at locus A and 12 and 14 repeats at locus B. It matches the number of repeats at the same loci for suspect 2. Thus suspect 2 cannot be excluded as a potential source of the evidence sample, while suspect 1 can be excluded.

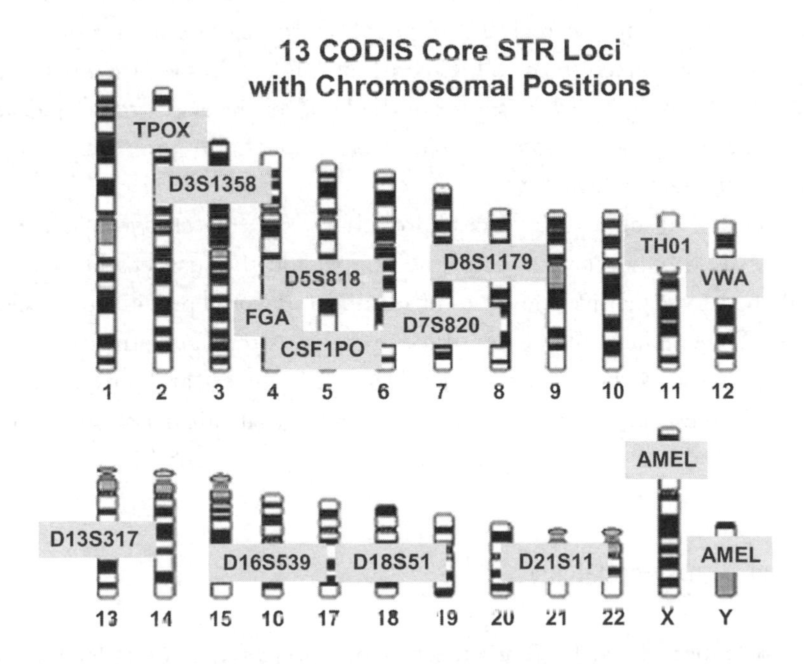

FIGURE 1.7. The alphanumeric names of the DNA sites used by U.S criminal justice authorities for forensic DNA analysis. *Source*: National Institute of Standards and Technology, "13 Core CODIS STR Loci with Chromosomal Positions," http://www.cstl.nist.gov/biotech/strbase/images/codis.jpg (accessed May 23, 2010).

TABLE 1.1 Short Tandem Repeats for Two Alleles

	Locus A	Locus B
Evidence sample	13, 15	12, 14
Suspect 1	12, 16	12, 17
Suspect 2	13, 15	12, 14

Source: Authors' construct.

STRs with more repeats are longer. The lengths of the STRs are compared in a DNA analyzer in a process similar to that of electrophoresis described earlier. The DNA segments are exposed to an electric current in a narrow capillary tube instead of a flat gel. As with electrophoresis, the shorter segments move more quickly through the capillary tube, and thus the length of the segment can be determined by the analyzer.

In table 1.1 suspect 1's DNA does not match the evidence sample in two loci. All we need is one mismatch of an STR length in a single locus to declare that the samples are not from the same individual, assuming that no errors were made in the analysis. Suspect 2 matches the evidence sample in two loci. If the match in the STRs continued across 13 loci, then suspect 2 would not be excluded as the source of the DNA of the evidence sample, and the probability that suspect 2 carried that same DNA profile by chance would be extremely low.

To obtain absolute 100 percent probability of a match, we would have to compare 3 billion base pairs of the sample and the suspect's DNA. That would be very expensive and time consuming and would provide more scientific evidence than would be necessary to make a strong case. Instead, a very small portion of the genome is analyzed (26 alleles) and, in cases of a "match," an estimate is provided for the chances that this match might have occurred by chance. This is called the "random-match probability."

Random-Match Probability

On November 13–14, 1997, at a meeting of representatives of 21 laboratories throughout the United States, forensic scientists and the FBI reached an agreement on using 13 STR loci for submitting profiles to the national forensic DNA database, otherwise known as the Combined DNA Index System

(CODIS). What level of certainty does this system achieve, and on what grounds?

Let us suppose that we are comparing a crime-scene sample with the DNA of a suspect using 13 loci (see table 1.2). For each STR locus, forensic technicians determine the number of repeats for each of the two alleles. For example, suppose that at locus A (table 1.2) both the suspect and the crime sample have 10 and 6 repeats. The next logical question is: how many people in the population of the suspect (racial, ethnic, and/or ancestry group) have 10 and 6 repeats at that locus? At locus B we find that there are 12 and 8 repeats for both the crime-scene sample and the suspect. Again, how many people in the suspect's reference population have 12 and 8 repeats for locus B? The reliability of the statistics for estimating that a random person in the population has the same repeats (random-match probability) is based on the frequency of the number of repeats at a locus in various reference populations. This information is critical for establishing the likelihood that two random people have the same number of STRs for 13 loci. The chance that two people have identical repeats is increased the closer they are in their genetic lineage. Close family members are more likely than those who are not related to have the same number of repeats in many (but not all) loci.

Table 1.2 also shows the frequencies with which each specific short tandem DNA segment would be observed in 13 loci (26 alleles) for a hypothetical reference population. Some geneticists like L. A. Zhivotovsky argue that the reference population is critical in establishing the probability of a random match between two samples because the frequencies of matched STRs are likely to be higher and the random-match probability also higher for people in the same geographical region or of the same ethnic or racial ancestry.[9] If

TABLE 1.2 Calculating the Probability of a Match in Two DNA Samples for 13 Loci

Loci	A	B	C	D	E	F	G	H	I	J	K	L	M
Number of repeats	10 6	12 8	13 8	7 9	9 7	14 4	11 8	12 10	10 6	12 9	14 8	8 8	12 5
Probability of repeats in a peer population	.01	.06	.03	.08	.02	.06	.07	.03	.04	.06	.08	.07	.10

Source: Authors.

the suspect is Asian, the chance of finding a random match in the European population is less than it would be for the Asian population. Because the suspect is presumed innocent, it is incumbent on our criminal justice system to consider seriously the conditions under which someone else in the population who has a forensic DNA profile identical to that of the suspect left his or her DNA at the crime scene. For locus A, the frequency of 10/6 STRs is .01, or 1 in 100. Similarly, for loci B, C, and D, the frequencies within the reference population are .06, .03, and .08, respectively.

The next step is to determine the rarity of the forensic DNA profile in the population. If the STRs at the loci are truly independent (as they are assumed to be), the chance that two people in a reference population will have the same number of repeats is obtained by taking the product of the frequencies for each locus. This is analogous to determining the chance of getting two heads when one simultaneously flips two different coins (which are independent events). The chance of getting a head on the first flip is .5, and similarly for the second. The chance of getting two heads on flipping both coins is $.5 \times .5 = .25$. Thus, if one flips both coins enough times, then one-quarter of the simultaneous flips will yield two heads.

For the first four loci in our example, the frequency of the specific STRs in a reference population would be determined by $(A = .01) \times (B = .06) \times (C = .03) \times (D = .08) = 1.44 \times 10^{-6}$, or about one chance in a million people. Thus in a population of 6 million people, approximately 6 people would be expected to have the same eight STRs on four loci.

If a crime were committed on an isolated island that had only 500 people, then a random-match probability of one person in nearly 1 million might be high enough to convince a jury that the suspect is the source of the crime-scene DNA, but few places are that isolated. If we completed the calculations for all 26 alleles, the probability of getting a random match in STRs for two people would be quite small—typically less than one in many billion.

Criminal justice agencies may differ in what they choose as the match probability that eliminates any doubt they have that a suspect was the source of the crime-scene DNA. Whatever number they choose as evidence of a "definitive match," they should not neglect to consider that there is a much greater likelihood that a close relative would have a matching profile.

The Laboratory Process for DNA Analysis

When evidence is received by a facility, it is characterized and coded, and a determination is made whether it is appropriate for DNA testing (see figure 1.8). If so, it is sent to a section of the laboratory for DNA extraction. Whether a garment, blood, hair, or saliva is involved, the extraction process removes proteins and other non-DNA components, leaving purified DNA residue for analysis. The extraction process may also involve the separation of sperm-cell and non-sperm-cell mixtures, such as sperm cells from vaginal epithelial cells. Different methods and reagents are used for the extraction of DNA depending on the nature and homogeneity of the sample. Once the DNA is extracted and purified from the sample, it is sent for DNA quantification, a process of measuring and setting the proper amount of DNA so it falls within the optimum ranges for analysis.

As an example, in a rape case the sample might consist of a stain, which is expected to contain a mixture of the perpetrator's sperm and the victim's epithelial cells (cells on the outer layer of the skin or organs). A portion of the sample is removed and exposed to a group of chemical reagents that degrade the proteins and break open the epithelial cells to release the nuclear DNA. The sperm pellets remain intact, and the freed epithelial DNA in solution is removed and analyzed. The original sample has undergone "differential extraction" when sperm and nonsperm fractions of the sample have been separated. The sperm pellets can then be exposed to another set of reagents to isolate the pure DNA. The DNA analyzer is optimally designed to handle certain amounts of DNA. If the DNA quantities are too large, artifacts (any result which is caused by the DNA analyzer itself and not by the DNA entity being analyzed) and other problems will likely occur, such as split peaks (a peak is an indicator of an allele at a locus) or off-scale peaks. If there is too little DNA, then some peaks will not appear. This is called allele dropout or possibly locus dropout. The analysis of STRs works best for samples containing 0.25 to 2.0 ng of DNA. One nanogram of DNA is contained in about 160 to 168 cells.

When the profiling is complete, the DNA analyzer produces a graphic and numerical output (an electropherogram) to a computer. The output shows a continuous oscillating line with periodic peaks and numerical values next

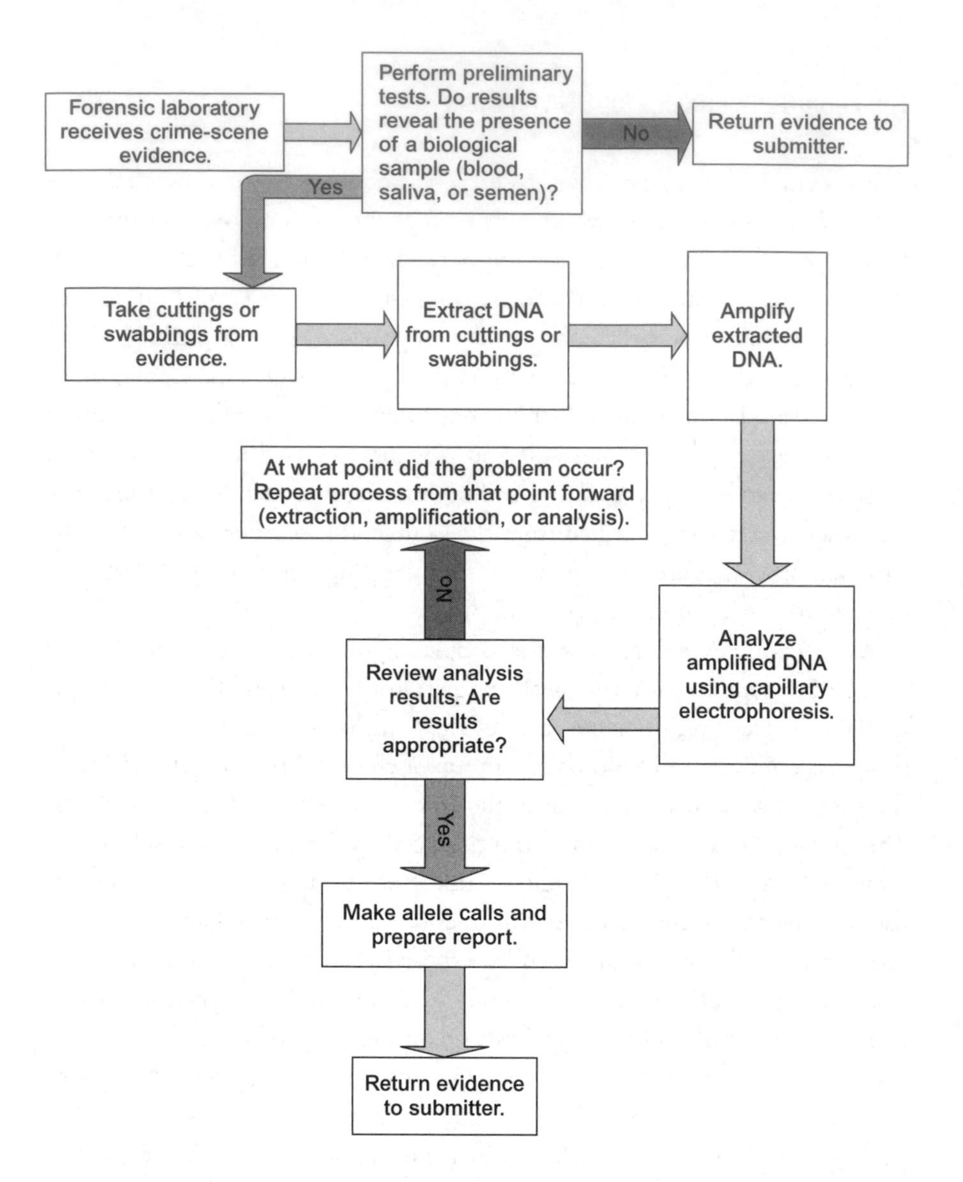

FIGURE 1.8. Flow diagram of how DNA evidence is handled in a forensic laboratory. *Source*: U.S. Department of Justice, http://www.justice.gov/oig/special/0405/chapter2.htm (accessed May 23, 2010).

to the peaks. The heights of the peaks represent the amount of DNA that was amplified for each allele segment (see figure 1.9), which is related to the amount of evidentiary DNA used for the analysis.

If the DNA comes from a single individual and is not degraded, two peaks should appear for each locus where the individual has inherited different alleles from each of his or her parents (heterozygous for that locus). In cases where an individual has inherited the same allele from each of his or her parents (homozygous for that locus) only one peak will appear (figure 1.9). If the DNA consists of a mixture of two individuals, there can be anywhere from one to four peaks at a particular locus, significantly complicating the picture. For example, if both individuals are heterozygous at a particular locus and have the same STR alleles at that locus, only two peaks will appear; if the two heterozygous individuals have no STR alleles in common, then four peaks will be observed; and if they share the same STRs for just one allele, then three peaks will be observed.

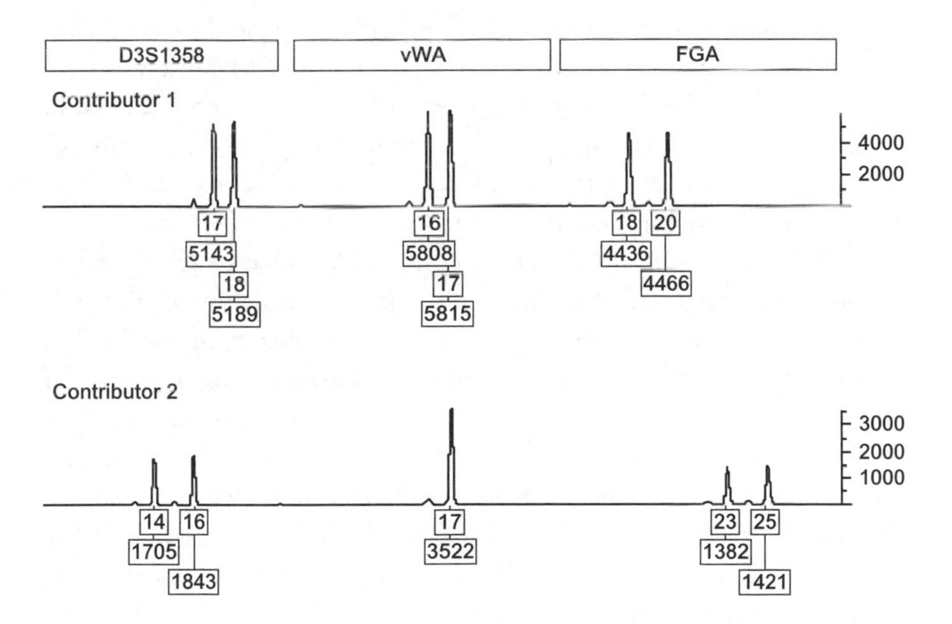

FIGURE 1.9. The output from a forensic DNA analyzer (called an electropherogram) showing the STR alleles for two contributors at three loci. The top number reports the number of STR repeats for the allele. The bottom number is a measure of the height of the peak (in RFUs), which represents the quantity of DNA (vertical axis). *Source*: Dan Krane.

The position of the peaks on the x-axis of the electropherogram indicates how long it takes the allele to pass through a capillary tube, which correlates with the length of the DNA segment, and therefore indicates the number of repeats. The height of the peaks in "relative fluorescence units" (RFUs) measures the amount of DNA present in the sample. Peaks from all the alleles in a profile coming from the same individual are expected to have about the same heights. The computer-generated numbers under each peak tell us how many repeats there are in the sample at that locus and the height of the peak relative to a baseline.

Quality-Control Procedures

The machines are calibrated and the reliability of the overall assay is monitored by analysis of samples of known DNA type as a control standard (like an official measuring rod). In all accredited forensic laboratories and others that follow good scientific practice, technicians are instructed to include a sample of known DNA sequence in every analysis sample they undertake. Any analysis where systematic errors have occurred (affected possibly by temperature, pressure, vibrations, or machine malfunctions) will be reflected in the control sample (see chapter 16). In addition, every step in the process, including technician interpretations, is required to be documented for purposes of troubleshooting and replication.

For example, the New York City forensic DNA lab has designed a number of control procedures that aim to minimize the possibility of false positives (false matches) and false negatives (false exclusions). The facility has an isolated laboratory that tests all reagents for purity and for any traces of DNA. The reagents are put through a DNA analyzer to validate that they are DNA free. Every sample of DNA that is put through the DNA analyzer is required to be checked against a DNA sample that has a known STR profile. If the profile of the quality-control standard does not match its expected values, the run is invalidated. There are strict protocols for how evidence moves through the different stages in the testing regime: The people who initially prepare the evidence for testing are different from those who do the extractions or place the sample in a DNA analyzer. The stages in the DNA analysis are segmented by physical spaces and are also separated by personnel who conduct different tasks in the process. In addition, those who handle the evidence during extraction, the PCR, or DNA analysis are required to be suited up from head to toe in protective garments designed to prevent cross-contamination of DNA. No one whose DNA is not on file is allowed to enter the high-security laboratory. In case there is a breach in the protective outerwear, the contaminant DNA from the technician can be traced.

Notwithstanding the strict quality-control procedures adopted by some laboratories and the crisp-looking output of electropherograms, there is still room for judgment calls in reading the output of the analyzer. In the words of William Thompson and colleagues, "Although many technical artifacts are clearly identifiable, standards for determining whether a peak is a true peak or a technical artifact are often rather subjective, leaving room for disagreement among experts."[10]

It is the policy of the FBI's forensic DNA laboratory that it will not report out a DNA profile with fewer than 10 complete STR markers. The New York City lab's policy is less stringent; it will accept 6 or more good STR markers. Also, the New York City lab's policy is that it will not enter a single peak at a locus (representing one allele) if the second peak at the locus (representing a second allele) is known to have been badly degraded and not found in the output. The computer program has a peak (height) threshold for scoring alleles. If the peak height is below 75 RFU, it is not reported. This avoids giving the forensic DNA analyst some discretion in calling the peak real or spurious. By the time it reaches 75 RFU, the peak is considered unambiguous. Other laboratories may use a different threshold below which peaks are ignored. Erin Murphy observes: "The FBI Protocol allows only peaks over 200 relative fluorescence units (RFU) to be considered conclusive for match purposes, though it recommends interpretations of any peak over 50 RFU including for exculpatory purposes."[11] Thus the FBI has ostensibly adopted 50 RFU as the detection threshold.

DNA analysis for use in criminal investigation is a highly complex process that involves collecting biological samples from individuals and crime scenes, extracting DNA from those samples, using the PCR to amplify the amount of DNA from subanalytical levels to analytical levels along noncoding regions of the DNA, and then using some method to determine the genetic variation of the sample. The result of this process is a DNA profile that can be stored electronically and used to identify an individual or to link that individual to a crime or exclude him or her from it by comparing that profile with others.

A new generation of DNA analyzers, genetic typing techniques, and forensic laboratory protocols has dramatically increased the efficiency of DNA testing and has helped address the most obvious sources of error of the first generation of DNA testing systems. Separation of samples, isolation of functions, proficiency testing, and protocols for interpreting mixed samples are among some of the quality-control innovations practiced, in theory, by all accredited laboratories. The shift from the autoradiograph with the dark bands generated by RFLP and gel electrophoresis to the PCR and the electropherogram has also reduced the likelihood of reading errors.

Nonetheless, no technology is errorproof, and DNA analysis is no exception to this rule. Artifacts in the output can be produced by the machine because of imperfections or environmental effects. Human error can also occur. Sample contamination, misinterpretations or mischaracterizations of data, and the reporting of inaccurate or misleading statistics are all sources of human error that can and have occurred in DNA cases, sometimes resulting in the wrongful conviction of an innocent person. The potential for error and abuse was made painfully clear in the case of the infamous Houston Police Department Crime Laboratory, where a final report by an independent investigator revealed that, of a total of 135 DNA cases reviewed, major issues were identified in 43—or approximately 32 percent—of those cases, including four death penalty cases. The report found that:

> The Crime Lab's historical DNA casework reflects a wide range of serious problems ranging from poor documentation to serious analytical and interpretive errors that resulted in highly questionable results being reported by the Lab. The profound weaknesses and flawed practices that were prevalent in the Crime Lab's DNA work include the absence of a quality assurance program, inadequately trained analysts, poor analytical technique, incorrect interpretations of data, the characterizing of results as "inconclusive" when that was not the case, and the lack of meaningful and competent technical reviews. Furthermore, the potential for the Crime Lab's analysis of biological evidence to result in a miscarriage of justice was amplified exponentially by the Lab's reported conclusions, frequently accompanied by inaccurate and misleading statistics that often suggested a strength of association between a

> suspect and the evidence that simply was not supported by the analyst's actual DNA results.[12]

The fallibility of DNA testing will be discussed in more detail in chapter 16.

Fallible or not, DNA is here to stay. Few would argue with Simon Cole's statement about the stature of forensic DNA in criminal justice: "New technology is changing the administration of criminal justice. Among the most prominent of such changes is the development of forensic DNA technology, which includes a forensic assay with potentially enormous discrimination and sensitivity and the development of large databases based on that assay."[13] However, as stated in the book *Truth Machine*, "This seemingly unassailable, transcendent form of criminal evidence remains bound up with stories that are infused with contingent judgments about the mundane meaning and significance of evidence."[14]

Chapter 2

The Network of U.S. DNA Data Banks

The purpose of DNA databases is to solve crimes that would otherwise be unsolvable.

—John M. Butler[1]

The power to assemble a permanent national DNA database of all offenders who have committed any of the crimes listed [in 18 U.S.C.] has catastrophic potential. If placed in the hands of an administration that chooses to "exalt order at the cost of liberty" . . . the database could be used to repress dissent or, quite literally, to eliminate political opposition.

—Judge Stephen R. Reinhardt in
United States v. Kincade (2004)[2]

In their original conception, forensic DNA data banks were designed to hold DNA profiles of violent felons and recidivist sex offenders. Because of their initial successes, over the last 15 years there has been an inexorable drive among law-enforcement agencies to push their use to their limit, stopping just short, at least for the present, of including every person's DNA in a national network.

By the new millennium every state law-enforcement agency and many local ones in the United States had become connected through a coordinated and federally managed forensic DNA data bank. At the heart of this system is a computer network overseen by the Federal Bureau of Investigation called the Combined DNA Index System, or CODIS. This network provides police with tools to solve "cold cases," that is, cases that have been dormant because of lack of leads, to link a current suspect to a crime scene, and to track suspects who may seek to change their identity.

The Origins and Workings of the Combined DNA Index System (CODIS)

CODIS was initiated in 1990 as a pilot program serving 14 state and local laboratories. The main idea behind the program was to collect, analyze, and store the DNA of individuals convicted of felony sex offenses and other violent crimes on the theory that these offenders were likely to be recidivists and frequently leave biological evidence at the scene of a crime.[3] By maintaining their DNA on file in a shared database system, police could develop investigative leads and solve crimes that might have been committed in the past or would be committed in the future by the same individual both within and across jurisdictions.

In 1994 Congress passed the DNA Identification Act as part of the Violent Crime Control and Law Enforcement Act (P.L. 103-322). The law called for the establishment of a forensic laboratory under the authority of the FBI with the capability to analyze DNA for identification purposes. In addition, it authorized the FBI to establish and maintain the CODIS software system, which would allow the sharing of DNA profiles uploaded at local, state, and federal levels.

By June 1998 all 50 states had authorized criminal DNA databases, and by 2004 all 50 state databases were connected through CODIS. The FBI officially launched CODIS on October 13, 1998, three years after England and Wales introduced their national forensic DNA databases (see chapter 9).

CODIS contains separate indexes of DNA profiles generated from four different sources: known persons convicted of crimes (the Convicted Offender Index); DNA samples whose source is unknown and that were recovered from crime scenes (the Forensic Index); DNA samples recovered from unidentified persons who may have been killed in a natural hazard or through criminal activity (the Unidentified Human Remains Index); and DNA samples voluntarily contributed by relatives of missing persons (the Relatives of Missing Persons Index). In this last case, if the remains of someone are found, his or her DNA profile can be submitted to the Relatives of Missing Persons Index in the hope that his or her identity can be ascertained because close relatives share similar DNA alleles. CODIS also includes a population file of anonymous contributors, whose DNA is used to determine the frequency of alleles in the short tandem repeats that are found in the general population or in racial and ethnic subpopulations.

The DNA Identification Act contained a set of privacy provisions applying to federal law-enforcement agencies regarding who has access to the DNA profile information. According to the act, the results of the DNA tests performed for law-enforcement purposes can be disclosed under only three conditions: to criminal justice agencies for law-enforcement identification purposes; in judicial proceedings; and for criminal defense purposes. In the last case defendants can have access to the information for exculpatory purposes. There is also a provision in the act for research on forensic DNA population databases to develop protocols and for quality-control purposes if personally identifiable information is removed.

Once CODIS was under way, state laws began to individuate their criteria for including DNA profiles in their data banks. Those criteria range from sex offenders to all convicted felons, and more recently to those convicted of white-collar crimes, juvenile offenders who are not legally considered felons, and arrestees, whether or not they are convicted. In 2001 the FBI's official recommendation to the states was that they include "all felony offenders and misdemeanor sex offenders within the scope of their database laws."[4]

Although states are allowed to set their own laws governing DNA collection, federal law determines the categories of DNA profiles that can be uploaded to the shared database system. In addition, each state has an FBI-approved laboratory that serves as the master site for all local and county laboratories that generate DNA profiles. All State and Local DNA Index Systems (SDIS and LDIS, respectively) are connected through the National DNA Index System (NDIS) (see figure 2.1). The NDIS is a system of DNA profile records compiled by criminal justice agencies (including federal, state, and local law enforcement agencies). CODIS is the automated DNA information processing and telecommunication system that supports NDIS. Sometimes CODIS is used interchangeably with NDIS.

The FBI also sets the standards that states must follow to upload DNA profiles and to run searches against the NDIS. For a state to upload a DNA profile to NDIS, it must contain at least 10 out of 13 loci. If DNA collected from a crime scene is too badly degraded or if mixtures are too complex to extract 10 readable loci, the NDIS will not accept the profile.

Anticipating a law establishing a national forensic DNA system, in 1991 the FBI issued *Legislative Guidelines for DNA Databases* as guidance to the

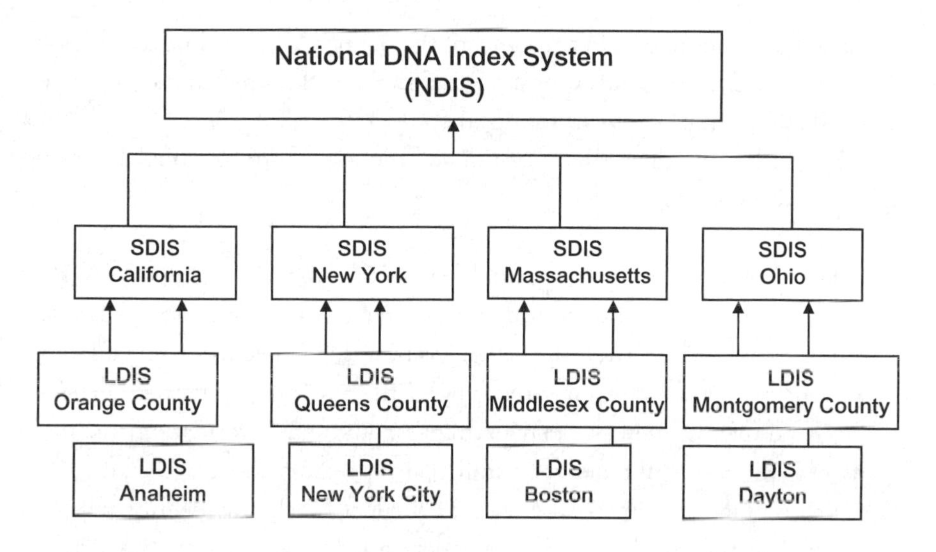

FIGURE 2.1. Three tiers of networks forming CODIS. The four-level structure of the National DNA Index System, federal, state, county, and local police agencies, illustrated by four states. SDIS=State DNA Index System; LDIS=Local DNA Index System. *Source*: Authors.

states for developing their statutes.[5] The guidelines included some limited privacy and civil liberties considerations. For example, the FBI recommended that only DNA records that relate to the identification of individuals, not records about physical characteristics, traits, or predispositions for diseases, should be collected. Personal information stored in the state DNA database, according to the guidelines, should be limited to the data necessary to generate investigative leads and to support a statistical interpretation of any test results. Finally, the guidelines recommended that access to the state DNA database system should be limited to duly constituted federal, state, and local law-enforcement agencies through their forensic laboratories.[6]

Currently the DNA profiles stored in the NDIS contain a specimen identifier, the sponsoring laboratory's identifier, the initials or names of the DNA personnel associated with the analysis, and the actual DNA profile characteristics. The NDIS does not store criminal history information, case-related information, Social Security numbers, or dates of birth. No personal identifying information other than a specimen identification number is stored in the NDIS. If a match (or association) is identified and later confirmed, a public forensic laboratory must initiate contact with other laboratories involved in the match in order to obtain the name of the offender.[7] Unauthorized disclosure of DNA information is subject to criminal penalties not to exceed a fine of $250,000 or imprisonment for up to one year.

John Butler describes the operational function of the national forensic DNA network as follows:

> When CODIS identifies a potential match, the laboratories responsible for the matching profiles are notified and they contact each other to validate or refuse the match. . . . After the match has been confirmed by qualified DNA analysis, which often involves retesting of the matching convicted offender DNA sample, laboratories may exchange additional information, such as names and phone numbers of criminal investigators and case details. If a match is obtained with the Convicted Offender Index, the identity and the location of the convicted offender is determined and an arrest warrant is procured.[8]

Federal Expansion of DNA Collection

Federal law stipulates which DNA profiles processed at state and local levels can be uploaded to CODIS. It also determines the categories of individuals from whom DNA can be collected and retained by federal authorities and uploaded to the NDIS ("qualifying Federal offenses"). As explained earlier, CODIS was initially limited to persons convicted of serious, violent crimes (namely, murder and felony sex offenses). Similarly, states only uploaded to CODIS DNA profiles taken from this narrow category of offenders, on the basis of guidelines issued by the FBI, which recommended that "states include all felony offenders and misdemeanor sex offenders within the scope of their database laws."[9]

Since CODIS was established in 1994, the federal law has been significantly expanded through the passage of three pieces of legislation (see table 2.1). First, the federal collection program was initiated through the passage of the DNA Analysis Backlog Elimination Act of 2000. This law required that DNA samples be collected from individuals in custody and those on probation, parole, or supervised release after being convicted of a "qualifying Federal offense."[10] Qualifying Federal offenses were limited to violent crimes and included murder, sexual abuse, peonage or slavery, kidnapping, and

TABLE 2.1 U.S. Legislative Authority for the Expansion of the DNA Data Banks

Legislation	Year	Persons subject to DNA collection	Regulations
DNA Identification Act, P.L. 103-322 (Title XXI, Subtitle C)	1994	Authorized the director of the FBI to establish an index of: (1) DNA identification records of persons convicted of crimes; (2) analyses of DNA samples recovered from crime scenes; and (3) analyses of DNA samples recovered from unidentified human remains.	61 FR 37495 (July 18, 1996)
DNA Analysis Backlog Elimination Act, P.L. 106-546	2000	Required collection of DNA from individuals in custody and those on probation, parole, or supervised release who have been convicted of any "qualifying Federal offense." Those include: murder or voluntary manslaughter; sexual abuse or sexual exploitation or other abuse of children; an offense related to peonage and slavery; kidnapping; and an offense involving robbery or burglary.	66 FR 34363 (June 28, 2001)
USA Patriot Act, P.L. 107-56 (Sec. 503)	2001	Expanded "qualifying Federal offenses" to include any Federal crime of terrorism (listed in U.S.C. 2332b(g)(5)(B) of Title 18) as well as "any crime of violence" and "any attempt or conspiracy to commit any of the above offenses."	68 FR 74855 (December 29, 2003)
Justice for All Act, P.L. 108-405, Sec. 203(b)	2004	Expanded "qualifying Federal offenses" to include any felony; any sex-abuse crime (under Chapter 109A or Title 18); any crime of violence; and any attempt to commit any of the above.	Interim Rule, 70 FR 4763 (January 31, 2005)
Violence Against Women Act, P.L. 109-162 (Sec. 1004)	2005	Authorized the attorney general to require, through regulation, DNA collection from anyone arrested or non-U.S. persons detained under federal authority.	73 FR 74932 (December 10, 2008)

Source: Authors.

offenses related to robbery or burglary. Although the federal collection program was limited to violent crimes, it allowed for the collection of DNA from those who had already served their sentences but were still in the system by way of being on probation or parole. The 2000 act also spoke to the conditions state law-enforcement authorities must meet in order to upload a DNA profile to CODIS. Specifically, DNA profiles could be submitted to CODIS when they were "taken from individuals convicted of a qualifying state offense."[11] So while federal authorities continued to collect DNA only from those convicted of serious, violent crimes, states were given the go-ahead to authorize the collection and uploading to CODIS of other categories of convicted individuals.

On October 3, 2004, President George W. Bush signed into law the Justice for All Act. Although the main purpose of this law was to provide funding to help states eliminate a substantial backlog of DNA samples collected from crime scenes and offenders, the law also significantly expanded CODIS. "Qualifying Federal offenses" were expanded to allow for the DNA testing of all persons convicted of a felony offense. In addition, the act allowed states to upload to CODIS the DNA profiles of "persons convicted of crimes" or "persons charged in an indictment or information with a crime," as well as "other persons whose DNA samples are collected under applicable legal authorities, provided that DNA profiles from arrestees who have not been charged in an indictment or information with a crime, and DNA samples that are voluntarily submitted solely for elimination purposes shall not be included in the National DNA Index System."[12]

By November 2005 CODIS contained 124,200 forensic profiles and 2.8 million offender profiles. On that date, according to the FBI's own account, CODIS had produced 27,700 matches and assisted 29,600 investigations.[13]

On January 5, 2006, President George W. Bush signed into law the Violence Against Women and Department of Justice Reauthorization Act of 2005. Attached to this broadly popular budgetary reauthorization was the DNA Fingerprint Act of 2005. With only 99 lines of text, this amendment, introduced initially by Senator Jon Kyl (R-AZ), marked a quantum leap in the power of police to collect DNA. The act authorized the U.S. attorney general to direct federal agencies to collect DNA from "individuals who are arrested or from non-U.S. persons who are detained under the authority of the United States."[14] In addition, the law allowed states to upload DNA profiles to CODIS from anyone whose DNA samples were collected under applicable legal authorities, so long as they were not voluntarily submitted. Therefore, it removed the requirement that prohibited the uploading of DNA profiles from arrestees who had not been charged. This change opened the floodgates for the states to submit names to CODIS of arrestees or detainees even if they were not charged with a crime. There is an expungement provision in the law that allows people who are arrested or detained but not charged or convicted to have their profiles removed from CODIS. Expungement is not automatic, however, and occurs only where the director of the FBI receives a certified copy of a final court order establishing that the charges were dismissed or resulted in acquittal. Therefore, the burden is on the individual arrested to request removal of his or her DNA from the system. Individuals who were convicted of a crime and later had their conviction overturned can also request that their DNA profiles be removed. Despite proper authorization, if the FBI fails to remove a profile, a subsequent cold match to it has been accepted in the courts so long as the FBI's failure to remove it was unintentional.

On December 10, 2008, the U.S. Department of Justice (DOJ) issued a final rule to implement the DNA Fingerprint Act of 2005.[15] The DOJ interpreted Congress's loosely framed mandate in the broadest way possible. Under the final rule, any federal agency with authority to take fingerprints ostensibly was given comparable authority to collect DNA. The proposed rule states quite unambiguously: "Agencies of the United States that arrest or detain individuals or supervise individuals facing charges will be required to collect DNA samples, if they collect fingerprints from individuals, subject to any limitations or exceptions the Attorney General may approve."[16] In addition, the regulation appears to allow the forcible taking of DNA even from those arrested for misdemeanors, such as trespassing on federal land during a demonstration.

The DOJ laid out a three-part justification for its broad interpretation of the law. First, by collecting DNA samples at the time of arrest or at some other early stage in the criminal justice process, the DOJ stated, it can deter subsequent criminal behavior. This deterrence function could be lost if law enforcement waits for conviction to collect DNA.[17] Second, the expanded database of arrestees "may show that the arrestee's DNA matches DNA in crime scene evidence" in crimes for which they were not arrested.[18] Third, the DNA sample may provide "an alternative means of directly ascertaining or verifying an arrestee's identity, where fingerprint records are unavailable, incomplete or inconclusive."[19] This so-called triple benefit, namely, criminal deterrence, crime solving, and true identity, provided the DOJ its justification for broadening the reach of federal authorities into the privacy of individuals who were not convicted of a felony crime.

The 2008 rule went into effect in January 2009 and applies to any federal agency that arrests or detains individuals or supervises individuals facing charges. The adoption of the final rule served as a signal to state legislatures that, like the federal government, they too could expand the criteria for inclusion in their DNA data banks by including arrestees, whether or not they were charged with or convicted of a crime. Before this ruling 13 states had expanded their DNA collection to arrestees. With a trend toward forensic DNA harmonization, a number of states, such as Missouri, Washington, New York, Oregon, and Vermont, which had chosen not to pass legislation expanding DNA collection to arrestees, have started to consider otherwise.

The DOJ estimates that under its new rule more than 1.2 million additional individuals will have their DNA involuntarily collected by multiple federal agencies, profiled, and maintained in CODIS each year. This represents a fifteenfold increase in the number of DNA samples that have been collected from federal offenders.[20] The full privacy and racial justice implications of the expansion of DNA databases to arrestees, as well as the question whether expanding databases to arrestees helps solve crimes, are discussed at length in part III of this book.

The 2008 federal rule on DNA collection by federal law-enforcement agencies is the first official government document to assert the exact analogy and the synergy between fingerprints and DNA profiles. It states that "the uses of DNA for law enforcement identification purposes are similar in general character to the uses of fingerprints, and those uses will be greatly enhanced as a practical matter if DNA is collected regularly in addition to fingerprints."[21] The government's decision to make DNA collection and fingerprints an exact analogy obviates any need to include any provisions for removing samples from the database that may have been obtained without court warrant, or without reasonable suspicion of a crime or illegal entry into the country, because fingerprints are viewed as a means of identification. Moreover, the rule permits agencies to use "such means as are reasonably necessary to detain, restrain, and collect a DNA sample from an individual . . . who refuses to cooperate in the collection of the sample."[22] In its comments on the proposed rule, the American Civil Liberties Union (ACLU) noted:

> The fingerprint analogy is misleading, because perhaps the most significant privacy concerns with DNA data banking are associated not with the DNA *profiles* that are retained electronically, but instead with the original *biological samples* that are stored indefinitely by forensic laboratories. Unlike fingerprints—two-dimensional representations of the physical attributes of our fingertips that can be used only for identification—DNA samples can provide insights into disease predisposition, physical attributes and ancestry.[23]

The full privacy implications of DNA testing for law-enforcement purposes are discussed in chapter 14.

Another significant change in the DNA database system occurred in 2006 and without congressional authorization. Until July 2006 the federal overseer of NDIS would not release any identifier information about a profile unless the match between the source DNA profile and the profile in NDIS was exact, except in cases where any discrepancy between the two profiles could be explained by degradation or mixtures that could compromise the crime-scene profile. On July 20, 2006, the FBI issued an interim plan for the release of information in the event of a "partial match" of a DNA profile. The memorandum stating this interim plan defined a partial match as "a moderate stringency candidate match between two single source profiles having at each locus [there are 13 loci] at least one allele in common."[24] This means that a match of 13 out of 26 alleles (one at each locus) can trigger an investigation of the partially matched individual. The purpose of this policy was for state law-enforcement officials to follow DNA leads to the family members of suspects whose DNA profile did not match exactly but had some common properties with the crime-scene evidence. The interim plan stated:

> With the documented concurrence of the prosecutor, the Casework Laboratory that identifies a "partial match" shall provide the NDIS Custodian with a written request for the release of the offender's identifying information. Such written request should include the statistical analysis used to conclude that there may be a potential limited relationship between the suspected perpetrator and the offender. Each request will be reviewed by the F.B.I.'s Office of the General Counsel and the NDIS Custodian for approval.[25]

This change in FBI policy has encouraged a number of states to consider opening their databases to varying degrees of "familial searching." Proposals to conduct full-scale searches in the database for possible relatives of the crime-scene source represent an entirely new use of forensic DNA database and a de facto expansion of the database to all close relatives (e.g., parents, children, and siblings) of the individuals in the database. Familial searching is discussed in detail in chapter 4.

State DNA Database Expansion

The Commonwealth of Virginia was the first state in the United States to develop a forensic DNA data bank in 1989. By the close of the next decade all 50 states had statutes authorizing the creation of DNA data banks. Initially, as with the federal data bank, state data banks included samples from felony offenders with allegedly high recidivism rates, such as sex offenders and violent felons. But state legislators, riding the wave of DNA-philia and taking advantage of the silent encouragement offered by changes in federal law, began to expand the criteria for inclusion soon after their database systems were up and running. Some justified the expansion by arguing that persons who commit lesser crimes have high recidivist rates and graduate to more violent offenses.

Table 2.2 shows how the qualifying offenses for which DNA can be collected and retained vary among the states. As of June 2009, 47 states had authorized DNA collection from all felons (all states except for Idaho, Nebraska, and New Hampshire); 37 from at least some categories of individuals who have been convicted of a misdemeanor; 4 from numerous non-sex-crime misdemeanors; and 32 from juvenile offenders. Included under "misdemeanors" in many states are voyeurism (Florida), failure to register as a convicted person (Nevada), many types of sexual offenses (Arkansas, Illinois), and controlled-substance offenses (New York, Nevada, and Kansas). According to one survey, burglary by itself was a justification for DNA collection in 40 states, and drug offenses authorize DNA collection in 28.[26]

Juveniles in America have increasingly become part of the DNA-collection system. By June 2009, 32 states had laws permitting police to obtain biological specimens from juveniles for DNA profiles, including those who had committed nonviolent crimes. Some legal scholars argue that juveniles should be treated differently than adults with respect to DNA collection. Juveniles are often protected by sealing their criminal records. There is a

TABLE 2.2 State Variations in DNA-Collection Statutes

Qualifying offenses	Number of states[a]
All convicted felons	47
Misdemeanors	37
Juvenile adjudications	32
Arrestees	21[b]
Postconviction DNA testing	38[c]
Other provisions	
Expungement criteria and procedures	38[d]
Destruction of offender samples after DNA profile analysis	1 (Wisconsin)[c]
Provisions for handling voluntary samples	17[f]
Criminal penalties for improper disclosure of DNA samples and records	40[c]
Allows database use for humanitarian purposes	13[e]

[a] State qualifying offenses as recorded by DNAResource.com (November 2008), http://www.dnaresource.com/documents/statequalifyingoffenses2008.pdf.
[b] Includes Minnesota legislation, which was subsequently overturned by the Minnesota Court of Appeals. See In re Welfare of C.T.L. Minn. App., October 10, 2006, http://caselaw.findlaw.com/mn-court-of-appeals/1297648.html (accessed May 23, 2010) and http://www.aslme.org/dna_04/grid/statute_grid.html (accessed May 23, 2010).
[c] American Society of Law, Medicine and Ethics, http://www.aslme.org/dna_04/grid/statute_grid_4_52006.html (accessed August 7, 2008).
[d] Seth Axelrod, "Survey of State DNA Database Statutes" (2004), a report for the American Society of Law, Medicine and Ethics (ASLME). Forty states include provisions for expungement of profiles for individuals who have been convicted of a qualifying offense but have successfully appealed. Thirty-four states require that the offender initiate the process.
[e] "The United States and the Development of DNA Data Banks," *Privacy International*, February 20, 2006, http://www.privacyinternational.org/article.shtml?cmd%5B347%5D=x-347-528471 (accessed March 30, 2010).
[f] National Conference of State Legislatures, "State Laws on DNA Data Banks, Qualifying Offenses, Others Who Must Provide a Sample," February 2010, http://www.ncsl.org/default.aspx?tabid=12737 (accessed April 17, 2010).
Source: Authors.

greater emphasis on rehabilitation and less on punishment for youthful offenders. Steven Messner writes in the *Journal of Juvenile Law*: "A national criminal DNA database, which does not distinguish between adult and juvenile, could undermine the purpose of the separate juvenile justice system," for example, if the DNA profile were maintained after the juvenile records were sealed. "In order to preserve the purposes and integrity of the juvenile justice system, additional protections need to be afforded juveniles whose DNA profiles are obtained. . . . A juvenile's DNA profile should be expunged from the database if the juvenile's record is sealed."[27]

In 2003 only 4 states took DNA from arrestees whether or not they were convicted of a crime. Six years later an additional 17 states had passed legislation that authorized the collection of DNA from arrestees. Some states have authorized collection of DNA from all felons, while others have authorized more limited arrestee collection, such as from individuals arrested for murder or rape. One of these, Minnesota's statute, was declared unconstitutional by the Minnesota Court of Appeals (see table 2.3). Although the FBI continues to refer to its "Convicted Offender Index," this is no longer accurate, since this index necessarily includes individuals who have not been convicted of crimes.

TABLE 2.3 State Legislation Expanding DNA Collection to Arrestees, 2001–2008

State	Year	Arrestees included
Alaska	2007	Anyone arrested for violent felonies and domestic abuse.
Arizona	2007	Anyone arrested for homicide manslaughter, first- or second-degree murder, indecent exposure, public sexual indecency, sexual abuse, sexual contact with a minor, sexual assault, child molestation, bestiality, first- or second-degree burglary, keeping or residing in a house of prostitution, prostitution, portraying an adult as a minor, incest, or soliciting abortion through a weapon.
California	2004	All adults arrested for murder or rape; starting in 2009, all adults arrested for any felony offense.
Kansas	2006	Arrests for any felony or drug crime of severity levels 1 or 2.
Louisiana	2003	Persons arrested for a felony or other specified offense (includes battery, unlawful use of a laser on a police officer, simple assault, assault on a schoolteacher, stalking, misdemeanor carnal knowledge of a juvenile, prostitution, soliciting for prostitutes, prostitution by massage, letting premises for prostitution, and Peeping Tom offenses). Includes juveniles.
Maryland	2008	Arrests for violent crimes, burglary, and breaking and entering of a motor vehicle.
New Mexico	2006	All adults arrested for certain felony offenses.
North Dakota	2007	Any adult arrested for a felony crime. Effective January 2009.
South Dakota	2008	Any person arrested for a felony offense punishable by five years or more.
Tennessee	2007	Any person arrested for a violent felony.
Texas	2001	Individuals indicted for certain sex crimes, certain crimes against children, and burglary.
Virginia	2002	Persons arrested for commission or attempted commission of a violent felony.

Source: Authors.

Virginia was the first state to authorize DNA collection from arrestees. In 2002 the Virginia state legislature enacted an amendment to its state data bank law allowing the seizure and analysis of DNA samples from "every person arrested for the commission of a violent felony." Arrestees are not necessarily charged with or convicted of a crime. False arrests or charges dropped are commonplace in police work, sometimes because of lack of evidence, because of mistaken identities, or because the police simply get it wrong. For example, arrests are made during the tumult of mass demonstrations because people find themselves in the middle of a rowdy crowd or in the sight line of overzealous police.

Louisiana was also among the first states to broaden DNA collection to arrestees. In 2003 Louisiana enacted legislation that expanded its DNA data-bank criteria to include simple assault, stalking, prostitution, and Peeping Tom offenses. Under this law juveniles arrested for these offenses, as well as adults soliciting prostitution, are subject to having their DNA included in the state data bank.

In 2004 California became the fourth state to authorize DNA collection from arrestees through passage of Proposition 69, also known as the DNA Fingerprint, Unsolved Crime, and Innocence Protection Act. This ballot proposition mandated the collection of DNA samples from adults and juveniles convicted of felonies, sex offenses, and arson. In addition, it set January 2009 as the date for requiring a DNA sample from every adult arrested for or charged with a felony. At the time of its passage, California's DNA law contained the most sweeping DNA collection requirements in the country. Its DNA database is currently the third largest in the world, and, with more than 330,000 individuals arrested for a felony offense every year in California,[28] it promises to continue to retain that position so long as the arrestee provisions remain in effect. In October 2009 the ACLU of Northern California filed a class-action lawsuit, challenging the arrestee provisions of the law.[29] The case is ongoing at the time of this writing.

As of June 2009, 21 states had provisions for collecting DNA from some categories of arrestees. Eleven states followed California's lead by enacting provisions for taking DNA from all felony arrestees, while the other 10 states adopted a more limited approach (for example, allowing arrestee DNA collection only for those arrested for murder and sex crimes, and in some cases burglary).

The extraordinary expansion of DNA collection has resulted in a near doubling of the number of so-called offender profiles in CODIS every two years, as can be seen in figure 2.2. The impetus for expanding the scope of contributors to DNA data banks is coming almost exclusively from law enforcement. The assumption has been that the larger the data banks, the more crimes will be solved. Rarely, however, is a crime solved by DNA alone. At any crime scene there may be multiple DNA fragments left on the scene by passersby. Also, the fact that some suspect's DNA was not found at a crime scene does not mean that the suspect was not at the scene or did not commit the crime. Nevertheless, the successes of some highly publicized cold-hit cases of DNA collected from a crime scene have inspired lawmakers to broaden the inclusion criteria for taking DNA samples.

Not all attempts to expand DNA databases have been successful, especially in the highly contested arena of arrestee testing. In 2007 both houses of the South Carolina legislature passed a bill requiring that law-enforcement authorities collect DNA samples from anyone arrested or indicted for any crime that is punishable by a prison sentence of five years or more.[30] Governor Mark Sanford vetoed the bill on June 18, 2007. In his veto Governor Sanford argued that this bill to expand the state's DNA data bank was an overreach by government and an erosion of personal privacy. In explaining his veto, Sanford wrote:[31]

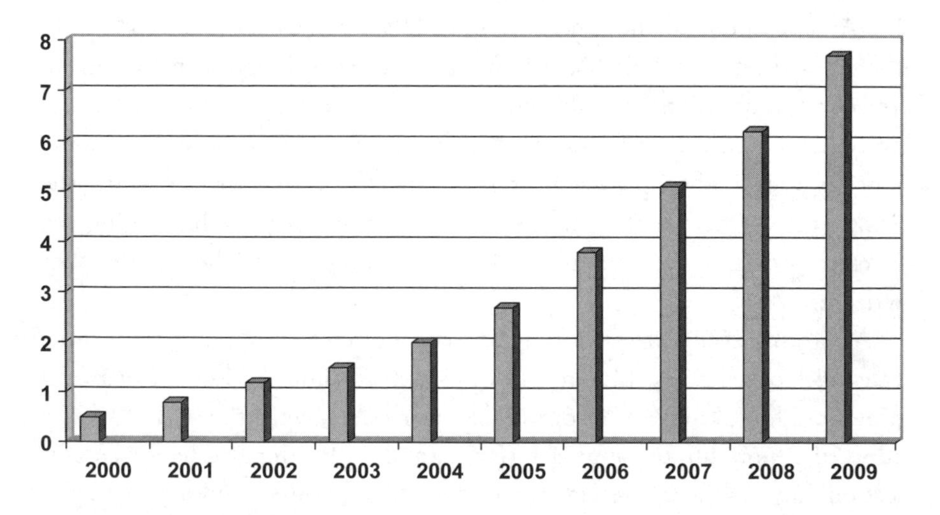

FIGURE 2.2. Growth of CODIS Offender Profiles, 2000–2009 (in millions). *Source:* Authors.

> Though American society values personal liberties, we are the first to recognize that persons *convicted* of a crime must give up some of those liberties, including the protection against search and seizure. By limiting DNA collection to those who have been convicted of a crime, we ensure that no DNA is collected unless that person has been granted due process of rights and has experienced a full vetting by the judicial system. If DNA collection were expanded to include custodial arrest for felonies, why stop there? Law enforcement could inevitably stop other crimes as well with an even further expanded database. We think the clear divide created with conviction has served us well because one of the central tenets of American law is that one is presumed innocent until proven guilty.

Sanford emphasized that DNA samples contain sensitive personal information such as disease predisposition, physical attributes, ancestry, and familial relationships. Only when there is a court order or conviction, he argued, should people be required to submit their DNA to a state database. Moreover, by weakening the standard under which DNA is collected, the state would be crossing the threshold from using personal information for criminal investigation to surveillance of its citizenry.

There have also been voluntary submissions of DNA to state data banks. These have come about through law-enforcement dragnets. When DNA left at a scene of a violent crime does not match any profile in the state or federal data bank, law-enforcement officers have on occasion asked local residents to submit their DNA voluntarily to eliminate themselves as suspects. Their DNA can remain in the data bank unless the state has a provision for removal of the profile and the contributors avail themselves of that right.

In what surely will be seen as an ironic state of affairs, on the one hand, states are expanding their DNA collections by broadening criteria to include innocent people; on the other hand, they are destroying crime-scene DNA evidence of those currently serving prison sentences. According to a report in *USA Today*, half the states do not have requirements for retaining crime-scene DNA evidence that either played a role in convicting or could play a role in exonerating an incarcerated felon.[32] States incur heavy costs in laboratory sequencing and in storage facilities for preserving biological samples when they expand their databases to include arrestees. The expansion of the databases is purportedly to catch future criminals. Law enforcement is dubious about the social benefits over costs for retaining crime-scene DNA for as long as a felon is in prison. The focus is on resources for obtaining new convictions rather than on protecting and preserving evidence that could be used to exonerate falsely convicted individuals. Bucking this trend is the state of Virginia. In the summer of 2008 the Virginia Forensic Science Board planned to send out letters to about 1,000 felons convicted in the 1970s and 1980s before DNA evidence was used in criminal cases to notify them that physical evidence pertaining to their case existed. On November 1, 556 letters went out, and by December another 528 were mailed. A little over 200 of those on the notification list were deceased. By late 2009 there was no confirmation that 545 of those individuals had been notified.[33] The letters inform the felons that biological evidence in their cases exists and that they qualify to have the DNA evidence tested. It was reported in 2008 that there were eight cases in which the defendant's DNA profile did not match the profile of the crime-scene samples.[34] While locating individuals and fulfilling postconviction testing are expected to continue for years, there has thus far been one posthumous exoneration, that of Curtis Jasper Moore, who was convicted on rape-murder charges.[35]

The Internet, consisting of tens of millions of personal computers linked seamlessly through thousands of transfer stations, has revolutionized human communication. Nothing since the creation of an international postal system has afforded so many individuals the power and opportunity to share information with people in distant places. Computer networks also provide new instruments for government bureaucracies whose mission is to maintain public records and keep public order. For both inter- and intragovernmental law-enforcement agencies, digital information in the form of photos and data has created new methods of surveillance. Centralized digital DNA profiles that provide a highly reliable means of establishing personal identification have made forensic DNA the most talked-about tool of forensic investigation since the development of the thumbprint more than 100 years ago.

The network of forensic DNA data banks is fast maturing in the modern industrial state, while debates continue over whether innocent people and juveniles or those convicted of minor offenses should be forced to have their personal identity held in a centralized computer that is linked to a biological specimen stored in some law-enforcement outpost. Efforts are currently under way to link the forensic DNA data banks of nation-states with Interpol, the international police agency. Currently there are some rules in place in the

United States about what information can be drawn from biological specimens for entry into criminal justice data banks, and what linkages can be made with other national data banks such as Social Security, Medicare, and the Internal Revenue Service. Nonetheless, the possibilities for abuse in combining criminal justice data banks with private databases, such as health records, credit accounts, books that people borrow from public libraries, phone calls, or shopping profiles held by national chains, are daunting. Whether for commercial interests or bureaucratic goals, there is a tendency for centralized digital information to expand and for cross talk to take place among databases. If personal privacy is to be protected as a constitutional right, then the current trend of rapid DNA database expansion will have to be reexamined in terms of the long-term and systemic trends toward greater governmental surveillance of law-abiding citizens.

Chapter 3

Community DNA Dragnets

> The decision was made, and announced the day after New Year's 1987, that the murder inquiry was about to embark on a "revolutionary step" in the hunt for the killer of Lynda Mann and Dawn Ashworth. All unalibied male residents in the villages between the ages of seventeen and thirty-four years would be asked to submit blood and saliva samples voluntarily in order to "eliminate them" as subjects in the footpath murders.
>
> —Joseph Wambaugh[1]

Floyd Wagster Jr., an African American living in Baton Rouge, Louisiana, in 2002, was on probation for marijuana possession. The Baton Rouge police telephoned Wagster in August 2002 to inform him that he was not in trouble but that they wanted him to come to the station for questioning. Wagster told the deputy that he wished to talk to his lawyer before going to the police station. About an hour later the deputy called Wagster again and inquired if he had spoken to his lawyer. Wagster responded that he had not yet gotten a call back from his attorney. He then left his home to run some errands. No more than half a mile from his home Wagster was stopped by a local deputy. From his account, he was ordered to exit his van and was handcuffed, taken to the station house, interrogated, and asked to provide a DNA sample.[2] He gave the police a sample under what he described as conditions of duress. According to documents Wagster filed in court, once he was at the station, Baton Rouge police denied his repeated requests to speak to an attorney and threatened to take him to drug court and jail unless he provided them with a DNA sample.[3]

Police selected Wagster as part of a manhunt to find a serial killer who they believed had struck between 2001 and 2002 in the Baton Rouge area.

Without probable cause, court warrants, or reasonable suspicion, police on the Baton Rouge Serial Killer Task Force requested DNA samples from 600 men. On July 31, 2004, Wagster filed suit in the U.S. District Court for the Middle District of Louisiana in Baton Rouge on behalf of the men who, while not required under state law, allegedly were coerced to give DNA samples. The judge threw out the charges Wagster filed against the police and the attorney general of Louisiana on the technicality that the complaint was filed too late.[4]

Evolution of DNA Dragnets

The police tactic used in Baton Rouge of rounding up large numbers of suspicionless individuals for a sample of their DNA is referred to as a DNA dragnet. The term "dragnet" entered the American popular lexicon in the 1950s as the title of a popular television police show that starred an imperturbable, street-smart detective named Joe Friday (played by Jack Webb) whose lack of emotion and devotion to facts were his calling cards. Law enforcement's use of the term "dragnet" is derived more closely from commercial fishing, where nets are dragged along the bottom of lakes and rivers to capture schools of fish. Like a shrimp trawl, long abhorred by some environmentalists for its high levels of bycatch (where many other sea creatures are unintentionally caught in the nets), law-enforcement dragnets are often employed with a low degree of specificity. Although a dictionary defines a police dragnet as an organized system or network for gathering or catching people wanted by the authorities,[5] in practice, dragnets are usually launched in search of a suspect. By soliciting DNA from scores of individuals fitting a loose physical description (e.g., "tall, black male"), if any, the dragnet seeks to generate a suspect when there is none. A dragnet is usually geographically focused, although the concept of a "national dragnet" is also used. The term "DNA sweep" is synonymous with "DNA dragnet." Samuel Walker characterizes a DNA sweep as "a situation where the police ask a number of individuals to give voluntary DNA samples in an effort to identify the perpetrator of a crime or a series of crimes."[6]

In recent years the DNA dragnet has been increasingly employed as a method for investigating crimes. When a crime scene has telltale forensic DNA evidence (blood, semen, hair follicles, and/or saliva) that is suspected to have been left by the perpetrator, and when the police do not have any viable leads on suspects, the DNA dragnet seeks to obtain biological samples from large numbers of people (sometimes the entire adult male or female population of a town) who either live or work in the vicinity of the crime scene and could conceivably have been at the location.

The first documented DNA dragnet was undertaken by police in Leicestershire Township in the United Kingdom, where they were investigating a double rape-murder of two 15-year-old girls. The story is fully documented in Joseph Wambaugh's 1989 book *The Blooding*,[7] a "nonfiction novel" that falls in the literary genre of Truman Capote's *In Cold Blood*.

The police investigation of the double rape-murder came to a dead end until one inspector heard about a scientific breakthrough in DNA identification. Alec Jeffreys, a geneticist from the University of Leicester, developed a method in 1983 that used the sequences of certain variable segments of chromosomal DNA as a means of establishing personal identity (see chapter 1). The first success of his technique came when it proved that a French teenager was the father of an English divorcée's child. Word of Jeffreys's success had gotten around to an inspector at the Leicestershire Constabulary, who had a suspect for one of the two teenage rape-murders in Narborough. He asked Jeffreys to compare the DNA left in both the crimes with that of the suspect to determine whether their suspect was responsible for the double murder. Jeffreys's technique showed that the DNA of the police's suspect did not match that left at the crime scene, but he did determine that the DNA at both crime scenes came from the same person.

In early 1987 the chief inspector at Leicestershire attempted something that neither his department nor any other police department previously had undertaken. The chief inspector initiated a program in which police would obtain voluntary blood and saliva samples from all unalibied male residents between the ages of 17 and 34 in the villages of Narborough (where the murders had taken place) and two nearby villages, Littlethorpe and Enderby. It was estimated that as many as 4,000 men gave their blood and saliva samples to police.

Months passed, but no DNA matches occurred. By the fall of 1987 the police were still collecting blood samples. Lab technicians continued to process a mounting backlog of DNA samples. Ultimately the case did not turn on a DNA match at all. Instead, a woman overheard an individual talking to a friend at a pub about how he had paid a coworker to provide a DNA sample under his name. She reported this information to the police, who followed up with the individual, Colin Pitchfork, who had a police record as a "sexual flasher." When the inspectors confronted Pitchfork with his scheme to substitute someone else's blood for his own, he confessed to both crimes. The DNA dragnet, though not successful, strictly speaking, in this particular case, had come of age.

The largest dragnet on record took place in the northern town of Cloppenburg, Germany, in 1998. It was reported that as many as 12,000 of the 18,000 men between the ages of 18 and 30 in Cloppenburg and 35 surrounding villages had voluntarily taken part in the saliva testing.[8] Police, hunting for a serial rapist who had raped and murdered an 11-year-old girl, indicated that they were prepared to collect DNA from up to 100,000 men to catch the perpetrator. The first group of 3,108 men aged between 25 and 45 lived near the scene of the rapes. *Science* magazine reported that another group of over 3,000 men was told to appear at the police station to provide samples.[9] The largest DNA dragnet to date in the United States took place in Miami in 1994. Police sampled 2,300 men in connection with a search for a serial killer.[10] The largest DNA collection in the United Kingdom consisted of a total of 4,500 local men sampled after a body was found in a quarry in Chipping Sodbury, England, in 1996.[11] The United Kingdom has carried out the highest number of dragnets of any single country; between April 1995 and January 2005 there were 292 DNA sweeps across England and Wales.

Civil Liberties Concerns

Mirroring the experience of the British police, a number of cities and towns in the United States, including ones in Louisiana, Florida, Virginia, Massachusetts, Illinois, and Michigan, have undertaken DNA dragnets. The imposition of these dragnets can conflict with the privacy rights of individuals and raises a number of additional concerns. The following are some of the core ethical and legal issues pertaining to DNA dragnets.

Dragnets are conducted without a warrant, probable cause, or individual suspicion. Should there be protections against coercion to obtain DNA samples? The Fourth Amendment of the U.S. Constitution protects individuals against unwarranted searches and seizure of property. When police take a nonvoluntary DNA sample from a suspect, it is considered a "search" and therefore generally requires a court warrant (see chapter 14).

In comparison, taking a fingerprint does not rise to the same level of privacy invasion and judicial involvement. Courts have generally ruled that law-enforcement officials are required to demonstrate that they have probable cause or that the fingerprint will establish the person's connection to an offense, but a distinction is made between acquiring fingerprints from a person to determine guilt or innocence and gathering fingerprints for identification purposes. Increasingly, states allow police to take fingerprints routinely even when a probable cause or suspicion is not established.

According to Aaron Chapin, "DNA dragnets inherently rely on suspicionless searches. Rarely do police have even reasonable suspicion, let alone probable cause in testing DNA in mass sweeps."[12] Other sorts of "sweeps" have been upheld by the courts, although probable cause is usually required. For example, the courts have upheld the right of police to stop cars on a highway in search of individuals involved in a robbery in cases where a description of the perpetrators is available. In addition, if the driver of a car has committed a vehicular infraction, the police are given authority to search the automobile for other unrelated infractions. Adam M. Gershowitz, writing in the *UCLA Law Review*, describes the incident-to-arrest doctrine:

> Imagine that Defendant Dan is stopped by the police for driving through a stop sign. The officer thinks that Dan looks suspicious, but has no probable cause to believe Dan has done anything illegal, other than driving recklessly. Because running a stop sign is an arrestable offense and the officer is suspicious that Dan might be involved in more serious criminal activity, the officer arrests Dan for the traffic violation.
>
> Under the search incident to arrest doctrine, officers are entitled to search the body of the arrestee to ensure that he does not have weapons and to prevent him from destroying evidence. The search incident to arrest is automatic and allows officers to open containers found on the person, even when there is no probable cause to believe anything illegal is inside.[13]

Police have been able to conduct DNA dragnets because the provision of samples in a dragnet is considered "voluntary." There is a legal precedent for rendering constitutional under the Fourth Amendment a person's voluntary consent in submitting evidence. The police may request that an individual against whom law enforcement has neither particularized suspicion nor probable cause submit a biological sample containing his or her DNA in a dragnet, so long as that individual is properly informed about the scope and the purpose of the request. When DNA is given voluntarily, it is considered a consent search. Consent by coercion, however subtle, is not voluntary consent. The Fourth Amendment of the Constitution was designed to protect people from a "police state" where a government becomes "powercentric" and personal privacy can be compromised at whim in a police investigation. Thus many of the core legal and ethical issues surrounding DNA dragnets hinge on the question whether the dragnet is truly voluntary.

Can DNA dragnets be truly consensual, or are they inherently coercive? A policeman who comes to someone's home and asks a person (face-to face) for a voluntary DNA sample has a more coercive effect and is more likely to draw a sample than if the request were made on local television or placed in a newspaper. The uniform worn and words used by the police soliciting the DNA can also affect how coercive it feels when one is asked to provide a sample. A policeman in uniform coming to a person's door to collect donations for a "police benevolent association" feels more coercive than if the request were made by mail.

In pursuit of a serial rapist-murderer, police in Baton Rouge, Louisiana, undertook a DNA dragnet in 2002. They targeted men in southern Louisiana and asked them to provide a DNA sample in order to exclude themselves as suspects. Fifteen men, including Shannon Kohler, refused to give police a buccal swab (cheek saliva).[14] Kohler claimed that police used coercive methods to get a sample from him, which allegedly included issuing threats to obtain a court warrant and to give his name to the press. The Baton Rouge Police Department obtained a court seizure warrant in November 2002 to force Kohler to submit a DNA sample. Kohler was subsequently cleared of the crimes.

Kohler filed suit against the Baton Rouge Police arguing that the seizure warrant lacked probable cause and was an invasion of his privacy. After he failed to obtain relief from the district court, on November 21, 2006, the Fifth Circuit Court of Appeals reversed the lower court and ruled that the DNA search warrant for Kohler lacked probable cause. To justify the warrant, police argued that probable cause was based on (1) two anonymous tips and (2) the fact that Kohler's physical characteristics were consistent with certain aspects of the FBI profile of the perpetrator. The majority of the Fifth Circuit judges wrote:

> These two traits are so generalized in nature that hundreds, if not thousands, of men in the Baton Rouge area could have possessed them, and they are, therefore, insufficient to warrant the belief that Kohler was the serial killer. . . . We conclude that the District Court erred in finding that the seizure warrant was supported by probable cause.[15]

In a well-publicized DNA dragnet in Truro, Massachusetts, coercion was evident in the fact that "both the police and the District Attorney Michael O'Keefe have said that men who refuse to be tested will be considered suspects, even though there had been no evidence linking them to the murder."[16]

When police investigators undertook a DNA sweep in 1999–2000 in Chicago, they were carefully advised against using intimidating tactics in communicating with potential donors of interest. The Chicago police used a consent form, which stated that the donors of DNA agreed that their samples "may be used for this investigation or any other investigation or any other legitimate law enforcement purpose."[17] Those whom the police targeted for a DNA sample were also told that they had the right to refuse the request. Thus it appears that the people approached by the police were sufficiently well informed about the use of their DNA. The police targeted a specific group of men to provide a DNA sample or to make an appointment at their own convenience. They were not told that they had to accompany the officer to the police station. Despite the apparent precautions taken by the police, there remains the question whether their approach was coercive. Jeffrey Grand notes, "In the context of the DNA dragnet, whether a donor is asked to accompany police to the station or is simply subject to a brief detention for field-testing, he still may reasonably feel that he is not free to terminate his encounter with the police."[18]

Do consent procedures currently used in DNA dragnets provide sufficient information to individuals to allow them to make a reasonable decision about whether to provide their DNA? Procedures for obtaining consent for DNA collection have varied considerably from one DNA dragnet to the next. Some have relied on oral consent, others on written consent. In some cases no consent was obtained at all.

Do dragnet consent procedures inform individuals of the potential uses of their DNA? Some police departments present a potential contributor with a written consent form that states how their DNA will be used. The consent form might also describe whether the source of the DNA (cheek swab or blood sample) will be destroyed after the DNA profile is processed. In the absence of state laws to the contrary, some police departments have held on to the voluntarily provided DNA samples and have maintained the profiles in separate DNA databases when the profiles do not qualify to be entered into the state or federal CODIS data bank. States differ on whether the consent forms for voluntary DNA samples stipulate that the DNA profile shall be used exclusively in a specific investigation and for nothing else and whether the sample is removed from the database when the case is closed.

Should police be permitted to retain voluntarily submitted samples beyond the close of the investigation for which they were collected? Should police be allowed to access those samples and profiles for use in future investigations? Law-enforcement personnel have argued that they need to retain samples as evidence for situations in which a case is reopened in the future. But DNA dragnets are crime specific and are not meant to be established as a general data bank for "suspicionless suspects" in the event that someone might commit a crime at some time in the future. DNA databases created for use in future investigations have been created under the CODIS system and are bound by state and federal laws. The federal government does not allow the profiles obtained from a dragnet to be placed in CODIS. The retention of DNA samples outside the scope of the data-bank statutes raises serious concerns.

Florida law allows the state DNA database to hold volunteer samples, as well as those taken from those convicted of a crime. In a Miami, Florida, dragnet, where more than 120 Hispanic males were asked to submit DNA samples to exclude them as a suspect in a serial rape investigation, samples from the volunteers were not destroyed when police found the rapist.[19] It has been reported that by 2003 Florida's DNA database contained more than 5,800 voluntary samples.[20]

On Christmas Day 1995, 18 year old Louise Smith was murdered in Yate, a town outside Bristol in the United Kingdom, on her way home from a nightclub. Police visited 10,500 homes, interviewed more than 14,800 people, and carried out DNA screening of 4,500 male volunteers without finding a suspect.[21]

David Frost, a 22-year-old who was staying with his parents at Yate during the Christmas holiday, was asked by police and agreed to give a DNA sample but failed to show up for the test. In August 1996, after Frost reported to police that he was leaving for South Africa, he became a prime suspect. The U.K. police arranged with South African authorities to obtain a DNA test from Frost, which matched the DNA taken from Smith's body.[22] The police reported that the likelihood of the match coming from another unrelated male was 1 in 35 million. On February 5, 1999, Frost pleaded guilty to the murder of Louise Smith. What remains of the 4,500 DNA samples collected by the police?

Should police be able to tell people that the only way they can clear themselves as a suspect is to provide a DNA sample? Is there language that could be used in a truly voluntary DNA dragnet that would not be coercive? According to Sepideh Esmaili, writing in the *Kent Law Review*, courts consider both objective and subjective factors in deciding whether duress was used by police in requesting a person's DNA sample. Threatening individuals with a court order or threatening to drag their names through the media if they refuse to submit a DNA sample is a form of coercion and voids any consent.[23] Some constitutional scholars argue that the police cannot conduct a noncoercive dragnet of people's DNA, and thus no dragnet should be conducted without a court warrant.[24]

In the United Kingdom, where there is no comparable constitutional guarantee of privacy, as there is in the United States, mandatory DNA profiling is becoming increasingly commonplace. A serial gerontophile rapist known as the Minstead Rapist committed his first offense in 1992 in the South East London area of England. In March 2004 Operation Minstead detectives hand-delivered letters to hundreds of black men in South London, asking that they provide a DNA sample for elimination purposes. Volunteers were told that their DNA sample would be destroyed as soon as police confirmed that it failed to match the rapist's DNA. On the basis of a visual description, 1,000 black men in South London were DNA profiled during a hunt for the serial rapist. However, 125 men initially refused to provide a sample, believing that it was discriminatory and breached their human rights. Police brought pressure to bear on those who refused, explaining that their behavior could be construed as suspicious. They also received intimidating letters from the police. Five were arrested, their DNA was taken, and then they were cleared.

Notwithstanding America's civil liberties tradition, similar coercive methods have been reported in collecting DNA samples. For example, Mark

Rothstein and Meghan Talbott have noted: "Although consent to participate [in a dragnet] is normally voluntary, such requests from law enforcement officers are inherently coercive. . . . In Oklahoma in 2001, people who refused to consent to DNA testing were served with search warrants and treated as suspects, thereby suffering public humiliation."[25] In Louisiana, police requested DNA samples from nearly 1,000 men in the search for a serial killer. Those who refused to provide a sample were threatened that a court order would be issued to get a swab of their saliva.[26] When police targeted young African American males for voluntary DNA samples to solve a murder case many refused to be tested. By their mere refusal, they became suspects (see box 3.1).

BOX 3.1 Refusal to "Voluntarily" Submit DNA in a Dragnet

In 1994 police in Ann Arbor, Michigan, conducted an unsuccessful dragnet that included data-retention problems and racial profiling. Police asked more than 600 African American men to submit DNA samples during the investigation of a serial rapist. Detectives decided to target African American men on the basis of a vague description that the perpetrator of the underlying crime was black. Approximately 160 men "voluntarily" submitted DNA samples and were excluded from suspicion. More than 400 men refused and were not tested. The police chief in charge of the dragnet said that anyone who did not volunteer DNA became a suspect. The perpetrator of the crimes, who was not among those initially tested in the DNA sweep, was eventually caught while attacking a fourth woman. A class-action suit was filed by some of the 160 innocent men who "voluntarily" submitted samples in the search. One of the litigants alleges that he lost his job after detectives informed his coworkers that they wanted to interview him. Police had sought to retain the DNA samples for 30 years but agreed to destroy or return them and paid monetary damages to plaintiffs in the suit.

Source: Electronic Privacy Information Center, "Kohler v. Englade: The Unsuccessful Use of DNA Dragnets to Fight Crime," http://www.epic.org/privacy/kohler (accessed October 28, 2007).

Refusing to Provide a DNA Sample Is Not an Admission of Guilt

It is sometimes assumed that for voluntary DNA dragnets the only persons who refuse to comply are those who have something to hide regarding the investigation in question. However, this is not borne out from the experience police have had with such DNA sweeps. Many factors other than protecting one's guilt may explain noncompliance with voluntary DNA submission.

Individuals who are approached for a DNA sample might be afraid that police will hold on to the sample even if the profile does not match the crime-scene DNA. They might feel stigmatized in knowing that their DNA profile and biological sample remain in a local or state database. They might feel generally that requesting their DNA is an unnecessary intrusion on the part of the government. Others might refuse to give a DNA sample because they believe that police might use their profile to investigate other crimes. In other words, their DNA will be subject to suspicionless searches without a time limit. Cases such as this have actually occurred. Police have on occasion excluded a suspect in one case and used his voluntary DNA sample to trawl the victim database for a cold hit. From the police perspective, the more DNA samples, the more cases will be solved. But there are also instances in which the match of DNA from secondary searches of nonsuspect samples yielded an exact match but did not solve the crime. Rather, the match from one rape case was from consensual sex of the cold-hit suspect with the victim before the rape.[27]

Potential voluntary DNA donors might also refuse to participate in a voluntary dragnet because they understand that mistakes in DNA analysis can and have occurred and that their DNA profile might result in a false positive match with crime-scene evidence (see chapter 16). Finally, although people might be convinced that their DNA evidence would not match the crime-scene DNA, while trusting the analytic laboratories, they may be concerned that the profile will be used in a familial search (see chapter 4).

In Charlottesville, Virginia, police asked 197 African American men to submit to cheek swabs as part of a search for a serial rapist who struck six times between 1997 and 2003. Police chose the subjects on the basis of their resemblance to a composite sketch of the perpetrator or because they behaved "strangely" when approached about the case. About 5 percent (10 men) of the 197 men from whom DNA samples were requested refused to comply.[28] One of these men was Steven Turner, 27, a University of Virginia graduate student. Turner was stopped by police in August 2003 while he was riding his bike in a neighborhood near the university. He allegedly fit the police profile of a 6-foot black man in his early twenties with an athletic build and "unnaturally white, bulging eyes."[29] Turner refused to give police a cheek swab for his DNA profile, even when two officers showed up at his home. Turner had a sense that his rights were being violated because this was a suspicionless search without a court warrant. He felt that he was being coerced to comply and that this practice was a form of racial profiling based on a vague description of the suspect. In response to significant opposition from the African American community, the police chief placed a hold on testing. Ultimately, the perpetrator was caught. He was African American, but he was neither in the criminal database nor in the group chosen for DNA swabbing.

Special-Needs Exception to Privacy

Is it possible that DNA dragnets might become mandatory and a standard part of police investigation? Is it possible that law enforcement will move away from any consent at all in DNA searches? Could the legislature or the courts allow mandatory DNA sweeps without warrants?

Courts have traditionally protected individuals from searches when there is neither probable cause nor suspicion. There is a broad consensus among legal scholars that without an act of Congress "the general interest in crime control served by DNA dragnets is not strong enough to outweigh the even stronger interest of free persons to be free from forced DNA testing and to keep their genetic information private."[30] One can only surmise whether the courts will issue DNA warrants when the prime suspects in a crime with DNA evidence consist of a dozen men, with no evidence other than their physical proximity to the crime scene.

On the other hand, searches in airports have become commonplace and do not require court warrants. Courts have referred to airport searches as an exception to the Fourth Amendment through a legal distinction called the "special-needs exception" where the government's need in conducting the search is considered a "special need, beyond the normal needs of law enforcement." Nations throughout the world have an overriding interest to protect their citizenry in air travel from terrorist activity or other human threats.

Thus U.S. courts have developed this special need of airport searches as a justified exception to the Fourth Amendment's protection from unreasonable searches, as long as the searches are not designed for general criminal investigation. It is possible that DNA dragnets could be employed in these sorts of situations, especially with advances in DNA technology that might allow for rapid DNA typing (see chapter 5).

It is possible that either the legislature or the courts could approve mandatory DNA searches without warrants. However, it seems unlikely that mandatory DNA searches would hold up under a constitutional challenge. Jeffrey Grand argues that if

> the sole purpose of the DNA dragnet is that of criminal apprehension . . . the dragnet procedure could not be considered a search "beyond the needs of law enforcement." Thus, the "special needs" exception is arguably inapplicable to the DNA dragnet. Thus, unless the government can demonstrate a compelling interest that outweighs the Fourth Amendment concerns of the individual, the collection of biological samples for DNA testing is in violation of the Fourth Amendment.[31]

Special-needs provisions for searches have been approved by the courts for administrative actions designed to prevent accidents, such as drug testing of railroad engineers, but have not been applied to crimes that have been committed. Once law enforcement is in search of a perpetrator, the Fourth Amendment of the Constitution applies in full force. Mandatory dragnets would be considered police and not administrative functions and therefore would require a warrant.

On the other hand, surreptitious DNA collection has been upheld in at least one state supreme court decision (see chapter 6). If the law continues to move in this direction, it seems more likely that known, mandatory DNA searches would be allowed when DNA collections that occur without consent or even knowledge that they are happening are considered lawful.

Effectiveness of Dragnets in Solving Crimes

How effective are dragnets in catching criminals? According to a 2004 report issued by the Department of Criminal Justice at the University of

Nebraska and conducted by Samuel Walker, a criminal justice professor, dragnets are entirely ineffective in solving crimes.[32] As of 2008 there had been at least 20 DNA dragnets conducted throughout the United States since 1990. All the dragnets were prompted by unsolved murders and rape cases and included dragnets that took place in Miami; San Diego; Ann Arbor, Michigan; Cheverly, Maryland; Lawrence, Massachusetts; and Truro, Massachusetts. Only one of these dragnets resulted in the resolution of a crime.

Upon closer inspection the single successful dragnet was not much of a dragnet at all. The case involved the rape of a resident in a nursing home in Lawrence, Massachusetts, in 1988. The victim was a 24-year-old comatose patient who became pregnant from the rape and gave birth to a premature baby. Thus investigators requested DNA samples from each of the 25 employees who had access to the patient. One of those samples from a nurse's assistant matched the DNA found in the rape victim. According to the University of Nebraska study, this was the only DNA dragnet out of 18 reported to have been carried out in the United States that resulted in the arrest and conviction of a perpetrator.[33] For each of the remaining situations, the case remained unsolved or was solved through other means.[34] It is important to note that the Lawrence dragnet was directed at workers in a nursing home, where the number of suspects was relatively small. According to Walker, because the crime was a sexual assault in a nursing home, and the requests for DNA samples were limited to employees known to have access to the victim, it was not a DNA "sweep," as in the other cases that involved many more and entirely suspicionless individuals.[35] At best, it was a dragnet of very limited scope.

Aaron Chapin argues that dragnets are an inefficient way to solve crimes. Crime solving by elimination of suspicionless individuals is also costly and unproductive.[36] In a letter published in the *Provincetown Banner*, Frederick Bieber and David Lazer wrote:

> Traditional DNA dragnets have had little success, as true perpetrators usually don't rush forward to volunteer a blood or cheek swab sample for comparison to crime scene evidence. Thus, sampling from elderly hobbling men in Truro would, at first blush, seem to be a distinctly inefficient way to look for new leads.[37]

In the case of the Truro, Massachusetts dragnet, police began collecting DNA from the nearly 800 male residents of the Cape Cod community three years after the 2002 brutal murder of former fashion writer Christa Worthington. The DNA samples remained unanalyzed for months. Ultimately a DNA sample was found to match the DNA from the crime, but this sample had not been obtained as part of the dragnet. In fact, the DNA belonged to Christopher McCowen, a 33-year-old trash collector who made weekly visits to the victim's home and was considered a possible suspect from early on in the investigation. When questioned, McCowen agreed to provide a DNA sample, but it took police two years to collect it and another year to process it. The cause for the delay in processing had been attributed to inadequate resources and a DNA testing backlog, no doubt exacerbated by the DNA dragnet.[38]

Uniform Standards and Good Practices

Currently there are no uniform standards or best practices according to which law enforcement collects DNA on a voluntary basis. People in the United States who are asked for their DNA are not read a statement like the Miranda Clause that would assert their rights not to give up their DNA without a warrant. The Electronic Privacy Information Center (EPIC), a public interest research center located in Washington and focusing on civil liberties, issued the following list of best practices in its amicus brief on behalf of the Kohler litigation:[39]

1. DNA dragnets should only be used by police as a last resort, when all other investigative avenues have been exhausted. In Walker and Harrington's Model Policy for DNA Sweeps, the authors argue that requests for voluntary DNA samples should be limited only to cases where "police officers have specific credible evidence linking a person or a very small number of people with a crime."[40]
2. DNA dragnets should be limited in scope to those who match the description of the perpetrator, or to those who had access to the victim. The dragnet should not serve as an excuse to build a DNA database of certain groups of individuals. These databases, sometimes called "rogue databases," are kept out of CODIS for the use of local law enforcement.
3. Potential donors of DNA samples should be informed of their right of refusal when asked by police to voluntarily submit their DNA to exclude them as a suspect in a crime.

4. Individuals approached for "voluntarily" submitting a DNA sample should not be subject to coercive techniques such as threats, additional scrutiny, or legal action if they refuse to comply.
5. Samples gathered from voluntary donors who are exculpated by their DNA profile from being a suspect in a dragnet should have their profile expunged and biological sample destroyed and not retained in either a separate file or in state or federal databases. A similar recommendation was noted in the Nuffield Council on Bioethics report on the forensic use of bioinformation: "Consent given by a volunteer to retain their biological samples and resulting profile on the NDNAD must be revocable at any time and without any requirement to give a reason."[41]
6. The consent form used to obtain voluntary DNA samples should disclose how the samples will be used and whether they will be destroyed if the sample does not match the crime-scene DNA.
7. Police must protect the privacy of the donors of DNA samples, as well as those who exercise their right to opt out. In Omaha, Nebraska, police investigating a series of rapes conducted a dragnet to obtain DNA samples of more than 36 men. On the basis of a rough witness description of the perpetrator, police went to people's homes and requested DNA in front of family members.[42]

While the best practices issued by EPIC on DNA dragnets may not be sufficient if one considers DNA dragnets to be inherently coercive, they provide a sensible balance between police efforts to investigate a high-priority crime where all other leads have been exhausted and the protection of civil liberties of suspicionless individuals brought into a search by virtue of their geographical proximity to the crime.

The original use of dragnets in police investigations before forensic DNA involved a type of blitzkrieg sweep through an area in an effort to pick up a suspect before too much time elapsed that would afford the perpetrator of a crime the opportunity to escape police interrogation. Both spatial and temporal elements were critical in traditional dragnets.

For DNA dragnets, time plays a different role. There is no immediacy to the search. The potential evidence retrieved from a DNA search will not disappear after a certain time period. What the police want and need is a match of the crime-scene DNA with the DNA profile of a person who is within the geographical area where the crime was committed. DNA dragnets are not of much value when the perpetrator of a crime is traveling through an area or when the number of people brought under the dragnet search becomes very large. For example, a crime committed in a rest stop at an interstate highway would not lend itself to a dragnet of people in the area.

As was demonstrated by the Truro case, DNA dragnets can be highly inefficient, and too much emphasis on DNA can inadvertently turn police away from more traditional methods of criminal investigation. In the meantime, the costs to community trust in law enforcement can be high. On June 19, 2008, the American Civil Liberties Union (ACLU) of Massachusetts filed suit on behalf of approximately 100 of the men who had voluntarily provided DNA samples to the Truro police in 2002. In August 2008 Massachusetts State Police returned the biological sample of one of the men who volunteered his DNA. By the winter of 2008 the other samples collected by Truro police were still in the database and awaiting a court decision on the expungement of the profiles and the destruction of the biological samples.

Dragnets are an inefficient means for solving a crime because they involve solving crimes by exclusion. From 1995 to 2005 about 7,000 people were tested in DNA dragnets in the United States. In only one case was a suspect identified where he gave his DNA voluntarily, and the list of possible suspects in this case was sufficiently narrow to begin with. For a small fixed number of suspects, exclusion can work efficiently. When the dragnet targets hundreds and thousands of people, and many people refuse to submit, experience has shown that it is not an efficient way to solve the crime. Moreover, the areawide data banking of innocent persons' DNA can undermine the willingness of individuals to cooperate and result in considerable hostility toward law enforcement. More broadly, the means by which police execute dragnets has been shown to be coercive, and there is little accountability, such as uniform standards of informed consent, to counteract this effect.

Requiring DNA from individuals picked up in a dragnet flies in the face of constitutional precedent that has protected suspicionless individuals from searches that are designed to solve crimes. Mass collections of DNA can have a chilling effect on society that could suppress constitutional speech and freedom of assembly. As one judge noted, a permanent national

DNA database "could be used to repress dissent or, quite literally, to eliminate political opposition."[43] People who know that they are in a database might hesitate to participate fully in civil society as activists. The creation of a database of innocents can be regarded as a silent way of suppressing dissent.

Chapter 4

Familial DNA Searches

> Familial testing, which uses biological relatedness as the trigger for criminal investigation, ensures that groups with more children and large families relative to other groups will be at higher risk for genetic surveillance.
>
> —Daniel Grimm[1]

In March 2003 motorist Michael Little suffered fatal injuries when a drunken individual hurled a brick from a footbridge in Surrey County in the southeast of England. The brick pierced the windshield of his moving vehicle, hit him in the chest, and caused him to suffer a fatal heart attack. Blood traces left on the brick were believed to be those of the perpetrator. The blood, it turned out, was from a wound the perpetrator had received during an earlier attempt to steal a car. Police profiled the DNA in the blood and compared it with 2.35 million profiles it had stored on the national U.K. database at that time. There was no match. They also conducted a DNA dragnet in the area, collecting DNA samples from more than 350 young men. Six months after the crime had been committed and still with no leads, the police decided to go back to the database, but this time to run a search not for a perfect match, but for one where 11 or more of the 20 alleles used in the British system matched. The search turned up 3,525 potential siblings to the source DNA.[2] The police narrowed this list to 25 white males who lived in the geographical region of the crime. One of these matched at 16 out of 20 alleles. This near match led the police to Craig Harman, a brother of the source profile and a prime suspect who admitted to and was convicted of the crime.[3] Harman was the first person in the world to be convicted following identification through a familial search of a DNA database.[4]

"Familial searching" of databases represents a new method of creating suspects in the absence of an immediate cold hit. These searches are premised on the notion that siblings and other closely related individuals share more common genetic material than unrelated individuals. Therefore, a DNA profile that is a near match to the DNA found at a crime scene may be that of a first-degree relative of the actual perpetrator (i.e., a parent, an offspring, or a full sibling). The source of the near match, then, may lead the police to the criminal suspect and, in some cases, provide valuable information about that individual.[5]

The most common method of familial searching involves generating a list of possible relatives of the unknown person (whose DNA was picked up at a crime scene) by performing a profile search of a forensic DNA database that is designed to find "partial matches" between crime-scene evidence and offender profiles. In the Combined DNA Index System (CODIS) the term "search stringencies" describes one of three search modalities. A "high-stringency" search means that all alleles of the loci that are present in both DNA profiles (crime-scene DNA and database profiles) are identical. This represents the standard search for a cold hit, where the crime-scene profile matches exactly one and only one profile on the database. In a "moderate-stringency" search the alleles of a locus (among two DNA profiles) with the least number of distinct alleles must be present in the corresponding locus of the other DNA profile. For example, under "moderate stringency" one DNA profile with STR alleles 7, 10 will generate a match with another profile with STR alleles 7, 7. The heterozygote 7, 10 is deemed a "moderate-stringency" match with the homozygote 7, 7.

For a "low-stringency" match (used to find parent-child relationships), each locus that is compared between two DNA profiles must have at least one allele of that locus present in the other DNA profile. Thus a locus with STR 7, 10 would meet the standard of a "low-stringency" match with locus STR 3, 7.[6]

In a second familial search method police may conduct what is called a "rare-allele" search, in which they analyze the crime-scene DNA for highly unusual DNA signatures. Close relatives of those matches are then tracked down and asked to "voluntarily" provide a DNA sample.[7] A third method of DNA profile searching is a variant of the second. It considers both the number of matching alleles resulting from a low-stringency search and their frequencies in a population. It is designed to reduce the number of false positives, that is, identification of individuals who are not relatives of the perpetrator.

Although this relatively new use of forensic DNA databases has led to a handful of somewhat remarkable success stories, familial searching, when practiced routinely, effectively expands DNA databases to all the close blood relatives of the individuals in the database, subjecting entire families (and perhaps even neighborhoods or even ethnic populations) to lifelong genetic surveillance. This gives rise to a number of social and ethical questions that deserve serious consideration.

Techniques and Practices

The premises behind familial searching are twofold. First, close matches in DNA profiles are more likely to indicate that the sources of the DNA are close family members rather than two unrelated individuals. Second, the closer the DNA match is, the higher the likelihood that the individuals are related, particularly when the matching alleles are rare in the general population. The most common and simplest technique used to identify potential relatives of suspects of a crime in the absence of a complete match with the database is allele counting, where forensic investigators compare the overall number of alleles shared between the crime-scene evidence and the database profiles. Generally speaking, the larger the number of shared alleles, the greater the probability of family ties. Siblings, on average, have about 18 of 26 possible alleles in common, while unrelated individuals average about 8 out of 26 in common.

BOX 4.1 Serial Killer Found from Sister's DNA

In 2006, 49-year-old James Lloyd was convicted of four rapes and two attempted rapes that had occurred between 1983 and 1986. A father of two and described as "a wealthy businessman," Lloyd was discovered to be the notorious "Dearne Valley Shoe Rapist," who had tied his women victims up with tights and stolen their shoes. The police got to Lloyd through his sister, who was in the database for a drunken-driving conviction. Forensic scientists narrowed an initial partial-match list and provided the South Yorkshire Police the names of 43 people in the database who were possible relatives of the sex offender. In following up on the 43 names, police knocked on the door of Lloyd's sister, who told the police that her brother roughly fit the age and height of the wanted individual and agreed to contact him. The sister called her brother and warned him that the police were after him. Upon hearing that he had become a suspect, Lloyd attempted suicide by trying to hang himself in his home but was saved by his 17-year-old son. While arresting him, police found more than 100 pairs of stiletto-heeled shoes hidden behind a secret trapdoor on the premises of Lloyd's printing firm, where he worked. Lloyd pled guilty to four rapes and two attempted rapes and was sentenced to life imprisonment.

Sources: Tony Lake, chief constable, Lincolnshire Police, presentation before the FBI Symposium on Familial Searching and Genetic Privacy, Arlington, VA, March 17–18, 2008; Paul Sims, "20 Years After His Evil Reign, Shoe Rapist Is Unmasked by His Sister's DNA," *Daily Mail* (London), July 18, 2006; Andrew Norfolk, "Genetic Bar Code That Reopened the Case," *The Times* (London), July 18, 2006.

Allele Counting: High, Moderate, and Low Stringency

As discussed in chapter 2, in the U.S. DNA data-banking system a complete match of two DNA samples occurs when all 26 alleles in the 13 loci (two alleles per locus) are identical, that is, each of the alleles at the locus has the same number of short tandem repeats (STRs). The FBI's definition of a "partial match" requires that the crime-scene DNA and the offender DNA profiles share at least one allele at each genetic location tested, but the definition was not intended to address familial (or kinship) relations.[8] In the hypothetical example in table 4.1, four samples, A, B, C, and D, are compared at locus number 1. The number of STRs for alleles at locus 1 is identical for samples A and B. If all 13 loci show the same concordance, we have an exact match for the samples.

In the event of a partial match, STR values will not be identical for all 26 alleles. Sample C matches the crime sample at moderate stringency, while sample D matches A at low stringency.

TABLE 4.1 High, Moderate, and Low Stringency for a Single Locus

	Crime sample A	Crime sample B	Crime sample C	Crime sample D
Locus 1	9, 13	9, 13	9, 9	9, 12
Stringency with respect to crime sample		High	Moderate	Low

Source: Authors.

The U.S. CODIS software system was designed to run moderate-stringency searches. This means that when a search is conducted against CODIS, the computer compares a given DNA profile with the database, looking not only for perfect matches but also cases where, for one or more loci, one sample contains only one allele (sample C, 9) and the other is heterozygote (sample A, 9, 13), with one of the alleles the same as the single one. This search criterion was designed not for purposes of conducting familial searches, but instead for capturing cases where DNA from a crime scene is partially degraded (so that the crime-scene DNA profile is a partial profile), or where the crime-scene sample may contain a mixture of two or more DNA profiles, or where "allelic dropout" (failure to detect an allele during sampling or failure to amplify an allele during the polymerase chain reaction) or a mistyping may have occurred. In other words, the intent of the moderate-stringency search was to build in a safety factor so that suspects who do have a profile that is identical to that of the crime stain but who would be overlooked by a high-stringency match search will in fact be identified. The CODIS software system was not set up to run low-stringency searches routinely, although it presumably would not be difficult to change the search parameters of the software.

For familial searches, forensic investigators must decide the criteria for a partial match. Some states use 13 alleles out of 26 to define a partial or low-stringency match. The state of Florida requires 21 out of 26. Investigators can examine the allelic similarities in partial matches and make predictions about how closely related the donors of the DNA samples are. They do this by making use of statistical information from large databases of the allelic homology of the DNA from family relations.

Henry Greely and colleagues reported on the allelic matching statistics for first-, second-, and third-degree relatives.[9] A close or first-degree relative is

TABLE 4.2 DNA Allele Matches Among Relatives

Relationship	Examples of relations	Allelic similarities
Parental	Father and son/daughter; mother and son/daughter	16 of 26 alleles in Caucasian population
First-degree relatives: siblings	Sister and brother	13–16 alleles
Second-degree relatives	Nephew	6–7 alleles
Third-degree relatives	Great-grandchild	3–4 alleles

Source: Authors.

a parent or sibling. On average they are expected to share about 50 percent of one another's DNA variants (between 13 and 16 alleles). A father or mother and child match at no fewer than 13 alleles. Second-degree relatives include uncles, aunts, nephews, nieces, grandparents, grandchildren, and half brothers and sisters. These relatives share about one-quarter of their DNA variations. Finally, third-degree relatives, who consist of great-grandparents and great-grandchildren, share about one-eighth of their DNA variations (see table 4.2).

Greely and colleagues have estimated the probability of unrelated people matching a certain number of alleles on the basis of different scenarios. These calculations are useful benchmarks for understanding what the chances are that a low-stringency match will yield a relative of the perpetrator of a crime in the offender's database. They conclude:

> On average, the chance that an unrelated person's genotype will match the genotype from crime scene DNA of 13 or more of the 26 alleles, allowing for all possible ways of distributing the matches across the markers, is around three percent. However, the chance that two unrelated people match at thirteen or more sites with every marker having at least one match (as will occur for parent-child pairs) is about one in two thousand.[10]

Matches from these criteria, then, are likely to produce close family members. However, they are also likely to miss most family members. In a memorandum sent to Attorney General Jerry Brown of California, Michael Chamberlain, head of the state's DNA legal unit, wrote: "Under the FBI's definition of partial match, it would preclude detection of 99.9% of brothers, many of whom have no alleles in common at a genetic location."[11]

BOX 4.2 Familial Search Reveals a Posthumous Suspect

On September 16, 1973, two 16-year-old girls were raped and strangled in southern Wales, United Kingdom. Some months later a third 16-year-old girl met the same fate. The similarity of these heinous murders suggested that they could have been committed by the same man. During the initial investigation the police focused on interviewing 30,000 individuals. Subsequently, using techniques such as psychological profiling and an intelligence-led screen, police targeted 500 potential suspects, including a man named Joseph Kappen, but there was insufficient evidence to charge any of these individuals. In 2000, 27 years after the crimes were committed, investigators in the Forensic Science Service obtained DNA profiles from clothing stains of two of the girls and submitted the profiles to the United Kingdom's National DNA Database (NDNAD). Although no exact match was found, a low-stringency analysis indicated that the DNA partially matched the DNA profile of a man named Paul Kappen. Police surmised that someone in Kappen's family was the murderer, and this led them back to Paul Kappen's father, Joseph, who had since died in 1990 at age 49. Meanwhile, in 2002 a comparison of the crime-scene DNA of the third girl's murder showed that the three crimes were linked. British law-enforcement authorities obtained DNA samples from the Kappen family, including Paul Kappen's mother and his siblings. The close matches between the crime-scene DNA and family DNA profiles were sufficiently credible for the police to obtain a warrant to exhume the body of Joseph Kappen. After his body was exhumed on May 17, 2002, it was learned that his DNA was an exact match with the crime-scene DNA from the three murders. Forensic investigators found the murderer by familial searching, albeit posthumously.

Source: Robin Williams and Paul Johnson, "Inclusiveness, Effectiveness and Intrusiveness: Issues in the Developing Uses of DNA Profiling in Support of Criminal Investigations," *Journal of Law, Medicine and Ethics* 33, no. 3 (2005): 545–558.

Limitations of Allele Counting

Partial-match searches give rise to a number of questions that are technical on the surface but quickly become ethical. First, there is the question whether partial-match searches will miss potential relatives (false negatives). The second question is whether these searches will falsely identify unrelated individuals (false positives) and send investigators off on wild-goose chases. As was seen in the Craig Harman case, partial-match searches can yield results in the thousands. Ultimately, the question becomes, "How close is close enough?" In other words, what is the appropriate threshold for determining whether a partial match is likely to have revealed a relative of the actual perpetrator, warranting further investigation?

The false-negative and false-positive problems are compounded by the search criteria employed by the CODIS software system, which is not only used in the United States but also has been exported to at least one-fifth of the European countries using forensic DNA databases.[12] Canada also employs CODIS software. The FBI gave the system to the Royal Canadian Mounted Police for free.[13] As noted earlier, the system was designed to perform moderate-stringency searches for the 13 CODIS markers. The original purpose behind its use was to account for instances where crime-scene samples are compromised—for example, where they are partially degraded or in cases where there is a mixture, so that it is difficult to tell which alleles come from the alleged perpetrator. As applied to looking for family members, moderate- and low-stringency searches will pick up most parent-offspring relationships, since a child necessarily inherits one allele from each of his or her parents. However, as pointed out by researchers at the University of Washington, a moderate-stringency search will miss the overwhelming majority of full as well as half sibling relationships.[14] This is because for each of the 13 pairs of alleles in a DNA profile, there is a 25 percent chance that siblings will not match at either allele. But if this occurs at any one of the 13 pairs of alleles, the familial relationship will be overlooked at the level of a moderate-stringency match. It turns out that the chance is only about 1 in 1,000 that a sibling will be identified at the level of a moderate-stringency search.[15]

The problem becomes more complicated, however, if we move to a low-stringency search. By relaxing the search criteria, we will increase the chances of picking up real siblings, or other related individuals, thereby improving the false-negative problem. However, we will greatly exacerbate the false-positive problem; the overwhelming number of individuals identified will not in fact be related at all. The reason this occurs is simply because there is a significant sharing of alleles within the general population. What we are looking for in a familial search is not this general sharing but, instead, sharing that results from two individuals with a recent common ancestor. The problem with allele counting, whether of low or moderate stringency, is that there is no way to distinguish whether these matching alleles are an indication of relatedness—echoes of common ancestry—or are simply coincidental.

In practice, both low- and moderate-stringency searches may very well send law enforcement off on wild-goose chases, especially in cases where they are conducted against large databases. As Greely and colleagues reported, although the chance that two unrelated people match at 13 or more sites with every marker having at least one matching allele is small (1 in 2,000), even this small percentage can yield a high number of false leads when one is dealing with a database of several million.[16] More recently, George Carmody, a population biologist at Carleton University and a member of the New York State Forensic Science Commission's DNA Subcommittee, estimated that a low-stringency search against the U.S. national database would generate on the order of 8,000 false partial matches for every real sibling match.[17] This number of false leads will necessarily grow as databases continue to expand.

The number of false positives will depend in part on the relative rarity of the alleles in the crime-scene sample. As Greely and colleagues have cautioned, "The partial match is only a lead—a relatively weak one for a common genotype though possibly a very strong one for a rare genotype."[18]

Similarly, David Paoletti and colleagues have pointed out that the value of a given partial match is dependent on a number of important parameters, including the relative frequencies of the matching alleles, the number of initial suspects considered, and the population size, or the number of potential alternative suspects. They conclude that it is not possible to come up with a single threshold (such as "number of shared alleles") for use in determining when a partial match warrants further investigation. A partial match of as few as 5 alleles out of 26 might be significant and sufficient grounds for follow-up investigation when all 5 of those alleles are rare and the alternative suspect pool is small. But when all the matching alleles are relatively common, as many as 15 alleles might need to be shared.[19]

Mathematical models (discussed in the next section) that predict kinship relationships from DNA can support a hypothesis that two people are related on the basis of a few shared alleles even when they are not related. In other words, aggressive police work can pursue family members of partially matched individuals on the basis of tenuous or tentative mathematical assumptions. In one study of 194 Caucasians profiled by the FBI on 13 loci, pairwise comparisons of the profiles were made. In this sample 1,654 pairs of individuals partially matched at 9 loci, and 797 pairs partially matched at over 9 loci.[20] Unless the mathematical models are validated and become canonical, their use in forensics will remain controversial.

BOX 4.3 The Case of the Grim Sleeper

A serial killer, colloquially named the Grim Sleeper, who had murdered African American women in South Los Angeles since 1985, left his saliva and other DNA at several killing sites. Los Angeles police connected the perpetrator to 10 victims. A genetic sample preserved from one of the crime scenes matched samples collected from a 14-year-old killed in 2002, the body of a woman killed in 2003, and another victim found in 2007. The Los Angeles police ran the DNA found at the crime scene against millions of genetic profiles of convicted criminals in the state DNA database. Police tried to locate the unknown killer's relatives in the hope that there would be similar DNA patterns by comparing the crime-scene DNA profile (under low-stringency conditions) with those in the state's DNA database of more than 1 million felons. Initially no suspects were found. Eventually the state forensic lab found a partial match in the Grim Sleeper case with Christopher Franklin, recently convicted of a felony, that suggested a father-son relationship. The trial led to Christopher's father, Lonnie D. Franklin Jr., 57, when police found a DNA match from his saliva on a discarded pizza slice. In July 2010 Lonnie Franklin was declared the Grim Sleeper by Los Angeles police and charged with 10 counts of murder and one count of attempted murder.

Sources: Joel Rubin and Maura Dolan, "DNA Search Fails to Find Relatives of Unknown Serial Killer," *Los Angeles Times*, December 3, 2008; Maura Dolan, Joel Rubin, and Mitchell Landsberg, "DNA Leads to Arrest in Grim Sleeper Killings," *Los Angeles Times*, July 8, 2010.

More Advanced Familial Searching Techniques

Population geneticists and biostatisticians have recommended using more advanced statistical techniques for conducting familial searches as a way of improving both efficiency and accuracy. The most commonly agreed-upon approach is to use "likelihood ratios" (LRs). As applied to familial searching, this is a statistical method that goes beyond simple allele counting and instead evaluates the genetic evidence to support the likelihood that two individuals are related compared with the likelihood that they are not. In general, the LR is the ratio of two probabilities of the same event under different hypotheses. For DNA testing the LR is the probability of the observed genetic profiles given proposed familial relationships versus the probability of observing the genetic profiles if the donor of the evidence and the identified partially matching profile sources are unrelated.[21] Charles Brenner, a consultant on paternity testing and forensic DNA practices, has described an LR as follows: "[It is the] amount of coincidence that you have to swallow if you want to believe that two profiles are similar merely by chance."[22]

The benefit of using LRs is that they make better use of genetic information and produce a prioritized list of partial matches. In addition to comparing the amount of sharing between any two individuals with the amount of sharing that would be expected if those individuals were siblings or parent and offspring, the LR can take into account the relative frequencies of the matching alleles in the population. The rarer the alleles that match, the more likely it is that the match indicates a potential familial relationship.

Frederick Bieber and colleagues performed some mathematical simulations to investigate the chances of finding a true relative using this approach. Specifically, they ran a series of simulations by comparing an "unknown" sample with each registered offender in the database to determine the likelihood ratios of successfully identifying biological relatives of the offenders. Assuming a database size of 50,000, they argue that a parent of a child of a known offender would be identified 62 percent of the time as the very first lead (highest likelihood ratio) and 99 percent of the time among the first 100 leads (99 families would be investigated to find the one family related to the perpetrator). This assumes that the relative is in the database in the first place, and that the crime is committed in the state where that relative resides.[23] They conclude: "Our simulations demonstrate that kinship analysis would be valuable now for detecting potential suspects who are the parents, children, or siblings of those whose profiles are in forensic databases."[24] LRs can also weigh the significance of a partial match on the basis of the racial and ethnic populations from which the DNA samples were drawn. For example, in the Caucasian population fathers and sons share an average of 16 out of the 26 CODIS-designated alleles, while two completely unrelated Caucasian individuals will share an average of 9 alleles. These figures will vary slightly across racial and ethnic populations. In groups that are more isolated and where there has been more inbreeding (e.g., cousins having children with one another), the number of shared alleles will be greater, even for so-called unrelated individuals. On the other hand, populations where there has been a lot of breeding across races and cultures and that have exhibited more mobility will show more allelic diversity and thus less allele sharing.

Although Bieber and colleagues are optimistic about the value of conducting familial searches using current techniques, others have pointed out that even these more refined techniques will produce a high number of false positives, especially when used in the context of large databases. High numbers of false leads result in resource-intensive investigations. In addition, these techniques are not likely to turn up any significant investigatory leads beyond full siblings and parental relationships.[25]

One way to narrow potentially long lists of possible relatives of suspects is to subject the stored DNA samples identified in a partial match to additional genetic testing. Kristen Lewis and colleagues advocate the use of ancestry-testing techniques—specifically Y-chromosome analysis and mitochondrial DNA sequencing—to exclude individuals from investigation who cannot be related to the true perpetrator through either paternal or maternal lineage.[26] Y-chromosome (Y-STR) analysis involves examining genetic markers along the Y chromosome; since the Y chromosome is paternally inherited, it can be used to trace family relationships among males. Similarly, mitochondrial DNA (mtDNA) is inherited from the mother and can be used to trace back maternal lineages. Frederick Bieber and David Lazer suggest that Y-STR analysis alone could eliminate 99 percent of false leads.[27] Others have recommended expanding the scope of genetic markers used in the DNA profile as a way of narrowing long lists of potential relatives. All these techniques would require accessing the stored biological samples of individuals in the database and subjecting them to additional genetic tests.

U.S. Policy on Familial Searches

In the United States the practice of familial searching was limited by a policy adopted by the FBI that prohibited the release of any identifying information about an offender in one state's database to officials in another state unless the offender's DNA was an exact match with the DNA evidence found at the scene of the crime (with limited exceptions for cases where slight differences between the profiles could be explained by degradation). In the summer of 2006 the FBI changed its policy in response to a request from Denver authorities after a close match was found between evidence taken from the scene of a rape and a convicted felon in Oregon, indicating that he was a potential relative of the actual perpetrator.[28] The interim policy, which became effective on July 14, 2006, allows states to share information related to "partial matches" upon FBI approval.[29] Following FBI authorization, discretion whether to release the offender information is left to the state that holds the information. According to the FBI memorandum distributed on July 20, 2006, "For situations in which there is no other available investigative information, the FBI Laboratory has instituted an Interim Plan that may permit the release of the offender's identifying information."[30]

The FBI's interim plan defines a "partial match" as "a moderate stringency candidate match between two single-source profiles having at each locus at least one allele in common."[31] This plan did not allow information to be released between states as a result of low-stringency searches. In this way the FBI was insisting that the search parameters initially established for CODIS remain intact. Furthermore, the FBI distinguishes these partial matches from familial searching, claiming that these partial matches are "inadvertent," as opposed to purposeful searches for incomplete matches. According to Tom Callaghan, FBI CODIS unit director, "The FBI does not do familial searching."[32]

The FBI's distinction between an "inadvertent partial match" and familial searching seems tenuous; after all, the follow-up to such a partial match is to seek out family members of that individual. Nonetheless, the FBI's insistence that partial-match follow-ups do not constitute familial searching may indicate the agency's own recognition that full-scale trawling of the databases is not authorized under current law and that allowing this more aggressive searching would render the database operations vulnerable to legal challenge. Whether legislative authority would be needed to authorize familial searching was a central question posed by the FBI itself at its national symposium on familial searching in March 2008.[33]

To be fair, the DNA Identification Act of 1994 that established CODIS is silent on the issue of familial searching and partial matches and simply states that the database will be used "for law enforcement purposes."[34] Nonetheless, there is some indication in the legislative history that Congress certainly did not intend for the database to be used to trawl for individuals other than those in the database. For example, Senator Herb Kohl from Wisconsin stated on the Senate floor in support of the DNA Analysis and Backlog Elimination Act of 2000:

> Currently, all 50 states require DNA samples to be obtained from certain convicted offenders, and these samples increasingly can be shared through a national DNA database established by Federal law. This national database . . . enables law enforcement officials to link DNA evidence found at a crime scene with any suspect whose DNA is already on file. By identifying repeat offenders, this system does make a difference.[35]

In other words, the idea was to create a database of known, convicted offenders so that law-enforcement officials could link those offenders to other crimes they might have committed and to have a way of catching them if they acted again. It was not to use the database as an intelligence or surveillance tool to investigate other people who were not in the database.

Perhaps the more significant evidence that familial searching crosses a well-established line is in the design of CODIS itself. Familial searching of the database is not a new concept; it was well known during the time at which CODIS was established that the database could be used in this fashion. In 1992, two years before the establishment of CODIS, the National Academy of Sciences issued a comprehensive report that made a series of recommendations for DNA data banking and testing. The report addressed head-on the issue of familial searching:

> The ability of DNA to recognize relatedness poses a novel privacy issue for DNA databanks. . . . To put it succinctly, DNA databanks have the ability to point not just to individuals but to entire families—including relatives who have committed no crime. Clearly, this poses serious issues of privacy and fairness. . . . It is inappropriate, for reasons of privacy, to search databanks of DNA from

> convicted criminals in such a fashion. Such uses should be prevented both by limitations on the software for search and by statutory guarantees of privacy.[36]

In addition, the Privacy Act of 1974 requires that a formal rule making be issued for any new government database that defines clearly the categories of individuals who will be affected by the database.[37] Such rules were issued by the U.S. Department of Justice for the National DNA Index System (NDIS) at its inception in 1998. Those rules established four categories of individuals covered by the system: convicted offenders; missing persons and their close biological relatives; victims; and DNA personnel. In addition, the rules included the following safeguard: "NDIS will disclose to a criminal justice agency the DNA records of another criminal justice agency *only when there is a potential DNA match*."[38] Given that family members represent an entirely separate category of individuals affected by the database system, and that partial matches result in disclosure of information between agencies when a DNA match is not complete, it seems that, at the very least, a formal rule-making process is required under the Privacy Act in order for the database to be used to identify potential family members.

The agency's hesitation to cross the line was at one time shared by most state database administrators. At the national symposium on familial searching in March 2008, sentiments were nearly unanimous among the 50 state CODIS administrators and their legal representatives that they did not currently have authority to move forward with familial searching.[39]

Even so, it is not clear that the line the FBI is attempting to draw between "partial matches" and "familial searching" will hold as a practical matter. A memorandum dated August 2, 2007, from Tom Callaghan stated that a "Next Generation CODIS" was under development that would provide a new search engine capable of performing "joint pedigree likelihood ratio" analyses.[40] Although the stated intent of the upgrade is to conduct missing-person searches, it is hard to imagine how the states or the FBI will be prevented from using this for familial searches in criminal investigations. In the meantime, the interim policy appears to have encouraged some familial searching advocates to push harder at the state level to initiate statewide familial searching programs. Several states, including Florida, South Carolina, North Carolina, Colorado, Missouri, Oregon, Arizona, and Massachusetts, have agreed to disclose partial matches to law-enforcement agencies in accordance with the FBI interim policy. These states have generally focused

BOX 4.4 A 14-Year-Old Offender's DNA Leads Police to a Murder Suspect

A 20-year-old woman named Lynette White was fatally stabbed in southern Wales in 1988. It was one of the most brutal murders in Welsh history. The case went cold for 12 years, but the police were not ready to give it up. In 2000 forensic investigators completed a new DNA sweep of the victim's apartment, hoping to acquire new forensic evidence. The sweep turned up spots of blood on a baseboard that had been missed on the first search. They profiled the bloodstain DNA and compared it with profiles in their national database but did not find an exact match. However, an allele in the crime-scene DNA was found in only 1 to 2 percent of the profiles in the NDNAD. By using a low-stringency familial search combined with geographical constraints, police found approximately 70 potential relatives of the person who had left the crime-scene stain. Forensic investigators eventually found a person in the NDNAD who had a reasonably close DNA profile to the crime-scene DNA—a 14-year-old boy who was not alive when Lynette White was murdered. The boy's DNA was in the database because he had previously gotten into trouble with the police. The police began looking into the boy's family and focused attention on the boy's paternal uncle, Jeffrey Gafoor. Gafoor's DNA was an exact match with the bloodstain, and he eventually admitted to committing the crime.

Source: "How police found Gafoor," *BBC News,* July 4, 2003, http://news.bbc.co.uk/1/hi/wales/3038138.stm (accessed April 15, 2010).

almost exclusively on the technicalities of familial searching and very little on the political, legal, or ethical ramifications of such searches. In contrast, Maryland statutorily banned the use of familial DNA searches.[41]

California's Reversal on Familial Searching

The shift in the FBI policy also sparked disputes between states. Denver's district attorney, Mitchell Morrissey, applied the interim FBI policy in familial searches to a cold case involving a Denver rape. The crime-scene DNA did not yield an exact match in CODIS, but a moderate-stringency search found what Morrissey believed could be a close family member who had been convicted of a felony in California. He requested information about the felon from California attorney general Jerry Brown in July 2007. On August 3, 2007, Brown reportedly denied the request, citing the need to protect the privacy of California felons who are not exact matches in CODIS. Brown noted that reporting on DNA near matches was beyond the scope of court opinions that authorize DNA database searches and could prompt a lawsuit.[42]

Five months later Rockne Harmon, a former senior deputy district attorney from Alameda County, California, and a strong proponent of familial searching, announced at the New York State Forensic Science Commission meeting on familial searching that California was about to change its policy on the release of partial-match information and unveil a full-scale familial searching program. He repeated these comments two months later at the March 2008 FBI symposium, and when he was asked by a representative from the Louisiana state lab why he was moving forward without legislative guidance, given the potential privacy and state constitutional concerns, he simply responded, "We're doing it without legislation."[43]

On April 14, 2008, the American Civil Liberties Union of Northern California filed a Public Records Act (PRA) request with the California Department of Justice (DOJ), asking for all records relating to the department's policy regarding familial searching, as well as any plans to change its policy. Ten days later (the deadline for the California DOJ to issue a response) the agency released publicly a new "DNA Partial Match Policy." A complete turnaround from Jerry Brown's initial response against releasing partial-match information, the policy not only allows the release of information in the event of partial matches, in concert with the FBI's interim policy, but it goes much further in explicitly allowing low-stringency searches. Requests for these searches are to be considered on a case-by-case basis. The crime-scene profile must be single source (not a mixture), the search must produce a "manageable number of candidates," and Y-STR analysis is required.[44] A minimum of 15 shared alleles is required for partial-match follow-ups; no minimum threshold is given for low-stringency searches.

California is the first state to release an official policy directive on familial searching. Some states, such as Oregon, updated their CODIS operations manuals to incorporate the FBI's interim policy on partial matches. Massachusetts and New York have regulations that explicitly address the issue of low-stringency searches, requiring that a minimum of four loci be provided for a forensic search against the DNA database. The intent of these regulations seems to be to address the issue of crime-scene sample degradation or limited sample availability rather than familial searches. However, Massachusetts provides that "the laboratory . . . may, at its discretion, request that a search be performed using fewer loci if there are scientific reasons which support using fewer than four loci in a particular case, including but not limited to the apparent presence of mixtures, sample degradation, limited sample availability, *or the possible involvement of relatives*."[45] Where other states have permitted familial searches, the threshold of similarity required for allowing follow-up is ambiguously defined and described in terms of such matches needing to "be very, very close" (Virginia) or is set at an arbitrary number of alleles (for example, 21 out of 26 in Florida).

In the absence of both clear legislative authority and an established methodology for familial searching, the move by some states to rush forward with low-stringency searches is at best premature, if not irresponsible. The failures and complications of such a headstrong approach may not rise to the level of public discourse; in general, it is only the sensationalized success stories of forensic DNA that we tend to hear about, not the dead-end investigations, the wasted resources, or even the errors.

Just as the United Kingdom has enacted the most aggressive policies on DNA expansion, it has also introduced a number of new, controversial investigational techniques using DNA, including familial searching (see boxes 4.1, 4.2, and 4.4). Familial searching is currently used routinely in the United Kingdom for high-profile investigations. As of January 2008, 148 cases in the United Kingdom had been submitted for analysis using familial searching techniques. Seventy-nine of those cases were active at that time, and 15 of them had been cleared (i.e., someone is arrested or charged) or resolved (no convictions) through familial searching. Of those 15, 9 had resulted in convictions, 4 in no conviction, and 2 were still going through the criminal justice system in 2009.[46] On average, the cases investigated so far in the United Kingdom using familial searching techniques have generated 1,500 to 3,000 partial matches,[47] which are then narrowed using geographical parameters and Y-STR sampling.

Proponents of familial searching in the United Kingdom boast a 90 percent success rate for those cases where it has been employed.[48] In addition, they argue that it saves money: Richard Pinchin of the U.K. Forensic Science Service claims that between $.5 and $2 million are saved for each murder investigation where familial searching is employed, simply because of a reduction in investigation time.[49] Pinchin also argues that law enforcement should feel obligated to use DNA in any way it can to solve a crime, given the time and resources it has spent collecting DNA from crime scenes and establishing and building the national DNA database.

Unlike the United States, the United Kingdom has developed some procedural guidelines for familial searching. According to Tony Lake, chief constable of Lincolnshire Police and the former chair of the National DNA Database Strategy Board, familial searching is considered only for serious crimes where there has been no match with the database. Familial searching can occur only with approval by the custodian of the National DNA Database (NDNAD) and is approved as "proportionate and ethical" only if it is restricted to the most serious cases and if intrusion into the privacy of individuals is minimized. For each case, a series of search parameters is established, such as level of genetic similarity, allele rarity, age, ethnic appearance, surname, and geographical area. The United Kingdom routinely uses a number of techniques to narrow long lists of partial matches, including Y-STR analysis, mitochondrial DNA analysis, and, finally, taking swabs from relatives. "Swabbing teams" are trained to find a person's DNA in the event someone refuses to give a sample and to handle cases where a potential revelation occurs for a family member who, for example, finds out that he or she is adopted.

So although the United Kingdom has moved forward more quickly with familial searching than has the United States, some recognition, at least, has been given to public concerns about the way in which these searches are conducted and how they might be followed up. Even so, the U.K. practice is not without criticism. Hugh Whittall, director of the Nuffield Council on Bioethics, has commented that the U.K. system still lacks a clear and transparent framework for determining the circumstances under which and how familial searching should be used.[50] According to Robin Williams and Paul Johnson,

> Discussions between ACPO [the Association of Chief Police Officers], the Home Office, the Information Commissioner, and representatives from the Human Genetics Commission, have resulted in an agreement about the circumstances under which such [familial] searches will be carried out and their results integrated into existing investigative procedures. *However,* the agreement is operationally sensitive and has not been publicly disseminated.[51]

A public consultation initiated by the Nuffield Council indicates that the public remains seriously concerned about whether familial searching constitutes an unjustifiable intrusion into personal privacy. The Nuffield Council has recommended that detailed and independent research be conducted on the operational usefulness and practical consequences of familial searching before it is more widely deployed.[52]

Civil Liberties and Familial Searching

From a civil liberties perspective, familial searching is problematic at several levels. First, if it is practiced routinely, it effectively expands the database designed to identify known offenders to include millions of innocent people—those who happen to be relatives of convicted offenders. These individuals have never been suspected of any wrongdoing, yet they are being placed under lifelong genetic surveillance by way of their relation to individuals who are in the database.

George Washington University law professor Jeffrey Rosen has noted that familial searching is inconsistent with a basic pillar of American political thought: "The idea of holding people responsible for who they are rather than what they've done could challenge deep American principles of privacy and equality."[53] The United States has a troubling history of profiling individuals on the basis of their biology, and familial searching, in this sense, plays into this history. Rosen has also argued that familial searching is antithetical to our founding fathers' rejection of the "corruption of blood," which under the common law of England stripped the descendants of anyone convicted of a felony of their right to inherit the felon's estate or title. Section 3 of Article III of the U.S. Constitution states, "The Congress shall have power to declare the punishment of treason, but no attainder of treason shall work corruption of blood, or forfeiture except during the life of the person attainted."[54]

In his carefully crafted memorandum on familial searching to California attorney general Jerry Brown, Deputy Attorney General Michael Chamberlain identifies four possible consequences for individuals who are in the database and are identified as part of a familial search: (1) they may be contacted by detectives; (2) they may not even have relatives who could have committed the crime; (3) they may have relatives who are completely innocent; and (4) they may have relatives who are not themselves in the database, but who will be under suspicion nonetheless. He sums up: "The Databank Program, designed as an investigative scalpel, could be used instead as an indiscriminate investigative fishing net."[55]

This fundamental shift in the intent and purpose of the database—from one of investigation of known offenders to its use as an intelligence tool to investigate individuals outside the database without probable cause—may conflict with some state constitutional laws. For example, the right of privacy guaranteed by the California Constitution has been interpreted by the California Supreme Court to prevent the government (and business interests) "from misusing information gathered for one purpose in order to serve another purpose."[56]

The use of the database to target innocent people also appears to be inconsistent with the very protections that have been built into most, if not all, state database statutes. First, although some states have moved to include categories of innocent people—for example, arrestees—in their databases, most states have in fact rejected these proposals on privacy grounds. In addition, states are required as a condition of participating in CODIS to allow for expungement of DNA records in situations where a conviction is overturned or a case is dismissed, although this usually requires that the falsely convicted individual initiate the process. It seems puzzling to say, on the one hand, that innocent people have the right to have their DNA removed from the system, but, on the other hand, that it is okay to mine the DNA of other innocent people by way of their relatives.

What balances are there (or should there be) against the unfettered use of familial searching? Let us suppose that a familial search brings police to the family members of someone in the national DNA data bank. The person in the data bank—the "genetic informant"—may not have committed any crime. Increasingly, U.S. states are including arrestees; England and Wales include all detained, charged, or arrested individuals over 10 years old. Now this individual is being questioned by police to divulge information about family members who are not yet suspected of any crime. Even if the individual is in prison and considered to have a lower expectation of privacy, he or she nonetheless has no obligation to report on personal matters of members of his or her family.

There are also privacy considerations of the relatives of the individual who is in the national database. Other than the fact that they are relatives of an individual whose DNA shares some limited homology with the crime-scene DNA, there is no suspicion against them for the crime in question. Their privacy must be intact until there is reasonable suspicion. Can investigators prompt them to give a DNA sample? Should investigators be permitted to obtain a DNA sample surreptitiously from family members?

Bieber and colleagues note that through familial searching "genetic surveillance would shift from the individual to the family."[57] Individualized suspicion expands to group suspicion. Courts have been resistant to give police warrants for suspicion based on group properties. The fact that a witness saw a man with a black hat commit a crime does not enable police to invade the privacy of all men with black hats. Courts will not issue warrants for such a group dragnet. It is also unlikely that courts will issue warrants for DNA searches of all family members of a low- or moderate-stringency match in a national database. But what if there were only a handful of individuals identified in a smaller database search? And what if only two of those had relatives who were of a plausible age range for the crime in question? And suppose that one of those relatives has a criminal record? Would that be enough suspicion to yield a warrant for that person's DNA?

In a sense, familial searching creates a backdoor way for law enforcement to investigate people without their knowledge or consent, let alone a search warrant. The call to use Y-STR analysis and other techniques that require returning to the stored biological samples is especially troubling in this regard. These analyses can reveal sensitive information about individuals and their families—information that they themselves might not want to know. Such testing may run contrary to state genetic privacy laws. For example, Section 79-1 of the New York Civil Rights Law prohibits the performance of any genetic test (including DNA profile analysis) without the "written informed consent" of the individual and forbids any other unauthorized testing or dissemination or disclosure of test results.[58]

There are many other unanswered procedural questions associated with the follow up by criminal justice to partial match searches. As discussed earlier, low-stringency searches can generate thousands of partial matches, and these will continue to grow as the databases grow. A partial match indicates only that there is some *possibility* that a relative of that person could have

DNA that fully matches the crime-scene evidence; the probability that the partial match is useful depends, in part, both on the number of alleles that are found to match and on their relative rarity in the population. Even if the list can be winnowed by using a variety of the techniques discussed earlier, the police might be tempted to knock on the doors of tens if not hundreds of individuals, disrupting the private lives of many innocent people unnecessarily. These shotgun-style investigations result in personal and social harm in the form of distress and stigma. Another danger is that in following up with potential relatives or their family members, law enforcement may overstate the significance of the familial match and make it seem as though they "have the DNA." This could lead some individuals to wrongfully confess to a crime they did not commit. In addition, if a partial match is not sufficient evidence to compel a relative to provide a DNA sample via a court order, what happens if those individuals refuse to provide a sample? What is the fate of the samples collected? Will they be destroyed if that person is excluded from the crime? Will there be a temptation on the part of law enforcement to follow people around to get their DNA surreptitiously when a court warrant cannot be obtained because there is insufficient evidence of probable cause?

Family searches may also reveal family secrets. Low-stringency database searches can bring nonsuspects into police inquiries that demand that they reveal intimate and personal information that falls well beyond establishing their identity. It is possible that some of this information could be sensitive and involve paternity, incest, immigration eligibility, human immunodeficiency virus (HIV) status, or fertility. Familial searches can also reveal unknown genetic relationships or an absence of a presumed genetic relationship. For example, a partially matching individual might name someone as a parent or child who turns out not to be genetically related to him or her, or a family member who does not know that he or she has a relative in prison could be contacted by the police and asked to provide a DNA sample. Sonia Suter, a professor at the George Washington School of Law, suggests that the risks to the family associated with these revealed secrets depend on why the information was hidden and how it is disclosed and to whom.[59]

Private information about the family, once it gets into police hands, can change family dynamics. A low-stringency search can create a large pool of suspects from the family. Erica Haimes puts an anthropological perspective on such searches:

> Realizing that one is part of such a large pool of potential relatives could change an individual's or family's self-perception. Individuals might gain or lose social capital through that membership and the collectivity might experience an enhanced self-consciousness of themselves as a group. They might also, for the first time, define themselves as being "related" to each other.[60]

Frederick Bieber uses a license-plate analogy to explain why police seek to justify law enforcement's investigation of partial matches. "Not investigating such leads would be like getting a partial license number in a getaway car and saying 'well, you didn't get the whole plate so we're not going to investigate the crime.' "[61] But the license-plate analogy is problematic. When you buy a car, you agree to have a license plate and to make that license plate visible. The license plate protects you, as well as others. If your car is stolen, the license plate can be used to locate it. If you drive recklessly, someone can report you to the police. By all means, the requirement to have a license plate on your car interferes with your privacy, but this is a reasonable condition for owning a car that we all agree to and are all aware of. The same, it seems, cannot be said of our DNA. Following up a partial DNA match means potentially learning a lot of personal information about a person. It is not a requirement—at least not so far—that we turn over our DNA as a condition of living in this society.

Racial and Ethnic Disparities

Daniel Grimm questions familial searches on grounds of privacy and racial disparities. In regard to the latter, certain minority groups with larger-than-average families will be disproportionately affected by familial searches. More innocent people in large families will become suspects in familial searches. They will be subjected to harassment, surveillance, and DNA sample collection and possibly even permanent storage of the family's DNA. On the basis of the demographics of Hispanic populations in the United States, Grimm concludes, "A partial match between a crime scene sample and an index sample from a Hispanic defendant will, on average, lead investigators to more biological relatives than if the sample had been from another group."[62] Grimm's conclusions were already anticipated by Bieber and colleagues in a 2006 *Science* article. Although the authors argue that by using a combination of cold matches and familial searches, more crimes will be solved, they also conclude that familial searching potentially will exacerbate disparities among racial and ethnic groups that exist in the criminal justice system.[63] In resonance with concerns about racial disparity associated with familial searching, Barry Scheck, cofounder of the Innocence Project, noted, "The genetic surveillance of innocents would be along racial lines. . . . I think it is a troublesome idea."[64]

In practice, further honing in the lens of the criminal justice system on particular populations means that racial minorities who have committed no crime are far more likely to be targeted than white people. It also means that white criminals are more likely to get away with crimes than blacks or Hispanics. These gross imbalances, combined with the increasing use of familial searches, may have the deeper impact of reiterating the faulty and highly dangerous notion that criminal propensity is genetic and racialized.

We have seen how familial searching uses statistical methods to generate suspects in a crime where there is no prime suspect. Under this method forensic DNA has been transformed from a precise tool for identification into a blunt instrument for using DNA similarities to troll for family members of a person who happens to be in a DNA data bank—even while there are no independent grounds of suspicion of those family members. Anyone who has his or her DNA profiled in a state DNA data bank, whether or not that person has been convicted of a crime, brings his or her entire family unit under DNA surveillance. Although police, by tradition and law, have had the right to generate suspects to a crime, there has also been a legal tradition of protecting the privacy and rights of suspicionless individuals. The courts and state legislatures have yet to set boundaries or prohibitions on trolling data banks for suspects on the basis of the fact that the crime-scene DNA is statistically more likely than a random DNA profile to be from a member of the extended family of the so-called near match. Currently, familial searching is carried out without special protections for juveniles, people who voluntarily donate their DNA, surreptitious DNA profiling, and the threats that some families may face by becoming identified as "crime families" because they are more frequently brought up in so-called close matches.

Chapter 5

Forensic DNA Phenotyping

Scientists now have the ability to identify an indefinite number of physical traits including height, eye color, sex and race from a trace of DNA material. Recent breakthroughs in the Human Genome Project (HGP) mandate an expansion of DNA evidence as an investigative tool.

—Lindsy Elkins[1]

A few years from now we're going to have figured out so many traits that a criminal might as well leave his driver's license at the scene of the crime.

—Tony Frudakis[2]

An unidentified girl, presumed to be 3 or 4 years old, was found decapitated in Kansas City, Missouri, in 2001. Local residents had given her the name "Precious Doe" in order to humanize her tragic fate and draw public interest to her case. The police sent her DNA to be analyzed for clues to her ethnic heritage. Results of her DNA test indicated that Precious Doe was of mixed ancestry, approximately 40 percent Caucasian and 60 percent African American. From the tests, forensic specialists in DNA analysis estimated that Precious Doe had a white grandparent. Police also had a tip from an Oklahoma man who said that he was a relative of the slain girl and knew who killed her. They also sought out people who had failed to report the disappearance of a child. From the tips they narrowed the suspects to one woman who had not reported her child as missing and had one black and one white parent. When they ran a DNA test of the woman, they discovered that she was the biological mother of Precious Doe. Both she and her husband, the girl's stepfather, were charged with her homicide. Writing about the case for *USA Today*, Richard Willing noted, "Advances in DNA testing are allowing investigators to learn more about suspects whose profiles are not in the databases. Tests that can identify a suspect's ancestry are being used not to identify the suspect by name, but rather to give police the idea of what he or she looks like."[3]

The term "DNA profiling" first came into use when Alec Jeffreys discovered how to use DNA samples to determine identity or paternity. "DNA profiling" became synonymous with terms like "DNA typing," "DNA fingerprinting," "DNA identification," and "forensic DNA testing." In this sense DNA profiling means sequencing the DNA loci that can be used to compare two samples to determine if they came from the same person. The *DNA profile* is the set of 26 numbers that characterize the short tandem repeats (STRs) at 13 loci in the human genome (see chapter 1).

However, the genetic age has brought with it a new usage of DNA profiling that is similar to visual profiling of crime suspects. In a visual profile an eyewitness identifies some characteristics of the alleged perpetrator of a crime, such as height, hair color, distinguishing marks, skin color, and body size and shape. A composite sketch (profile) is created from bits of eyewitness information. In *forensic DNA phenotyping* forensic scientists analyze DNA from a crime scene in an attempt to determine certain aspects of the individual's physical characteristics. This "second generation" of DNA profiling goes well beyond examining the 13 loci that are used and stored in the U.S. Combined DNA Index System (CODIS) database. Instead, one or more genetic tests are run to glean information from the DNA that might provide clues about the suspect's physical features. Lindsy Elkins describes DNA phenotyping as follows: "Scientists can now discern from DNA a virtually indefinite number of physical traits possessed by an individual from height, eye color, sex and race, down to the shapes of a person's toes. In addition, genetic typing permits inferences as to inherited disorders and may offer clues to facial or other bodily features."[4]

Forensic DNA phenotyping is not without socioethical concerns. These include the following: Does DNA phenotyping by police infringe on civil liberties? Will DNA phenotyping exacerbate racial or ethnic stereotypes? Will it help law enforcement solve crimes? Will DNA phenotyping reduce the use of prejudicial and subjective (based on eyewitness reports) profiling?[5]

How scientifically credible is phenotyping by DNA? Because the technology of DNA phenotyping is still in its early stages of development, the answers to these questions must be tentative.

Linking Genotype to Phenotype

Lindsy Elkins in her law review article on physical profiling based on genetics is far too optimistic about the relationship between the DNA and the phenotype ("down to the shapes of a person's toes").[6] There are significant uncertainties associated with predicting racial or ethnic origins or physical qualities on the basis of DNA. The remaining sections of this chapter discuss the current possibilities and limitations for "phenotyping" DNA. Can we draw inferences about physical appearance or the medical condition of an individual from his or her DNA? What are the possibilities of creating a physical profile of an individual exclusively from crime-scene data?

Three research programs in genetics have provided the scientific foundations behind forensic genotype-to-phenotype profiling: ancestral testing, behavioral genetics, and medical genetics. In each case scientists seek to find links between components of the human genome and an individual's physical characteristics, behavior, or response to environmental factors such as medication or antigens.

Ancestral Genotyping

The Human Genome Diversity Project, an outcome of the Human Genome initiative, has been used to identify genetic patterns that reveal one's biogeographical ancestry. Population-specific alleles (PSAs) enable geneticists to distinguish the genotypes of a select number of ethnic populations, such as Europeans, Africans, Native Americans, and Asians. And although there are more genetic variations within ethnic/racial groups than between groups, some researchers have found that there are sufficiently stable genetic segments (haplotypes) that can distinguish among several ancestral population groups. These are called ethnic geographical markers of ancestral origin, or Ancestry Informative Markers:

> Ancestry informative markers (AIMs) . . . are autosomal genetic markers that show substantial differences in allele frequency across population groups. . . . Factors such as isolation by distance, range expansions, land bridges, maritime technologies, ice ages, and cultural and linguistic barriers have all affected human migration and mating patterns in the past and have therefore shaped the present worldwide distribution of genetic variation.[7]

Current ancestry tests examine upwards of 176 genetic markers in which the DNA varies at only one position. These single-nucleotide polymorphisms (SNPs) have been found to occur more frequently in certain population groups than in others— the result of centuries of geographical separation and group intermarriage.[8]

Because most people have mixed biogeographical ancestral origins, the programs that yield information of this kind provide percentages (such as 80 percent African and 20 percent European). The composite percentages typically cannot yield definitive information about the phenotype of specific individuals falling within these biogeographical categories, but the question of probabilities for populations is left open.

According to Susanne Haga, assistant research professor at the Duke Institute for Genome Science and Policy, because admixture confounds the prediction of ancestral origins, "a substantial gap remains between ancestry and/or race and physical appearance. . . . Knowing an individual's race [percentage of racial origin] can be misleading if used to predict certain physical traits."[9] When someone is told they are 60 percent Asian and 40 percent European it is based on large segments of DNA. We do not know what the 60 percent Asian DNA will express itself as; therefore we cannot from those percentages say anything about the phenotype of the individual. But if we had probability figures, such as "90 percent of the people who are tested by DNA ancestry as 60 percent Asian have 'yellow' skin tone," then we may be able to draw some probabilistic predictions about phenotype.

Ancestral Genomics in Law Enforcement

Ancestral genomics is beginning to find a place in law enforcement. In March 2003 a company known as DNAPrint Genomics of Sarasota, Florida, analyzed the DNA of a serial killer in Louisiana using a genetic ancestry technique trademarked as DNA Witness. The company concluded that the killer's "biogeographical ancestry" was 85 percent sub-Saharan African and 15 percent Native American. Until then, police had been seeking a Caucasian male in a pickup truck. The analysis of biological evidence at the crime scene by DNA Witness concluded that the Louisiana task force's search was misguided, and that the individual they were looking for was more likely to be a "lighter skinned black man." This description was inferred from probabilistic ancestry percentages revealed in the perpetrator's DNA. The Louisiana police were dubious about the reliability of DNA Witness in profiling the serial killer. They sent the company 20 other DNA samples of individuals of whom they alone knew the ethnic or racial identification. According to the research director at DNAPrint Genomics, the company correctly identified the ancestry of all the samples.[10] As a result of the analysis, Louisiana police shifted the focus of their investigation and identified Derrick Todd Lee, an African American man, as a possible suspect in the crime. Lee became the first person in the United States to be identified as a possible murder suspect through the use of a DNA test that racially profiled his DNA. On August 11, 2004, Lee was convicted in the first of a series of murder and rape cases.

DNAPrint Genomics used its success in Louisiana to aggressively market its services to police departments, investigators, and agencies.[11] Toward the end of 2004 it started offering "RETINOME," a genetic test to infer eye color, to law-enforcement agencies in addition to ancestry testing.[12] Here is how the company described its forensic profiling services:

> Testing DNA to create a physical description from crime scene DNA and providing a photo database array of representatives closely matching the analyzed DNA, allows detectives a means of describing "persons of interest." This presumptive test method is a new market based on evolving DNAPrint™ Genomics technologies. Common hereditary traits such as skin pigmentation, eye color, hair color, facial geometry and even height can be predicted through analysis of DNA sequences. This can be done indirectly, through an extensive knowledge of ancestry admixture or for certain genetic traits, directly through knowledge of the underlying genes. At DNAPrint™ we use both methods. Our goal is to continue to lead the field of forensic presumptive testing using our DNA Witness™ line of products and services.[13]

The DNA Witness software technology was contracted out by the Boulder, Colorado, Police Department in 2004 to develop suspects in the highly publicized investigation of the rape-murder of a 23-year-old woman. In December 1997 Susannah Chase was walking home after getting a pizza when she was savagely raped and beaten to death with a baseball bat. When her murder investigation failed to lead to any suspects, it was put in the cold-case file. Six years later, in December 2003, the Boulder Police Department contacted DNAPrint Genomics to acquire some phenotypic information from the male sperm preserved from the victim's body. In its January 2004 report the company stated that the source of the DNA was a person of Hispanic or Native American background. According to Boulder police chief Mark Beckner, "This technology gave our detectives a focus and direction that turned out to be right on the mark."[14]

Police arrested a prime suspect in January 2008 by the name of Diego Olmos-Alcalde, a Chilean native who had an uncertain U.S. immigration status. DNAPrint Genomics appeared to have correctly identified the Hispanic ancestry of the person whose sperm was found at the crime scene. But what role did the ancestry information have in solving the crime? Did the phenotype "Hispanic or Native American" help police narrow the lead to the prime suspect?

As it turned out, it was not the phenotyping of the DNA that solved the case, but a match on CODIS, albeit a delayed one. The suspect was known to both Colorado and Wyoming police authorities for other felony arrests. He was arrested in 1998 in Denver on charges of attempted sexual assault and carrying a concealed weapon. However, prosecutors dropped the sexual assault charge in exchange for a guilty plea to the weapon charge.[15] In May 2001, while in Wyoming, Olmos-Alcalde was given a 12- to 20-year prison term for kidnapping a Cheyenne woman in a parking lot near her apartment. Because of errors made by the trial judge, the Wyoming Supreme Court overturned the conviction and ordered a new trial, whereupon Olmos-Alcalde was resentenced in September 2004 to 7 to 10 years, with credit for time served.[16] The defendant was paroled in the summer of 2007.

When the Denver police got around to loading the crime-scene DNA profile into CODIS, there was no match, that is, not until six months after Olmos-Alcalde was released from prison. If Olmos-Alcalde's DNA profile had been uploaded to CODIS as soon as he entered prison, the police would not have needed the services of DNAPrint Genomics. There are two likely explanations for the delay. It is possible that his DNA was not profiled when he entered the Wyoming penal system but only after he was released. Alternatively, the profile of his DNA was not loaded into CODIS until he got out of prison or after some combination of delays. One report stated that officials were not sure when his DNA was profiled: "Melinda Brazalle, spokesperson for the Wyoming Department of Corrections, said the Department's policy calls for collecting DNA from inmates during the intake process. But she wasn't certain when Alcalde's genetic information was obtained."[17] Colorado loaded the DNA sample taken from Chase's body into CODIS as early as 2002 without a match, while Wyoming entered Olmos-Alcalde's profile into CODIS in January 2008.[18] The reason for the delay was the backlog of samples awaiting DNA processing. At the time there was a backlog of about 180,000 federal convicted-offender samples awaiting DNA processing and about 50,000 samples that were already processed but were waiting to be entered into CODIS.

There are two points to this story. First, DNAPrint Genomics's ancestry analysis did not help solve this case; rather, it was solved from a DNA profile match in CODIS. Second, had there been no backlog in processing convicted felon DNA, police would have had no reason to consult DNAPrint Genomics.[19]

When the National DNA Advisory Board was considering the alleles it would recommend for use in forensic testing of crime-scene samples, its members made a deliberate choice to avoid any sequence that had known phenotypic properties or that disclosed ancestral origins. According to Ranajit Chakraborty, a board member and director of the University of Cincinnati's Center for Genome Information, "In 1997, when members of the national DNA Advisory Board officially selected the gene markers for DNA evidence matching, they could have included a few markers associated with ancestral geographical origins (European, East Asian, Sub-Saharan African), which are a good indicator of race and ethnicity." But the board instead decided against using racial markers because of the political sensitivity they represented.[20] During the period in which CODIS was being developed, there was a heightened sensitivity both in law enforcement and among scientists that there were privacy issues involved in decoding people's DNA that law enforcement should steer clear of. Although it was the board's noble intention to use DNA exclusively for identification purposes, scientists continue to investigate methods that link allele variations in the STRs with racial and ethnic ancestry.[21]

Behavioral Genetics and Profiling

Research in behavioral genetics seeks to find links between genotype and certain human behaviors, specifically, but not exclusively, criminal behaviors.[22] Thus, if genes could be strongly correlated with a person's "fits of anger" or one's pedophiliac tendencies, police could use genetic screening to narrow—or generate—a list of suspects in the search for a perpetrator of a violent and/or sexually deviant crime. Claims made by behavioral geneticists have sparked vigorous debates. Some scientists have critiqued the underlying science, pointing out the inherent limitations in studies of simple correlations that occur between genetic factors and complex behaviors.[23] Others have cautioned that the miscommunications and misapplication of behavioral genetic research to policy could result in grave social consequences, especially in the area of criminal justice.[24] Still others view the new field of behavioral genetics as a reincarnation of widely disavowed beliefs in genetic determinism and eugenics.[25]

In 1965 a study published in *Nature* found that a significantly higher number of inmates in a prison hospital in Edinburgh, Scotland, described as "dangerously violent" had an extra Y chromosome (XYY males), compared with the general population.[26] The authors hypothesized that the presence of an extra Y chromosome produces extra aggressiveness and concluded that this condition increased the chances that an individual would be institutionalized. In 1970 President Richard Nixon's personal medical adviser suggested that the country make use of this science, proposing a massive program of genetic screening for every 6-year-old to detect the "criminal potential" of preadolescents. Moreover, he suggested that every "hard-core 6-year-old" be sent to "therapeutic" camps where they could learn to be "good social animals."[27]

The XYY study and other similar early studies lacked a satisfactory control group. When the studies were replicated with the proper methodology, scientists learned that XYY males tended to be taller, less intelligent, and more hyperactive but not necessarily more violent than their XY counterparts.[28] The search for biological markers for social deviance has been a central theme in behavioral genetics. If such markers were discovered and their reliability were demonstrated, their forensic use would be nearly impossible to prevent so long as the entire human genome remains within the reach of law enforcement.

Another genetic marker called MAOA deficiency has been linked to violent behavior. A 2002 study published in *Science* by Avshalom Caspi and his colleagues looked at the DNA of 1,037 children who had participated in a 23-year survey of health and development conducted in New Zealand. Caspi and colleagues observed that a particular polymorphism—a gene on the X chromosome that regulates the production of the enzyme monoamine oxidase A (MAOA)—tended to moderate the effect of childhood maltreatment. That is, children with a version of the gene that caused greater production of MAOA were less likely to respond to maltreatment by lashing out and developing antisocial tendencies, while children with the low-MAOA producing version of the gene had a higher rate of violent behavior. The MAOA gene oxidizes a chemical tryptophan into serotonin, one of several neurotransmitters (with dopamine and norepinephrine) that it uses to send signals. Mutations in MAOA can alter the amount of serotonin produced in the brain and biochemically affect emotion and behavior. Caspi and colleagues' study makes no broad claims about genetic determinism but rather asserts that genetics may lend a partial explanation for differing behavioral responses to childhood maltreatment and concludes with the conjecture that "these findings could inform the development of future pharmacological treatments."[29]

The MAOA gene could play several roles in criminal justice. Defense lawyers may seek to use MAOA testing results to support their client's claim of mitigating circumstances in a violent crime (i.e., genetic defense against culpability). Alternatively, detection of an MAOA mutation in a crime-scene sample can lead police on a search for individuals with the defect, culling medical information for the match or using the result to profile individuals who may have a history of aggressive behavior. MAOA screening could become a factor in sentencing determinations or a requirement for consideration for early parole. Correctional facilities might prescribe routine MAOA testing for prisoners and forcibly administer medication to those presumed to be prone to violence on the basis of that testing. A 2002 Consensus Report by the Council of State Governments found that "staff at many correctional facilities have overrelied on the use of psychotropic medications and, in many cases, sedative-hypnotic medications, simply to pacify and to control inmates with mental illness and others believed to be disruptive."[30] Medication is often administered without adequate psychiatric evaluation and follow-up counseling and is often the sole treatment for mental illness and even nonclinical behavioral problems.[31]

The potential applications of behavioral genetic research to the criminal justice system raise profound questions for individual liberty and social justice. Past attempts to apply behavioral research to social policy—as evidenced most poignantly by the eugenics policies of the early to middle parts of the twentieth century—have been dismissive of the rights of individuals and vulnerable groups. Despite the increasing precision of molecular biology and repeated calls for caution about making social claims based on such research, there is a recurrent tendency to "biologize" crime and antisocial behavior and to use behavioral science as a justification for inequality and the promotion of new measures of social control. Therefore, we should be wary of any applications of behavioral genetics to the criminal justice system, whether for the narrow purpose of testing crime-scene samples on a case-by-case basis to predict behavioral traits of individual suspects or for the far broader purpose of characterizing the future dangerousness of the prison population or the general citizenry.

Medical Phenotyping

Medical geneticists seek to identify genes or multiple genetic loci (genetic markers) that correlate with individual disease states, predisposition to disease, or drug sensitivity. According to Wojciech and colleagues, "Studies developing genotype to phenotype correlations have advanced rapidly in medicine."[32] According to the National Institutes of Health, genetic tests are currently available for more than 1,700 diseases.[33] In 2004 Hui Huang and colleagues utilized nearly 1,200 human disease sequences in a study.[34] Some of these genetic mutations are asymptomatic, such as the Tay-Sachs trait, where only one copy of the mutation will not trigger disease manifestations. Others, such as Duchenne muscular dystrophy, which afflicts boys whose mothers are carriers, are severely disabling and prevent a person from living an independent life. Between these extremes are many cases where the genetic mutations may be connected to a disease of late adult onset where there are different intensities of disease expression. If DNA at a crime scene were analyzed for some or all known disease-related genetic mutations and there were a positive result for one that required medical treatment, police could track down potential suspects from hospital or pharmacy records of individuals treated for the relatively rare condition.

Consider the following scenario:

> A series of burglaries take place in a small town in Virginia. DNA evidence in the form of a blood sample is collected from a window ledge of one of the homes. The Virginia authorities run the DNA against the state database; no match is found. They then send the sample to an outside laboratory for further testing. The lab runs a series of genetic screens on the sample and finds that the sample contains all four of the mutations that are among those most commonly associated with Gaucher's disease. Upon receiving this analysis, Virginia authorities contact the National Gaucher Foundation and learn that most people living with Gaucher's disease receive biweekly enzyme-replacement treatments. Furthermore, these treatments are administered at only two locations in Virginia, the Children's Hospital and the University of Virginia. They contact the treatment centers and request a list of all individuals who have been treated for Gaucher's disease over the last five years. The Children's Hospital complies and turns over the records, but the University of Virginia refuses, claiming that this is private medical information that cannot be released.

To our knowledge, this scenario has not occurred. However, the facts about Gaucher's disease and the locations of the treatment centers are accurate. In their pursuit of a felon, can and should the police be permitted to explore medical information from the genome of crime-scene evidence? Currently there is nothing to stop law enforcement from using whatever means it chooses to identify and analyze evidence obtained from the scene of a crime, including blood, tissue, hairs, and semen—that is, any biological materials. Crime-scene evidence does not possess any rights or privacy. The person who left that evidence at the scene or who is the source of the biological materials found at the scene has legal rights and privacy rights under the Fourth Amendment, but those rights do not include a protection against police acquiring information from the shed DNA. Therefore, criminal investigators can use whatever techniques they choose to find suspects from the evidence left at the crime scene. Some law-enforcement agencies have even resorted to hiring psychics who claim to have extrasensory powers that can provide information from objects found at the crime scene.[35]

If investigators are allowed to obtain scientific evidence in the form of mutations that are strongly correlated with Gaucher's disease from discarded

DNA at a crime scene, they then have a phenotypic clue to narrow the field of possible suspects. Can police gain access to medical records from hospitals, physicians, or medical centers for an undesignated and unspecified number of individuals who meet a specific medical criterion? Should this be considered a normal part of police investigation, or is this an intrusion on medical privacy that must be accompanied by a court warrant? Is this a type of "medical dragnet" without informed consent?

When police suspect that a fleeing suspect has been shot, they contact hospitals and seek information on whether someone was recently treated for gunshot wounds. Why would contacting local hospitals for information about people who are treated for Gaucher's disease be any different? One answer is that when a person has a gunshot wound, he cannot be expected to have an expectation of privacy because he has entered the hospital in full view of everyone in his direct sight and is expecting emergency attention, perhaps to save his life from loss of blood or from damage to an organ. The situation is different for his medical information that may reside in hospital records.

Federal law does protect individual privacy in regard to personal medical records, but those protections are not absolute. If there is probable cause that an individual may have committed a crime, police can usually obtain a court warrant to gain access to medical records of an individual suspect. Confidentiality of medical records is established under the Health Insurance Portability and Accountability Act (HIPAA, passed on August 21, 1996). Applied to public health information, confidentiality means that information or data are not made available or disclosed to unauthorized persons and without proper cause. However, HIPAA contains a broad exception that allows disclosure of protected health information to law-enforcement officials, not only in compliance with a court order or grand-jury subpoena but also in response to an administrative subpoena, summons, or civil investigative demand—all legal instruments issued without judicial review.[36] Broad administrative discretion is given to those with stewardship over health information at the hospitals in determining how to respond to written requests from law enforcement for patient records. HIPAA also allows health-care providers to disclose to law enforcement, on request, a broad array of identification information, including name, address, Social Security number, blood type, date of treatment, and a physical description.

When there is individualized suspicion and probable cause, police generally have no problem obtaining a court warrant for a suspect's medical records. But in the case outlined, there is no individualized suspicion and therefore no probable cause against any single Gaucher's patient. Should law enforcement be able to acquire information about all people of a certain geographical region who are treated for Gaucher's disease because the crime-scene DNA has been medically screened and shown to have the alleles for the disease?

Judges do not typically give police investigators warrants when they are on a fishing expedition without probable cause that an individual or some small number of individuals is suspect. So our investigators may not be successful in getting a "wide-net" warrant from the courts to obtain the medical records of all Gaucher's patients in an area. Should police be legally permitted to access medical records at Children's Hospital and the University of Virginia without court warrants on the basis of the finding that DNA at the crime scene (which may or may not be the perpetrator's) has the Gaucher mutation? Will the privacy of medical records keep the police from obtaining the information about males treated for Gaucher's disease without a warrant? Will health centers turn over the information under current federal medical privacy statutes?

In our scenario the police are undertaking a kind of medical dragnet. Is a medical dragnet without informed consent ethically and/or legally justifiable? Should the same informed-consent principles hold for a medical dragnet that we expect to be in place for a DNA dragnet? Suppose that Children's Hospital turns over to the police the names of the male patients who have been treated there for Gaucher's disease over a period of 10 years. If police then narrow their suspects to three males on the basis of treatment for Gaucher's disease and a suspected age range, is that reasonable suspicion to get a warrant for their DNA? These questions remain to be answered by the courts.

Predicting Sex, Hair Color, Eye Color, and Skin Color from DNA

Companies like DNAPrint Genomics that began providing products and services to the criminal justice sector for correlating phenotype with segments of DNA focused on a few genes and chromosomes where there had already been published studies that linked DNA alleles to physical characteristics. The most definitive trait prediction from DNA is sex. This can be done by chromosome analysis because the male has an X and a Y chromosome, while the female has two X chromosomes. The sex of a DNA donor can also be determined by a gene called the amelogenin sequence, which is present on both the X and Y chromosomes. But the genes on these chromosomes have different sizes, and those sizes can be read off a DNA analyzer.[37] There are also male-specific genes (e.g., SRY) and female-specific genes (AR), which can be analyzed by PCR techniques.[38] The amelogenin genetic analysis has been used in forensics and prenatal diagnosis.

According to Elkins, "Genetically-derived trait information may be superior to human-derived trait information because, unlike humans, machines cannot be fooled by changes in physical appearance."[39] This may be true, but the practical value of predicting appearance is less clear. For example, let us suppose that a witness described a possible suspect in a crime as having blond hair. Let us also imagine that the perpetrator of the crime left DNA at the crime scene, and a DNA test revealed that the suspect in fact had brown hair. Indeed, such a test might have greater reliability for natural hair color than an eyewitness report. But how helpful would this information be in tracking down the suspect if in fact the individual had dyed his hair?

In terms of inferring human nondisease physical characteristics from DNA left at a crime scene, we probably cannot do much better than the phenotype red hair/light skin. Red hair has been linked to variants of a single gene called MC1R, which encodes the melanocortin-1 receptor. Receptors reside at the surface or in the nucleus of cells and are acted on by specific proteins (such as hormones) to produce other proteins through the mechanism of DNA transcription. The action of a stimulating hormone called alpha-melanocyte (aMSH) on the receptor MC1R controls the switch between a red/yellow substance (phaeomelanin) and the black/brown substance (eumelanin). People with red hair produce more phaeomelanin than brown- or black-haired people. One study found 12 variants of MC1R. Of those, 8 were associated with red hair. The authors found that for 96 percent of individuals they studied, those with 2 of the 8 red-hair-causing mutations had red hair. Two of their subjects who had the mutations but were not red haired (either blond or light brown) described themselves as having had red hair in their youth.[40]

Writing in the *Journal of Forensic Sciences*, Branicki Wojciech and colleagues noted that "determination of one of the [relevant] MC1R variants in the homozygous state or heterozygous combination can be considered a strong indicator that the sample donor has red or strawberry-blond hair and fair skin."[41] The relationship between certain genetic variants in MC1R has been reproduced and validated.[42] From a forensic viewpoint, if a DNA sample shows two MC1R sequence mutations, it is a good bet that the donor of the sample has or had red hair. If the DNA shows that the individual is homozygous (has two copies) of the allele that is not associated with red hair, then it is a good bet that the donor of the DNA does not have red hair. The presence of one mutation would be inconclusive.[43]

Iris color is largely a genetic rather than an environmental trait. Studies of twins show high correlations (about 85 percent) of iris color between homozygote twins. How much pigment a person expresses in his or her iris is linked to three SNPs, also known as single-letter variations, in a DNA sequence near a gene known as OCA2. According to a 2007 study in the journal *Human Genetics*, OCA2 is the major human iris-color gene, and SNPs within this gene can accurately predict melanin content from DNA.[44] Among the individuals carrying the same SNP sequence in all three locations on both copies of the gene, 62 percent were blue eyed.[45] Because iris color does not change with age or sunlight, it is viewed as a stable and predictable trait that could be useful for crime-scene DNA profiling. However, with only slightly more than 50 percent genotype-to-phenotype predictability, phenotyping for iris color has limited reliability.

Skin color is considered a polygenic trait. This means that many genes in different loci of the human genome are responsible for pigmentation of the individual. There may be 30 to 40 genes responsible, which may act both additively and nonadditively in producing a spectrum of pigmentation colors that are found within the human family. Others have speculated that there are more than 120 genes that play some role in skin pigmentation. Variations of the color of human hair and skin are determined by the amount, density, and distribution of the two components making up the pigment melanin, which is produced in specialized cells known as melanocytes. The synthesis of two organic polymers, namely, eumelanin (dark brown/purple/black pigment) and phaeomelanin (yellow to reddish brown pigment), determines the amount of melanin in the skin. The exact role of the genes responsible for producing the ratio of these pigments (eumelanin and phaeomelanin) is not fully understood. Much of what is known about human pigmentation has been learned from mouse studies.[46] Not only are the genetics and chemical pathways of melanin synthesis complex, but environmental factors in skin pigmentation also play a role.[47]

In his book *Molecular Photofitting* Tony Frudakis reports that an individual with each of nine major variants of a gene MC1R will more likely than not express a fair or pale skin phenotype relative to the average European. But he adds that "we need to be able to measure more than just the MC1R genotypes to make inferences and because skin color varies so much with biogeographical ancestry (on a continental level), the approach that could be useful is the admixture mapping approach."[48] Admixture mapping is a tool for uncovering genes that contribute to complex traits. It involves examining the gene frequencies of two or more genetically diverse intermating populations (admixture populations).

Given the complexity of the genetics and the gaps in our knowledge of human skin pigmentation, we cannot yet go directly from DNA to skin color. The indirect route to specifying skin color from genotype is through ancestry analysis, as previously discussed. However, ancestry analysis will yield mixtures of a person's biogeographical heritage; skin pigmentation cannot be predicted determinatively from that knowledge.

Troy Duster reported that the forensic science laboratory in Birmingham, England, claimed that its DNA test can distinguish between "Caucasians" and "Afro-Caribbeans" in nearly 85 percent of cases.[49] The original work cited for this result was published in 1993.[50] Bert-Jaap Koops and Maurice Schellekens note that the differential power of phenotyping can result in racial bias:

> The use of characteristics for a composite drawing or description of the perpetrator of a crime may lead to stigmatization of certain groups within society. . . . This is an even greater risk if one ethnic background can be better determined than others, as the FSS [Forensic Science Service, United Kingdom] indicates is the case with Afro-Caribbeans. This might lead to relatively more cases involving Afro-Caribbeans, simply because they can be determined where other DNA types would yield inconclusive information.[51]

This is the DNA counterpart to the old joke, "Why are you looking under the streetlight for your lost key?" The answer: "Because that's where the light is."

Tony Frudakis, formerly of DNAPrint Genomics, also supports the use of DNA for developing "facial geometry" of an individual. This might include facial bone structure as well as the size and shape of the face. In a project at University College London, scientists scanned the faces of hundreds of volunteers in order to find correlations between digitized facial geometry and genetic markers, but the project was abandoned in 2000 when it proved too complex. One German forensic biologist commented that we may never be able to fully reconstruct a suspect's face from genes alone.[52]

Policies on Phenotyping DNA

The debate over forensic DNA phenotyping has been more acute in the European Union than it has in the United States. Only three states, namely, Indiana, Rhode Island, and Wyoming, disallow by statute the use of DNA submitted to their data banks for purposes of obtaining phenotypic information about the DNA sources.[53] For example, the Wyoming law states that "information contained in the state DNA database shall not be collected or stored for the purpose of obtaining information about physical characteristics, traits or predispositions for disease."[54] But this does not restrict phenotyping crime-scene DNA to generate information about suspects before the DNA enters the data bank. Michelle Hibbert surmises that the vast majority of states permit DNA trawling in order to allow police investigators to trace unknown suspects or anonymous murder victims.[55]

At the federal level, as discussed earlier, HIPAA provides individuals privacy protection in regard to their medical information and data. However, HIPAA contains broad exceptions that provide medical administrators discretionary authority to disclose information to law-enforcement officers upon request. As a result, phenotypic testing of a bloodstain left at a crime scene that revealed that the suspect had a rare medical disorder could very well lead to a medical dragnet.

The European Union has issued a number of directives prohibiting the exchange of DNA information that can reveal physical traits. On June 25, 2001, a European Council resolution urged member states to exchange only the results of DNA analysis of noncoding chromosomal loci. Under the Prüm Convention of 2005, party states cannot make available DNA data drawn from coding segments of the DNA.[56] Belgium and Germany explicitly prohibit deriving physical traits other than gender from DNA. In contrast, a 2003 amendment to the Dutch Code of Criminal Procedures permits investigators to obtain phenotypic information from DNA found at the scene of a crime when the suspect is not known.[57] The Netherlands is the only country to explicitly approve the use of DNA forensic phenotyping.[58]

Increasingly, when law-enforcement investigators cannot get a match between the DNA found at a crime scene and the DNA profiles in their database, they use the services of specialized firms that construct phenotypic profiles of the suspected perpetrator from markers in the crime-scene DNA.[59] This trend is likely to continue with the advent of gene chips, or DNA microarrays, such as those that have been developed by the company Affymetrix.[60] These gene chips allow researchers to access information on thousands of genes simultaneously.

While at DNAPrint Genomics, Tony Frudakis predicted that in the future people's DNA will be used to describe their physical appearance.[61] Thus far there is little science to support that prophecy. Nonetheless, efforts to develop genotype-to-phenotype indicators will undoubtedly improve as financial investments in research, ancestry analysis, and forensic markets for crime-scene DNA profiles expand.

When a person's DNA is taken for phenotyping, information about genetic predispositions can be revealed unexpectedly. For example, a suspect could learn the information for the first time in a courtroom or during sentencing. This scenario breaches a basic principle in medical law. People have the right to know—or *not* to know—about their genome. This right cannot be protected if forensic authorities gain access to sensitive genetic information and are authorized to use it as evidence against a suspect.

Some traits are less sensitive than others. If direct phenotyping of DNA could be limited to externally perceptible traits, such as hair color or stature, and nonsensitive internal or behavioral traits, such as voice type, left-handedness, or absolute pitch, then an appropriate balance between preserving genomic privacy and assisting law enforcement might be achieved. These traits raise no serious objections based on the right not to know, privacy, or the risk of stigmatization.[62] Perhaps phenotyping the DNA from unknown subjects can fall under an ethical principle "the right not to know" when the proposed DNA test is for characteristics that are beyond the individual's physical identifying features. Unfortunately, no such distinction has been made by crime investigators to date, and in the meanwhile, the temptation on the part of law enforcement to mine crime-scene DNA to make predictions about the physical, behavioral, or medical conditions of the alleged perpetrator only increases. Already claims have been made that genetic factors have been found that are associated with sexual orientation, intelligence, addictive behavior, and aggression. Even if they are unsound, law enforcement will be tempted to use them to generate profiles of suspects from the DNA, such as "Likely to be a tall, African American homosexual male, with high intelligence, a propensity for addiction, and recessive for sickle cell anemia." Koops and Schellekens argue that "once phenotyping for criminal investigation is an accepted practice, phenotypical information will be used for other purposes as well, such as eugenics to redress bad genotypes related to aggression and pedophilia."[63]

Chapter 6

Surreptitious Biological Sampling

We can't go anywhere without leaving a bread-crumb trail of identifying DNA matter. If we have no legitimate expectation of privacy in such bodily material, what possible impediment can there be to having the government collect what we leave behind, extract its DNA signature and enhance CODIS to include everyone? Perhaps my colleagues in the plurality feel comfortable living in a world where the government can keep track of everyone's whereabouts, or perhaps they believe it's inevitable given the dangers of modern life. But I mourn the loss of anonymity such a regime will bring.

—Judge Alex Kozinski[1]

People constantly leave genetic material, fingerprints, footprints, or other evidence of their identity in public places. There is no subjective expectation of privacy in discarded genetic material just as there is no subjective expectation of privacy in fingerprints or footprints left in a public place.

—Justice Charles W. Johnson[2]

In 1974 Barbara Lloyd was raped and stabbed to death in her home in Buffalo, New York. Police had a suspect in Leon Chatt, the husband of the victim's stepsister at the time of the murder, but no evidence to connect him to the crime scene. Over three decades later the police revived the case with DNA retained from the forensic evidence of the crime scene. Although police did not have enough evidence to obtain a warrant for a search, they followed Chatt around, picked up his DNA after he spat on the sidewalk, and compared it with the 30-year-old crime-scene sample. He was charged with and convicted of one of Buffalo's oldest unsolved crimes in 2007.[3]

This is one example of a number of documented cases where police have collected DNA from individuals surreptitiously. Sometimes police have acquired these items by offering an individual a cigarette or a drink during an interrogation and then collecting the items afterwards; at other times police have simply followed individuals around without their knowledge, picking up items that they have discarded that might contain their DNA.

The presumption of law enforcement in these cases is that the practice of collecting and analyzing an individual's DNA without his or her knowledge or consent is legal. Law enforcement's primary argument in support of this position is that the DNA one leaves behind is "abandoned," and furthermore, an individual who "abandons" his or her DNA no longer has any privacy interest in it or in the information it holds about that individual.

In the handful of state court cases that have considered this issue to date, law enforcement's perspective on the use of so-called abandoned DNA has prevailed. The remainder of this chapter discusses one of these cases in detail and analyzes the implications for genetic privacy of law enforcement's prevailing view on DNA collected surreptitiously.

State of Washington v. Athan

On November 12, 1982, the Seattle police found the body of a 13-year-old named Kristen Sumstad. The teenager had been stuffed into a cardboard box in the Magnolia neighborhood of Seattle after she was sexually assaulted and strangled. In their search for the perpetrator, the police had questioned John Nicholas Athan, whose brother they claimed saw John transporting a large box on a grocery cart around the time and in the neighborhood of the crime. When questioned, Athan told police that he had been in the neighborhood stealing firewood the night before Sumstad's body was discovered. Police did not have enough evidence to charge Athan with the crime. The case went cold.

Twenty years later the Seattle Police Department's Cold Case Unit sent biological evidence preserved from the crime scene to the Washington State Patrol Crime Laboratory. The laboratory investigators sequenced and profiled the crime-scene DNA that had been taken from the body of Sumstad and compared it with DNA profiles in the state and federal DNA data banks, but no match was found. The police then turned their attention to living suspects from two decades earlier, including Athan.

The police located Athan in New Jersey. Concerned that he might flee the country if they approached him directly, they created a ruse to get the suspect to give up his DNA. The detectives posed as lawyers from a fictitious law firm and wrote Athan a letter inviting him to be their client in a fabricated class-action suit filed against municipalities that overcharged on parking tickets. Athan believed that he could share in the settlement costs if they prevailed in the suit. The letter he received contained the names of a fictitious firm and the names of fictitious attorneys, which in actuality were the real names of members of the Seattle Police Department. Athan signed, dated, and returned the fabricated class-action authorization form and then licked the envelope and mailed it. It was then retrieved by the police, who removed the flap, photographed the contents, and had it tested for DNA from the residual saliva that was used to seal the envelope, all without a warrant.

The DNA profile from the envelope matched the DNA profile from the semen found at the crime scene obtained from Sumstad's body. On the basis of these results the prosecutor obtained an arrest warrant for Athan. Once arrested, he was presented with a search warrant to obtain a sample of his DNA. When the sample was analyzed, police found that his DNA profile from the sample matched both the DNA profile obtained from the envelope and the profile taken from the semen left on the victim's body.

Athan's lawyers petitioned the court to suppress the DNA evidence and dismiss the case. They argued that the police had violated state law by taking his DNA without his consent or by court order. In addition, they argued that lawyers, not the police, were the intended recipients of the letter, and the police had violated state privacy law by opening a sealed envelope that was intended for another person. Finally, they argued that posing as a lawyer was a crime under state law, and therefore any information the police acquired under that ruse should be disqualified. The trial court rejected each of these motions, and Athan was found guilty of second-degree murder and sentenced to 10 to 20 years in prison.

The case was brought before the Supreme Court of the state of Washington. In its decision, dated May 10, 2007,[4] the court examined two questions: (1) whether Athan's DNA was collected in violation of either the state or the federal constitution; and (2) whether the actions of the police detectives were illegal and unfairly prejudiced his right to a fair trial.

In a majority opinion signed by five justices, the court concluded that the surreptitious taking of Athan's DNA violated neither the state nor the federal constitution. Washington State's Constitution, which guarantees that "no person shall be disturbed in his private affairs, or his home invaded, without authority of law," is known to provide greater protection than the Fourth Amendment. Although the state did not deny that one's DNA is normally part of one's "private affairs," it argued that Athan had no privacy interest in the DNA he left on the envelope both because he voluntarily abandoned it when he licked the envelope and mailed it to a third party and because DNA obtained from saliva is "voluntarily exposed" to the public just like appearance and physical description. The court agreed:

> We find there is no inherent privacy interest in saliva. Certainly the nonconsensual collection of blood or urine samples in some circumstances . . . invokes privacy concerns; however, obtaining the saliva sample in this case did not involve an invasive or involuntary procedure. . . . The facts of this situation are analogous to a person spitting on the sidewalk or leaving a cigarette butt in an ashtray. We hold under these circumstances, any privacy interest is lost. The envelope, and any saliva contained on it, becomes the property of the recipient.

Furthermore, in direct response to the argument put forth in an amicus brief of the American Civil Liberties Union of Washington,[5] the court found that although DNA has the potential to reveal a vast amount of personal information, the state's use of Athan's DNA was narrowly limited to identification purposes.

In examining whether Athan's rights were violated under the Fourth Amendment, the majority of the court wrote:

> There is no subjective expectation of privacy in discarded genetic material just as there is no subjective expectation of privacy in fingerprints or footprints left in a public place. . . . The analysis of DNA obtained without forcible compulsion and analyzed by the government for comparison to evidence at a crime scene is not a search under the Fourth Amendment.[6]

The opinion concluded in no uncertain terms: "No recognized privacy interest exists in voluntarily discarded saliva and a legitimate government purpose in collecting a suspect's discarded DNA exists for identification purposes."[7]

The court was also not particularly concerned about the approach used by the police to obtain Athan's DNA sample: "We find there is no absolute prohibition of police ruses involving detectives posing as attorneys in the State of Washington."[8] Furthermore, the use of the ruse was "not so outrageous as to offend a sense of justice or require dismissal of the case," in particular, because it was not used in an attempt to gain "confidential information" and because "public policy allows for some deceitful conduct and violation of criminal laws by police officers in order to detect and eliminate criminal activity."[9]

The 6–3 decision in *Athan* came as a shock to many privacy advocates. Washington State is known to have strong privacy protections. Article I, section 7, of its constitution is decidedly more protective of individual privacy than the Fourth Amendment. As described in the dissent in *Athan*, it protects objective expectations of privacy, as opposed to subjective privacy expectations that may be influenced by well-publicized advances in surveillance technology. Prior tests of Washington's privacy laws had rejected compelled testing of DNA to determine paternity, as well as compelled blood testing as a requirement for government employment. The court had also held that the government cannot, without a warrant, use a Global Positioning System (GPS) device to track a person's movement in a car,[10] use an infrared heat device to view a person's activities inside the home,[11] obtain long-distance telephone records by placing a pen register on a person's telephone,[12] or search the contents of garbage left on a curb for pickup.

In this last case, *State of Washington v. Boland*, the court found that an individual's private affairs were unreasonably intruded on by law-enforcement officers when they removed garbage from his trash can and transported it to a police station in order to examine its contents for evidence of drug-related activities.[13] This decision was in direct contrast to federal precedent. In *California v. Greenwood* the U.S. Supreme Court had held that under the Fourth Amendment no reasonable expectation of privacy exists in garbage that has been left on the curbside for collection. The Washington court found that the state constitution provided greater protection of the defendant's privacy interest in this context and went against federal precedent for two reasons:

(1) the fact that Boland had left his garbage at the curb (outside his home) had no bearing on whether the state had unreasonably intruded into his private affairs; and (2) the collection of garbage is necessary to the proper functioning of modern society and requires a person to reasonably expect that his garbage will be removed by a licensed trash collector, not indiscriminately rummaged through by the police; it would be improper to require that in order to maintain privacy in one's trash that the owner would have to forgo use of ordinary methods of trash collection.

Following the reasoning in *Boland*, the court could have found in *Athan* that the state had intruded into Athan's privacy in his DNA regardless of whether it was collected on his person or he had placed it in the mail or spat it onto the street. Furthermore, the court might have found that the "proper functioning of modern society" requires that an individual can walk about in the world or send letters through the mail without fear that his or her DNA might be picked up; it would be improper to require that we refrain from licking envelopes, walk around in a bubble suit, and stop disposing of personal items that might contain our DNA in order to protect our privacy.

How is it that instead, the court concluded that an individual has a greater expectation of privacy in his or her garbage than in his or her DNA? First, the court failed to distinguish the saliva sample from the DNA contained within the sample. The court found that an individual has "no inherent privacy interest in saliva." One can agree or disagree on this point, but the privacy interest of concern was Athan's *DNA information*, not his *saliva per se*. By focusing on the saliva and not the DNA in it, the court diminished Athan's privacy interest.

Indeed, there is at least a growing consensus, if not near unanimity, among bioethicists, medical professionals, and policy makers that an individual's DNA is a private matter. Our DNA can reveal both medical and nonmedical information, including inherited genetic disorders, predispositions (mutations that correlate with the onset of disease), familial disease patterns, environmental and drug sensitivities, parental linkages, ancestral identity, and sibling connections. Increasingly, scientists are hypothesizing about the possible and sometimes-dubious relationship between genes and behavior, for example, whether certain genetic markers make someone more prone to commit aggressive acts.[14] It is precisely this consensus (i.e., that we are entitled to privacy in our DNA) that allowed Congress to pass the Genetic

Information and Nondiscrimination Act (GINA) in 2008. This act prohibits insurance companies from requesting genetic information as a condition of hiring or enrollment, or using individual genetic information for hiring or enrollment decisions. (See chapter 14 for a detailed discussion of genetic privacy.)

The courts, too, have overwhelmingly agreed that the taking and analysis of DNA for purposes of identification is a search and therefore requires a warrant supported by probable cause. However, the courts have differed in why they have found this to be necessary. Some have focused entirely on the physical invasion associated with the forcible taking of the sample, either by drawing blood or swabbing the inside of a person's mouth. Others have considered another kind of invasiveness that relates to the informational aspect of DNA and distinguish the taking of a DNA sample from that of a fingerprint. Unlike fingerprints, which reveal very little information about an individual and cannot readily be used for anything beyond identification, a DNA sample provides an almost unlimited amount of sensitive information about a person. The potential to access sensitive personal genetic information remains as long as the biological sample is retained (see chapter 14).

In *Athan* the court acknowledged the potential for DNA to reveal a lot of private information but then dismissed this concern simply because there was no evidence to indicate that the police had used Athan's DNA for anything beyond identification. This seems shortsighted and contradictory. The issue is not what law enforcement did or did not do with Athan's DNA in this one instance; it is what law enforcement can do once it has a person's DNA in its possession. If one agrees that DNA can reveal highly private information about a person, then the only real way to protect against the potential misuse of DNA is to not collect it in the first place or to require that it be destroyed after an individual's identification has been confirmed.

The court also adopted law enforcement's argument that the DNA had been "voluntarily abandoned" or "discarded." Since Athan sent his sample in the mail, and it was not extracted forcibly from him, his privacy was not invaded. This argument is problematic on several counts. First, "abandoned" implies a knowing intent to part with an item. In Athan's case, although he may have knowingly abandoned his saliva in licking the envelope and sending it off, it is not at all clear that he knowingly abandoned the information contained within that saliva. Athan's intent was to return a letter that was to be read by attorneys, in whom he placed his trust. His intent was not to send them his DNA. Athan may not have known that his DNA was contained in his saliva sample on the envelope at a level sufficient for DNA testing, or even that his saliva contained DNA information in the first place.

DNA is not so much voluntarily abandoned as it is inadvertently, but naturally, released from our bodies in the form of skin cells, saliva, and hair samples. How "voluntary" is it that we discard DNA every time we sneeze, urinate, or bleed from a scratch? Under the court's reasoning in *Athan*, any and all DNA left behind by an individual is fair game for the police.

One justice recognized that regular shedding of DNA should not be considered an abandonment of privacy in it. However, this justice still concurred with the majority in concluding that Athan waived his privacy right when he placed the envelope in the mail. Athan's act of sending the saliva through the mail was an abandonment of privacy in the saliva. By his reasoning, DNA picked up from a cigarette butt or spit, sweat drops, or hair strands left behind on the sidewalk should not be fair game for testing without a warrant, whereas items that come into possession of the police through the use of trickery should. This approach seems highly problematic, since it encourages law-enforcement personnel to engage in ethically questionable behavior any time they want to get a person's DNA. If one has an expectation of privacy in his or her DNA, then it follows that the expectation will apply to any nonconsensual analysis of biological materials—whether acquired by entrapment or by following someone around. Either way, the analysis of the DNA is occurring without the knowledge or consent of the individual.

In a minority dissent, Washington Supreme Court justice Mary E. Fairhurst wrote a defense of Athan's DNA privacy that was signed by two other justices of the court:

> Because Athan's DNA provided the government vast amounts of intimate information beyond mere identity, I would conclude that Athan has privacy interests in his saliva and DNA. In stark contrast to saliva, fingerprints, and other physical characteristics, one never exposes one's DNA to the public. . . . The detectives intruded on Athan's expectation of privacy without authority of law when they collected his saliva from an envelope based on a ruse and tested his DNA. Athan did not voluntarily relinquish his privacy interests in

either his saliva or his DNA by licking the envelope and placing it in the mail.[15]

Beyond *Athan*: Revisiting Abandoned DNA

Whether it intended to or not, the Washington Supreme Court authorized a free-for-all with regard to the collection of DNA by law enforcement. By the court's reasoning, police could follow people around and take their DNA without their knowledge or consent, for more or less any reason. As Elizabeth Joh at the University of California, Davis, states, this collection technique "is a backdoor to population-wide data banking"—the banking of genetic information from virtually anyone. Furthermore, this backdoor method of DNA collection is "in stark distinction to the growing body of commentary on the collection of DNA samples from prisoners and parolees for state and federal DNA databases."[16] Indeed, state legislatures and the courts have gone to great lengths to debate and develop a complex array of policies concerning when and under what circumstances law enforcement can collect DNA and from whom (see chapter 2). If police can take DNA surreptitiously, why would they ever bother to get a warrant to obtain someone's DNA? And what would stop law enforcement from building a DNA data bank of suspicionless suspects for purposes of surveillance, completely outside the boundaries of the laws that currently govern our DNA data banks?

To date, no court has yet held police collection of "abandoned" DNA illegal. In the handful of other cases that have been brought to date, courts have adopted the notion that DNA collected from cigarette butts or coffee cups is "abandoned." According to Joh, "Once DNA is considered abandoned or knowingly exposed, the Fourth Amendment does not apply at all."[17]

If we wish to afford any protection to the genetic information contained in our DNA, we need to reframe the debate over surreptitious DNA collection and recognize that the trail of DNA that we leave everywhere we go is something other than "abandoned." We "abandon" items we no longer wish to own or carry around. In contrast, we have no choice but to leave our DNA pretty much everywhere we go. Short of walking around in the world in a plastic bubble suit, it would be virtually impossible to refrain from shedding our DNA (see box 6.1).

BOX 6.1 How Much DNA Do You Shed?

DNA is routinely released from humans. We shed our skin on an ongoing basis. DNA is also released through urination, bleeding, and blowing one's nose. As discussed in chapter 1, your DNA is found in each and every one of your skin cells. And only a few cells are needed to test your DNA.

Consider the following:

- The average human loses between 40 and 100 hairs each day.[a]
- Between 30,000 and 40,000 dead skin cells fall from the human body every minute.[b]
- From a single sneeze there are 3,000 droplets containing virus particles, bacteria, and both dead and live cells.[c]

[a] "ScienceFacts," http://www.science-facts.com/?page_id=16 (accessed May 23, 2010).
[b] Britannica Encyclopedia Online, "The Skin You're in," January 2009, http://www.britannica.com/bps/additionalcontent/18/36011874/The-Skin-Youre-In (accessed May 23, 2010).
[c] Diana Treece, "Tuberculosis," *InnovAiT* 3, no. 1 (2010): 20–27.
Source: Authors.

Law enforcement's treatment—and court acceptance—of DNA as "abandoned" could have spillover effects well beyond law enforcement. Currently there are no protections from private investigators or amateur investigators acquiring and analyzing a biological sample from an individual who is unaware of their motives.[18] Robert Green and George Annas, writing in the *New England Journal of Medicine*, imagine a scenario where someone obtains an abandoned hair or saliva sample from a presidential candidate with the purpose of engaging in a form of "genetic McCarthyism."[19] That is, the stealth DNA could be used to disclose that the candidate has an increased risk of a disease. This scenario can be generalized to any individual who is a cultural icon for whom there is a high paparazzi value on personal, especially embarrassing, information offered to the media.

Unlike in the court in *Athan*, some other courts that have considered this issue have likened so-called abandoned DNA to trash.[20] These cases have relied on the U.S. Supreme Court case *California v. Greenwood*, where, in contrast to the Washington State Supreme Court's decision in *Boland*, the Court ruled that an individual does not have an expectation of privacy in garbage left on the curb since it is "knowingly exposed" to the public.[21] But even if one accepts the Supreme Court's reasoning, garbage is not a very good analogy to our DNA. When we leave our garbage on the street, we may very well know that someone might rummage through it. Over four decades ago a self-styled member of the 1960s counterculture yippie movement named A. J. Weberman rummaged through the garbage of folk legend Bob Dylan. Weberman, who coined the term "garbologist" to describe his obsession with acquiring any shreds of information about his folk icon, viewed Dylan's garbage as a public, not a private, resource.

We expect and accept that the private information that might be contained in letters or bills can be accessed by virtually anyone who might come into contact with our garbage, which is why some people choose to shred personal papers and letters before disposal. But in contrast to letters tossed in the garbage, we cannot "shred" the DNA that continuously gets discharged from our bodies. While we can control what we choose to throw in the garbage, we cannot refrain from leaving our DNA everywhere we go.

The Supreme Court's decision in *California v. Greenwood* was also premised on the notion that garbage can be picked up by any passerby and deciphered (the "common knowledge doctrine"). But DNA found in the garbage or on other abandoned property does not satisfy the "common knowledge doctrine" because only specialists can decipher it with sophisticated, uncommon equipment. One can argue that the loss of privacy for our trash extends to any ordinary person who can read, but the loss of privacy for our DNA only extends to a small group of specialists who can extract our DNA and analyze it, revealing information that we ourselves might not wish to know.

Joh suggests that perhaps garbage is not a helpful analogy for thinking about DNA that we leave behind. She suggests that more helpful comparisons might be found with body parts or human waste, although even these comparisons are limited, since even a tiny amount of DNA has the potential to reveal far more information about a person. Perhaps most helpful, she notes, would be to change the terminology by which we refer to the DNA that is collected without our knowledge or consent. Perhaps if we thought of law enforcement's collection and use of this DNA as "covert involuntary DNA sampling," we might see a change in the current passivity toward this issue. In Victoria, Australia, residents have called for laws banning "covert DNA sampling" by Victorian police, and in response, the attorney general of

Victoria has promised to examine the "legal loophole" that currently allows police to collect DNA surreptitiously.[22]

Courts have been more protective of personal privacy when our private speech, objects hidden from view, or personal materials can be accessed only by individuals with special technology and are not open for everyone to see. In *Kyllo v. United States* the Supreme Court ruled 5–4 that the thermal imaging of the defendant's home constituted a search under the Fourth Amendment and thus required a warrant. The majority of the court found that a person's home is protected against warrantless searches by clever technologies that are not generally available to the public: "We think that obtaining by sense-enhancing technology any information regarding the interior of the home that could not otherwise have been obtained without physical 'intrusion into a constitutionally protected area' . . . constitutes a search—at least where (as here) the technology in question is not in general public use" (see box 6.2).[23]

BOX 6.2 *Kyllo v. United States*: Technological Intrusion into Home and Body

In the early 1990s Danny Kyllo, who lived in a triplex home in Florence, Oregon, was suspected by an agent of the U.S. Bureau of Land Management (BLM) of growing and distributing marijuana from his residence. The BLM agent enlisted help from a member of the Oregon National Guard to conduct surveillance of Kyllo's residence with a thermal imager, a device that can detect infrared heat emitted from the surface of an object. In this instance the thermal imager was used to measure infrared radiation being emitted from the walls of Kyllo's home. The surveillance was conducted in a car on a public street simply by pointing the thermal imager at the walls of Kyllo's home and comparing the output with the radiation emitted from neighboring homes.

In mid-January 1992 the BLM agent observed that higher levels of infrared heat were coming from Kyllo's walls than from those of his neighbors. He inferred that this was because Kyllo was using high-intensity lamps to grow marijuana. Citing tips from informants, utility bills, and

(*continued*)

the thermal-imaging data, the BLM agent sought and received a search warrant from a federal judge. When agents entered Kyllo's home, their suspicions were confirmed. The suspect had more than 100 marijuana plants. Kyllo was charged with and convicted of manufacturing marijuana.

Kyllo petitioned the court to overturn his conviction on the grounds that the warrant was granted under evidence from the thermal-imaging device that was unlawfully obtained and violated his privacy under the Fourth Amendment of the Constitution. The federal Ninth Circuit Court of Appeals rejected Kyllo's petition, stating that he had shown no subjective expectation of privacy because he made no effort to conceal the heat escaping from his home. In 1994 the Ninth Circuit Court of Appeals reviewed the *Kyllo* case, focusing on whether the warrant used to search the home of Kyllo was based on knowingly and recklessly false information. The court reversed and remanded the decision of the district court, requesting that the court hold an evidentiary hearing on the capabilities of the Afema 210 thermal imager. Once again, the district court denied Kyllo's motion to suppress the thermal-imaging evidence on the grounds that warrantless searches of homes with the thermal imager were permissible. Kyllo appealed again in 1998 to the Ninth Circuit, which found, in a 2–1 decision, that the use of thermal-imaging systems was unconstitutional. However, after the government petitioned for a rehearing, the case went back to the Ninth Circuit, where one judge retired and another replaced him. The decision this time was 2–1 against Kyllo on the grounds that the monitoring of heat emissions by a thermal-imaging system did not intrude on Kyllo's privacy.

The key point before the jurists was whether a technological device allowed police to gain confidential information from a suspect (namely, that he was using intense heat lamps and growing marijuana) that they could otherwise have obtained only by physical entry into the person's home. Does the use of this technology violate a person's privacy interests? Does it constitute an unreasonable search? Are police subject to the constraints of the Fourth Amendment when they avail themselves of a technological device that functions as a surrogate for physical entry into one's home? The case was argued before the Supreme Court on February 20, 2001, and decided on June 11 of that year. In a 5–4 decision the Court ruled that the thermal imaging of Kyllo's home constituted a search and thus required a warrant.

Source: Authors.

Like a person's home, one's body falls within the protection of the Fourth Amendment. Therefore, it is not a stretch to apply the reasoning in *Kyllo* to DNA. The privacy interest associated with our DNA comes into play not so much in the form of the materials that we leave behind us, namely, the macroscopic objects such as coffee cups, which ordinary people can observe and identify, but only when those objects are scientifically analyzed for the molecular information contained within.

Any information that cannot be obtained by direct observation of an individual can be considered private unless the individual decides to make it otherwise. If police were to use a technology that would reveal information about what is contained in one's wallet or pocket, it would follow from *Kyllo* that it should be considered a search as long as the technology is not available to everyone. The use of ordinary binoculars would not reach a threshold of breaching privacy under this standard. But subjecting a saliva sample to laboratory testing—which only those with specialized knowledge and with access to sophisticated equipment could carry out—perhaps should. As the use of a thermal imaging device invades the privacy of one's home, so too does the use of PCR to glean information from one's DNA breach the privacy of one's body.

When we throw a cup or personal items in the trash, we do not expect that people with sophisticated knowledge and access to a genetic laboratory will detect and analyze our DNA, possibly revealing intimate information. It might be reasonable to say that we have no expectation of privacy for the abandoned object, but certainly we do have such an expectation of privacy for the information in the genetic material that sits on that object.

What does it mean to live in a world where one has to assume that DNA, which people shed on a continual and involuntary or uncontrolled basis, might at any time be picked up from discarded objects, extracted, and analyzed for information? Some legal scholars have accepted the default position that the information on shed DNA is not protected: "There is even a strong indication that suspects possess no 'reasonable expectation of privacy' in shed DNA cells, and thus law enforcement can easily gather such probative evidence without worrying about the individual's Fourth Amendment rights."[24] Our examples and analogy call into question the default position that is the current practice.

Police point to individual success stories as a way of justifying surreptitious DNA collection as a "clever investigation technique." But by allowing police to take our DNA without our knowledge or consent, we are opening the door to mass DNA collections of individuals vaguely suspected or not at all suspected of a crime. This could extend to private detectives and amateurs who have some reason to use surreptitious methods to collect and analyze someone's DNA.[25] Individuals now have no way of contesting this collection or the use of their DNA. This scenario becomes increasingly worrisome when it is coupled with developments in behavioral genetics; for example, genetic markers for aggression or addiction could provide justification for identifying these individuals before crimes are committed and subjecting them to social control, or as a reason to mete out a more severe punishment when they are convicted of a crime.[26]

The current dominant framework that assumes that DNA collected from coffee cups, cigarette butts, and saliva samples is "abandoned" means that police can pick up DNA anywhere, from anyone, and at any time. The lack of meaningful consideration of the full implications of this assumption is threatening to erode any and all privacy interest in our DNA. If we take seriously the notion that our DNA is a private matter, then we need to ensure that our bodily materials in which that DNA resides cannot be tested and mined without our consent. As Joh states: "While Fourth Amendment law may not appear to protect a privacy interest in the human tissue left behind as the detritus of our daily lives, it is far from obvious that people do not harbor a privacy expectation in genetic information that 'society is prepared to recognize as reasonable'."[27] It is noteworthy that the United Kingdom has taken the first step in acknowledging DNA's special privacy protections by banning ordinary citizens from analyzing "abandoned" DNA without the consent of the person(s) from whom it originated.

Chapter 7

Exonerations: When the DNA Doesn't Match

So then after I have been on death row for twenty-two years, they find this DNA evidence, you know, and the prosecution says that this will be the final nail in Kerry Max Cook's coffin. "We'll show the world once and for all that he committed this murder." And then the results come in and it did just the opposite; it finally took the nail out of my coffin, told the world the truth—that that was Professor Whitfield's DNA they found on that girl. And he's still out. They never went after him. He's been walking around a free man, laughing at the system for twenty-two years.

—*The Exonerated* (play)[1]

Our legal system places great weight on the finality of criminal convictions. Courts and prosecutors are exceedingly reluctant to reverse judgments or reconsider closed cases; when they do—and it's rare—it's usually because of a compelling showing of error. Even so, some state officials continue to express doubts about the innocence of exonerated defendants, sometimes in the face of extraordinary evidence.

—Samuel R. Gross et al.[2]

Comparisons between scientific and judicial methods of fixing belief can be found in several recent studies.[3] Both the scientific and legal communities seek the truth, albeit in different contexts. The former focuses on the pursuit of knowledge of processes within the natural world, while the latter seeks guilt or innocence of human behavior in the social world. As noted by the late sociologist Robert Merton, science operates within a system of norms, including organized skepticism, which implies that scientists place a high burden on validating a claim.[4] Science is also deemed to be self-correcting. Either by retesting a result or by identifying bias and uncertainty, the gatekeepers of "certified knowledge," such as journal editors, reviewers, and professional societies, correct the mistakes of prepublication or published results.

The legal system pursues its truth through an adversary system that draws on experts and eyewitnesses, a process where legal advocates make their cases before the "triers of fact," either juries or judges. In contrast to the scientific culture, the legal system can correct mistakes through a process of appeals. Mistakes in law often involve procedural errors committed by the trial judge. For example, an appeals court might overturn a lower court's decision on the admissibility of evidence or whether an expert witness met the standards set under the Supreme Court's 1993 *Daubert* ruling.[5] Corrections by a higher court can lead to a mistrial or to exoneration in cases where the defendant's constitutional rights have been violated.

Although it may seem that there are stark differences between the pursuit of truth in science and in the judicial system, both are embedded in a social context, and as a consequence the methods they use share some common features. Fraud can be found both in science and in legal processes, where evidence can be cooked, fabricated, withheld, or discarded. Scientists who violate the standards of research integrity are usually publicly censured or ostracized from the community. Penalties imposed on police for the mishandling of evidence may not be as severe as those given to scientists charged with misconduct, but evidence mishandling can reverse a conviction and weaken the public's trust in a prosecutor. Just as there is scientific misconduct, there is prosecutorial misconduct, such as when a prosecutor knowingly uses false evidence. In the 1935 case *Berger v. United States* the Supreme Court outlined the prosecutor's legal and ethical role in pursuing justice: "It is as much his [the prosecutor's] duty to refrain from improper methods calculated to produce a wrongful conviction as it is to use every legitimate means to bring about a just one."[6] As an example, the California Penal Code requires prosecutors to disclose any exculpatory evidence, including notes compiled by the prosecution's expert witness containing exculpatory information.[7]

According to the philosopher Karl Popper, all scientific claims are tentative. They gain their standing as "scientific claims" by being falsifiable. Scientific claims (explanations, hypotheses, and theories) gain plausibility and validation (corroboration) when they are rigorously tested and resist falsification.[8] But no scientific claim is beyond falsifiability; otherwise it could not claim scientific (empirical) status.

Judicial decisions of guilt or innocence are also tentative, but there are some caveats. Innocence and guilt are not treated the same in legal epistemology. A person found innocent of a crime cannot be tried again in light of new evidence or procedural errors. Prosecutorial mistakes cannot be remediated in a second trial. The American judicial system is afforded one chance to prove guilt but offers multiple opportunities for correcting a guilty conviction that was in error.

In science the burden for demonstrating truth may be high, but negative results, such as experiments that do not corroborate a hypothesis, which also reveal truths, rarely get reported by journals. The burden, however, remains the same for positive or negative results. By contrast, legal theory suggests that the burden to demonstrate a positive finding of guilt is higher than to demonstrate innocence because "innocence" is the default position. Failure to demonstrate guilt to the "triers of fact" leaves innocence as the default, although it does not imply innocence in some objective sense. "No evidence of guilt" is not the same as "evidence of innocence." Similarly, failure to confirm a scientific hypothesis does not imply that the hypothesis is false. There are many cases in science where some tests are positive and some are negative.

Contemporary philosophers and historians of science have shown that science can too easily be idealized. There are examples that suggest that scientists are unwilling to give up hypotheses or theories in the face of falsifying evidence. This was the conclusion of the historian Thomas Kuhn in *The Structure of Scientific Revolutions* when he wrote that scientists hold on to a "paradigm" despite convincing evidence that it is false.[9] There is also increasing evidence that scientific results can be biased by the sponsor of the research, who may have a financial interest in certain outcomes.[10]

Science has always played a role in the judicial process in the form of expert testimony. The term "forensic science" is an established part of criminal justice. Juries, however, have not been a part of the certification of scientific knowledge. Periodically, proposals have been proffered for a science court in the form of an elite group of scientific experts who would reach a consensus on complex issues involving scientific causality that have a strong bearing on public welfare. Supreme Court justice Stephen Breyer suggested such a system to resolve the opinions of dueling experts in tort cases involving chemical exposures.[11] In a few cases judges have empaneled experts, notably, in the class-action case of silicone breast implants, to reach a decision on whether there was sufficient scientific evidence for the plaintiff's claim.

When DNA was first introduced into the courtroom in the 1980s, it created a sea change in forensic science. Microscopic traces of DNA at a crime scene could implicate a defendant with more precision and less uncertainty than traditional analysis of hair, blood, footprints, bite marks, and the like. By far the most revolutionary aspect of DNA evidence is its role in establishing innocence. The remainder of this chapter examines the impact of DNA evidence and DNA data banks on uncovering the truth about wrongfully convicted individuals.

The First DNA Exoneration

The first person exonerated in the United States by DNA evidence was Gary Dotson, a resident of a Chicago suburb convicted of rape and aggravated kidnapping. The story of Dotson's conviction and eventual exoneration has many twists and turns covering a period of 12 years. During this period Dotson was tried and convicted, was given a governor's commutation, was placed on parole, was charged with domestic battery, parole violation, and other minor offenses, had his original sentence reinstated, and served additional time in prison after his first release, all the while going through many appeals. Because it was the first case of DNA exoneration, it occurred while the science of DNA testing was in its infancy, and that added a level of complexity.

The case began on the evening of July 9, 1977, 10 years before forensic DNA was introduced. Cathleen Crowell, a 16-year-old cook and cashier in a Long John Silver's seafood restaurant, was standing on the side of a road in Homewood, Illinois, a Chicago superb. A police officer saw Crowell standing alone, vulnerable, agitated, and in tears. She told the officer that she had been grabbed by three men while crossing a mall parking lot, thrown into a car, and raped by one of the men in the back seat. The rapist, she reported, had scratched letters into her naked stomach with a broken beer bottle.

Crowell was taken to South Suburban Hospital, where she underwent a rape examination. The examination included a vaginal swab and observation of a seminal stain on her underpants. The emergency-room physician also made note of the marks on her stomach—a series of superficial scratches in a crosshatched pattern, which could have been an effort to spell out some words that were not decipherable.

Several days after the incident a sketch artist worked with Crowell to develop a drawing of her assailant. Subsequent to the artist's rendering of the rapist, Crowell was shown a series of mug shots. She picked out Gary Dotson as her rapist.

The trial of Dotson took place in May 1979. There were two witnesses for the prosecution. One was the alleged victim, a high-achieving student at Homewood-Flossman High School. She identified Dotson as her rapist in open court with no uncertainty expressed. The second witness was Timothy Dixon, a state police forensic scientist. Dixon testified for the prosecution that he detected type B blood antigens in the stain on Crowell's underpants. He noted that Dotson had type B blood. According to Dixon, B antigen secretions were found in about 10 percent of the white male population. He also testified that several loose hairs found on the victim were similar to hairs taken from Dotson.

The prosecutor used the forensic testimony and the victim's eyewitness account to gain a conviction of Dotson in 1981. He was sentenced to not less than 24 and not more than 50 years for rape and aggravated kidnapping. The Illinois Appellate Court upheld the conviction.[12]

The case against Dotson began unraveling in 1982. Crowell married a high-school classmate and moved to New Hampshire. The newlyweds joined the Pilgrim Baptist Church. In 1985 Crowell, then Cathleen Crowell Web, confided to her pastor that she had fabricated the rape allegation that sent Dotson to prison. Her motive was to protect herself from the embarrassment of becoming pregnant after she had consensual sex with the boyfriend she had at the time.

Cook County prosecutors were informed of the recantation of Crowell's testimony but were not interested in reexamining the case until the Chicago print media broke the story later that year. From there, further weaknesses and inconsistencies in the prosecutor's case against Dotson began to emerge. The state's forensic expert had misrepresented himself as having taken courses at the University of California at Berkeley. Also, he had failed to reveal that

Crowell was also a secretor of type B blood antigens and that the seminal content of the stain could have come from 60 percent of men in the white population. The forensic expert also erred in linking the loose hairs to Dotson. Moreover, the concentration of spermatozoa in the stain on Crowell's underpants was different than that on the vaginal swab, indicating that the two events had not occurred at the same time, as the victim had claimed.

News coverage created a public appeal for Dotson's release. On April 11, 1985, at a hearing on Dotson's motion for a new trial, the same judge who had originally heard the case refused to order a new trial on the grounds that he believed that the complainant (the rape victim) was more believable in her original testimony than in her recantation.[13]

After the governor of Illinois accepted authority for the case, a hearing was held before the Illinois Prisoner Review Board. The governor did not believe Crowell's recantation and refused to pardon Dotson. But on May 12, 1985, the governor commuted Dotson's sentence to six years that he had already served, under conditions of good behavior. In 1987, after Dotson was accused of spousal abuse, the governor revoked his parole. On Christmas Eve 1987 the governor granted Dotson a "last chance parole." Two days later Dotson was arrested in a barroom brawl, and his parole was once again revoked.

In 1988 Dotson's new attorney read a story in *Newsweek* magazine about a new technique for establishing identity by DNA analysis. He requested that the court have Dotson's DNA tested to determine if he was implicated in the rape. The decision was supported by the governor, and the testing was done by Alex Jeffreys, who had developed the first method for DNA fingerprinting. His process, called restriction fragment length polymorphism (RFLP), required a sizable quantity of DNA. Because there was degradation in the sample taken from Crowell's underpants, Jeffreys could not perform the test. The governor agreed to a new test using the latest DNA identification technology, polymerase chain reaction (PCR), patented by the Cetus Corporation, which had been shown to be effective in testing extremely small quantities of DNA. On August 15, 1988, the governor was notified that the PCR tests conclusively excluded Dotson and implicated Crowell's past boyfriend as the source of semen in the underpants.

Dotson was released from prison while serving additional time for a technical parole violation and was committed to a substance-abuse treatment center. His lawyer petitioned for a new trial on the rape charges on the basis of the PCR results. The presiding judge of the Criminal Division of the

Cook County Circuit Court granted the motion. The state attorney's office decided not to prosecute on the basis of the victim's credibility and the DNA test results. On August 14, 1989, the Cook County Circuit Court in Chicago, Illinois, vacated Gary Dotson's 1979 rape conviction and dismissed all charges. With the charges against Dotson dropped, he became the first convicted felon exonerated by DNA technology, after having served eight years in prison.

When the DNA Exclusion Is Not Enough

If DNA has such powerful probative value, then it would stand to reason that judges, juries, prosecutors, and the public would accept a negative finding as definitive evidence of innocence. In fact, DNA samples that do not match should hold more probative value for a case of innocence than DNA samples that do match should hold for evidence of guilt, since there is always a chance that a match might have occurred as a result of contamination (see chapter 16). But even the best evidence of innocence can be explained away by alternative hypotheses, however improbable. This is illustrated in the prosecution of Darryl Hunt. This case began on October 10, 1984, when the body of Deborah Sykes, a local reporter, was found in a field near her place of employment in Winston-Salem, North Carolina. The medical examiner reported that Sykes had suffered multiple stab wounds and that she had been sexually assaulted both vaginally and anally.

Two different juries found Darryl Hunt guilty of murdering Sykes. He was first tried and convicted in 1984 and sentenced to life imprisonment. The North Carolina Supreme Court overturned the conviction on direct appeal when it found that the state allowed a police officer to testify on the substance of unsworn statements by Hunt's girlfriend. Hunt was retried in 1990 and was convicted once again for felony murder and sentenced to life imprisonment. The second sentence was upheld on appeal.

During both of Hunt's trials the state offered no direct evidence linking Hunt to the kidnapping, robbery, sexual assault, or murder of Sykes. The state's evidence was based on the testimony of an eyewitness who placed Hunt near the scene of the crime. A hotel employee claimed that Hunt entered the hotel bathroom later that morning and exited, leaving bloody hand towels behind.

On the prompting of Hunt's lawyer, Mark Rabil, PCR DNA testing was performed on a fluid sample taken from Sykes's body after Hunt's trial. It showed that Hunt was not a contributor to the sperm contained in the sample. Rabil filed motions for a new trial on the basis of newly discovered evidence. His motion was denied by the superior court and the North Carolina Supreme Court. The court argued that the burden for a new trial on the basis of the appearance of postconviction evidence is very high and requires a "truly persuasive demonstration of actual innocence." According to the court, Hunt would have to show that, on the basis of the newly discovered evidence and the entire record of the case before the jury, "no rational trier of fact could [find] proof of guilt beyond a reasonable doubt."[14] In its decision the court wrote:

> Hunt has made no such showing. If we evaluate the PCR/DNA test results along with the entirety of the evidence introduced at both trials, we are unable to conclude that no rational jury would have convicted Hunt of the murder. The new evidence is simply not sufficiently exculpatory to warrant a new trial. Indeed, as the magistrate judge explained, the DNA results do nothing to discount a number of other possible scenarios reasonably implicating Hunt in the sexual assault. Hunt's sperm might have been present on a different untested sample; Hunt raped Ms. Sykes but did not ejaculate; or Hunt sodomized Ms. Sykes. Moreover, the DNA results do not exonerate Hunt from committing the murder, from committing the other underlying crimes of kidnapping or robbery, or from aiding and abetting any of these crimes, including the sexual assault.[15]

The North Carolina district attorney initially indicated that he would dismiss the case on grounds of the failed DNA match, but in the face of public attitudes and local politics he changed his position. When Hunt was not released after his DNA cleared him as a rapist in a case where murder and rape were interconnected, the members of his legal team began their own investigation of possible suspects. They figured that the only way Hunt would be freed would be if they found a match of the real killer-rapist. Mark Rabil said, "We began lining up suspects and did DNA tests secretly. We spoke to them and saved their cigarette butts. Then we had the butts tested for DNA."[16] Hunt's legal team even considered a suspect who had committed suicide years earlier. They traced his son, obtained a DNA sample, and ran a test on it. The test eliminated the deceased suspect.

Between 1996 and 2000 Rabil pursued a pardon for Hunt, which under law is under the governor's authority. But that proved futile. Having failed to find a DNA match with the suspects they could muster, the members of Hunt's legal team filed a motion in 2003 to have the crime-scene DNA checked against the North Carolina forensic DNA database. They received approval to determine whether there was a match between the DNA found on the rape victim and that of a person whose DNA profile was entered into the DNA felony profile database in North Carolina. There was no match. However, a partial match was found with the DNA of Anthony Brown, whose profile had been entered into the database around 2003. North Carolina FBI forensic DNA laboratory specialists told Rabil that there was no basis from the partial match to assume that the source of the DNA came from a close relative of Anthony Brown. Rabil received the opposite response from Alabama police forensic experts, who claimed that the close match constituted a familial connection.

The intent of seeking a match of the crime-scene DNA with the North Carolina forensic DNA data bank was to find a cold hit with a felon who had been convicted of other crimes. It was not a familial search, where one selects a predetermined number of alleles, less than the number used in a precise match, on the basis of the population genetics of family kinship. But the results of obtaining one match of seven out of nine alleles did open up the investigation to family members of Anthony Brown. One of those was Willard Brown.

Ironically, Willard Brown was a suspect in another case. A woman (Regina K.) was abducted, raped, and stabbed repeatedly on February 2, 1985. On April 1, 1986, she identified Willard Brown as her assailant from photographs. There was no physical evidence, and she was discouraged from pressing charges. The police dropped the case against Willard Brown even though they noted the similarities between the 1985 rape and the Sykes case. One detective specifically made the connection, but a typographical error in jail records led him to believe incorrectly that Brown was in custody at the time of Sykes's murder. According to a report by the city manager of Winston-Salem, "Police detectives should have connected the Sykes and Regina K. cases in the spring of 1986 and connected Willard Brown's blood group evidence with that from the Sykes case since the same investigator was working on both cases at the time."[17] Instead, Willard Brown was not reactivated as a suspect until the partial DNA match of his brother was made in 2003.

Police secured a discarded cigarette butt from Willard Brown and found a precise match with the crime-scene evidence in the Sykes case, and in

December 2003 he confessed to the 1984 rape and stabbing. Only after Brown confessed was Darryl Hunt released after having served about 18 years of a life sentence. On February 6, 2004, a Superior Court judge vacated Hunt's murder conviction. Despite the exonerating evidence and the admission by Brown, there remained people in Winston-Salem who refused to accept Hunt's innocence, including the victim's mother.

Frederick Bieber uses the case of Darryl Hunt to support familial DNA searching. However, it is worth noting that authorities would not have had to resort to a partial match to find Willard through his brother Anthony if Willard had been in the database or if the police investigation had been done thoroughly. The reference to familial matches in this case by Bieber overstates their value as an investigative tool.

Barriers to Postconviction DNA Testing

The progressive movement to use DNA for exculpatory purposes faces several obstacles. First, it requires considerable human resources to review cases involving claims of innocence, to undertake investigations, and to hire lawyers to reactivate a case that has, in the court's and prosecutor's mind, reached finality. Second, even when a claim of innocence looks promising, there may not be any crime-scene DNA to test against the DNA of the incarcerated individual. The crime-scene evidence may have been discarded or lost. Many states do not have laws requiring the preservation of DNA evidence. Without preservation of the biological sample there is no case for innocence when the critical issue in question is whether there is a match or mismatch between the crime-scene DNA and the suspect's DNA. Third, some states do not have laws that give convicted felons the right to postconviction DNA testing, although 44 states have some form of legislation. Fourth, even if the crime-scene DNA has been preserved, the convicted felon has a legal right to postconviction DNA testing, and the incarceree's DNA does not match that of the crime scene, the prosecutors and the courts may not agree to exoneration. They may come up with other theories to rationalize guilt that override their conception of the probative value of the DNA evidence.

Exonerations of wrongfully convicted individuals have been hampered by the premature disposal or misplacement of crime-scene DNA. Typically, prosecutors and police have little incentive to forage around in search of lost

DNA. In one notable case described below, members of the Innocence Project, a nonprofit legal clinic of the Benjamin N. Cardozo School of Law, by dint of persistence and luck, discovered critical DNA evidence that was believed to have been destroyed.[18]

In December 1982 an all-white jury in Hanover, Virginia, convicted Marvin Lamont Anderson, an African American, of kidnapping and raping a 24-year-old white woman.

> Marvin Lamont Anderson had a spotless record, but he lived near the victim, and he dated a white woman. Police included his photo in a [photo] spread. The victim picked out Anderson's picture. An hour later, she picked him out of a lineup. The photograph of the man who raped and tortured the petite, blue-eyed blond was [also] in the photo spread. But his name is John Otis Lincoln. It was Lincoln who dragged the woman into the woods the night of July 17, 1982. Anderson was at his mother's house washing his car.[19]

At the 1982 trial forensic DNA was not yet available. A state forensic scientist testified that she was unable to determine the rapist's blood type from the semen recovered from the victim. The jury found Anderson guilty, and he was sentenced to 210 years in prison on the basis of the mistaken identification. Just 19 years old, he cried when the verdicts were read. He spent the next 15 years in prison.

Six years after Anderson was convicted, another suspect admitted to the crime. Governor L. Douglas Wilder pardoned Anderson in 1993, but his felony conviction was not removed from his record. The Innocence Project took on the case to have his record cleared. Codirector of the Innocence Project Peter Neufeld noted: "The first thing that was so extraordinary about Marvin's case is that the only reason he was picked out for identification is because the perpetrator, who was a black man, said to the victim, 'I got me a white girlfriend.'"[20] Neufeld also noted: "It is the only case in America—of all the DNA exonerations—where the real perpetrator was shown to the victim, and instead she picked an innocent guy."[21]

The Innocence Project's investigators requested from the police a sample of the semen that had been obtained from the victim. At first it seemed like a lost cause. The police had thrown out the rape kit. Neufeld describes the serendipity of the misplaced DNA on which rested Anderson's exoneration:

> Well, Marvin's case is kind of special because we looked for the evidence for five years, and we were told it was destroyed. But fortunately, they took one last look back in the archives where the criminalist kept her notebooks, and they found that she had broken the rules in Virginia and instead of returning the evidence to the rape kit to be destroyed, she Scotch taped it into her lab notebook. Had she followed the rules, Marvin would still be convicted.[22]

Brandon Garrett, writing in the *Columbia Law Review*, examines how the U.S. criminal justice system handled the first 200 cases of convicted felons proved innocent and exonerated by DNA evidence. From his study of the background of these cases he concluded: "Even after DNA testing became available, courts and law enforcement imposed obstacles to conducting DNA testing and then denied relief even after DNA proved innocence. These data show how reluctant our criminal system remains to redress false convictions."[23]

According to the Innocence Project, 22 percent of the claims of innocence that its team investigated between the years 2004 and 2008 were terminated because the crime-scene DNA evidence was not available.[24] At least 24 states either have no statutes that require preservation of DNA evidence or, if they have statutes, do not require preservation of the DNA for all violent felonies, covering the full length of the incarceration. Without such a statute, wrongly convicted individuals do not have the opportunity to prove their innocence through DNA testing. Some states, like Arizona, have a preservation statute for sex offenses and homicides but not other felony offenses.[25] In Garrett's study of the first 200 exonerees he noted, "Even given its potential as exculpatory biological evidence, in a high percentage of cases DNA evidence is not preserved."[26]

Alan Newton was freed from prison in New York City on July 6, 2006, after 22 years of wrongful conviction for rape. Newton had requested DNA testing for 12 years. Both he and the courts were told by prosecutors that repeated searches of police storage facilities did not yield the rape kit from the case. The rape kit was last recorded to be at a police facility in 1985. A newly assigned prosecutor requested that the commanding officer of the police department renew the search for the rape kit. When a hand search of the specific bin was undertaken, the rape kit was discovered inside it. DNA testing of the sample conclusively proved Newton's innocence. He was finally exonerated by a joint state-defense motion.[27]

The next case is one of several documented by the Innocence Project in which, by some act of serendipity or pure coincidence, DNA evidence from a criminal case, which proved to be a critical piece of corroborating evidence for a claim of innocence, was not destroyed. The evidence could just as easily have been lost, removing the chance of exoneration. Calvin Johnson was convicted of rape in Clayton County, Georgia, in 1983. It was the practice in Clayton County for the court stenographer to keep all the evidence entered as exhibits after the trial was concluded. The stenographer who participated in the Johnson case had accumulated a closet full of evidence. When he was about to retire, he asked what he should do with the evidence. He was told to throw it away, which he did.

A district attorney saw the boxes of evidence in a parking-lot dumpster outside the courthouse and decided that the evidence should be saved. Because the DNA evidence was salvaged from the dumpster, Calvin Johnson was able to compare that DNA with his own and prove his innocence.[28] The Innocence Project has reported 10 cases to date in which evidence initially believed to have been lost or destroyed was subsequently found. In all these cases DNA testing exculpated a wrongfully convicted individual.

Arizona resident Larry Youngblood was picked up as a suspect in the sexual assault of a 10-year-old boy in 1983. Youngblood was placed in a police lineup because his description matched that given by the victim. Detectives had taken possession of the boy's underwear and T-shirt, which had stains of the molester's semen. However, police did not refrigerate the stains. As a consequence, the DNA degraded and could not be used to validate Youngblood's claim of innocence. The Arizona Court of Appeals reversed the conviction on the grounds that the state had failed to preserve the evidence.[29]

In 1989 the U.S. Supreme Court reinstated Youngblood's conviction, ruling that the plaintiff could not show that the police had acted in bad faith when they failed to preserve the biological evidence from the victim.[30] Justice Harry Blackmun, speaking in the minority, wrote: "The Constitution requires that criminal defendants be provided with a fair trial, not merely a 'good faith' try at a fair trial."[31] The Court set the precedent that negligence on the part of the police in handling evidence is not grounds for relief, even when the evidence has exculpatory potential. According to Barry Scheck of the Innocence Project, more than 40 percent of their cases go unsolved because of missing or discarded evidence that was in the possession of the police.[32]

When a convicted felon has confessed to a crime, prosecutors are dubious about putting any resources into claims of innocence. After all, as one often hears, "Why would an innocent person admit to the crime?" However, false confessions are not uncommon in the criminal justice system. Indigent suspects charged with homicide, who may face execution if they are convicted of a crime they did not commit, might choose a plea bargain that guarantees prison and avoids capital punishment. It has also been shown that police can force a confession from an individual by deploying heavy-handed interrogation and deprivation. In Garrett's study "Judging Innocence," false confessions were introduced in 31 out of 200 cases (15.5 percent).[33] Eleven out of 223 DNA exonerees (4.9 percent) investigated by the Innocence Project pled guilty to crimes they did not commit.

One of these is the case of Christopher Ochoa. The case began with the rape-murder of Nancy DePriest in 1988 at a Pizza Hut in Austin, Texas, where she had worked. Christopher Ochoa and his roommate Richard Danziger worked in another Austin-area Pizza Hut. They became prime suspects when a Pizza Hut waitress observed them drinking beer and toasting the victim. After being interviewed by a brutal police officer who threatened him with a needle, Ochoa confessed to the homicide to avoid the death penalty. He also implicated his friend Richard Danziger in the rape. Both Ochoa and Danziger were convicted and received a life sentence. Years after the rape-murder a private forensic laboratory in California cleared Ochoa and Danziger of the crime on the basis of DNA evidence. DNA evidence eventually implicated the real perpetrator of the crime, Achem Josef Marino, who, while serving three life sentences, declared that he had a religious conversion, and, while Ochoa and Danziger were serving life sentences, revealed details of the crime to police, including the location of items stolen from the Pizza Hut. In February 1996 Marino wrote letters to the Austin Police Department and to the *Austin American-Statesman* in which he claimed that he alone had killed DePriest and that Danziger and Ochoa were innocent. For two years there was no response to Marino's letters even though police found items he had described. It was only after a letter from Marino was made public that Ochoa and Danziger were able to obtain DNA tests, which excluded them and incriminated Marino. Danziger and Ochoa were officially exonerated in 2001. By then, Danziger had been attacked in prison by inmates, kicked in the head, and sustained serious brain damage. Ochoa was released from prison and entered Wisconsin Law School.[34] A report on false confessions in *Psychology Today* stated,

Although it is difficult, if not impossible, to estimate the number of false confessions nationwide, a review of one decade's worth of murder cases in a single Illinois county found 247 instances in which the defendants' self-incriminating statements were thrown out by the court or found by a jury to be insufficiently convincing for conviction. (*The Chicago Tribune* conducted the investigation.)[35]

According to the Innocence Project:

> In about 25% of DNA exoneration cases, innocent defendants made incriminating statements, delivered outright confessions or pled guilty. These cases show that confessions are not always prompted by internal knowledge or actual guilt, but are sometimes motivated by external influences. Why do innocent people confess? A variety of factors can contribute to a false confession during a police interrogation. . . . They include: duress, coercion, intoxication, diminished capacity, mental impairment, ignorance of the law, fear of violence, the actual infliction of harm, the threat of a harsh sentence, and misunderstanding the situation.[36]

As described in the case of Darryl Hunt, even where DNA evidence is available, subjected to testing, and definitely found to not match an individual convicted of a crime, prosecutors and the courts may go to great lengths to refrain from exoneration. An example that illustrates a prosecutor's use of hyperbolic rationalization in the form of an incredible theory in the face of a DNA mismatch is found in the case of Bruce Godschalk, who was convicted of a double rape in Philadelphia, Pennsylvania. He had falsely confessed to the two rapes, but the semen stains did not contain Godschalk's DNA. The Montgomery County district attorney, Bruce Castor Jr., believed that the initial confession trumped the DNA evidence. In a CNN interview Peter Neufeld explained the district attorney's theory why the DNA mismatch was not convincing: "He made the suggestion that perhaps someone could have sneezed in the one victim's apartment in July of '86 and then maybe sneezed in the second victim's apartment in September,"[37] implying that each of the crime scenes was contaminated by the sneezing individual. Given that the laboratory investigation showed that the DNA came from semen, the sneeze theory was inconsistent with sound science.

When the Defense Solves the Crime

Imagine that you are a defense attorney who has spent several years obtaining a postconviction DNA test for your client, who is serving a life sentence for a rape-murder conviction. Suppose that the test showed unambiguously that the DNA from the sperm found on the victim did not match the DNA of your client, but neither the prosecutor nor the courts will support the release of your client on the basis of the DNA evidence, even though no one questions the reliability of the test. The only option available to the defense attorney is to find the real perpetrator of the brutal crime. If there are other prime suspects whom the police overlooked, the defense attorney can obtain abandoned objects from these suspects to get a sample of their DNA. Alternatively, the defense can propose uploading the crime-scene DNA to CODIS and hope that there is a cold hit. In either case the defense attorney must obtain permission from the prosecutor or the police either to submit DNA to CODIS or to obtain a sample of the crime-scene DNA that he or she would compare with the DNA of the prime suspects.

In one of several cases handled by the Innocence Project, the lawyers investigating a wrongful conviction turned themselves into Perry Mason–like investigators solving the crimes in order to obtain justice for their client. The case in question began on August 1, 1982, in Dallas, Texas. Three armed African American males entered the home of a Dallas couple. One of the assailants robbed the male victim while the other two raped the female, who was five months pregnant. After the first rape all three men raped the woman repeatedly and eventually released her. The police took the rape victim to a hospital, where semen evidence was collected. She identified one of the three rapists (who was known by the couple) from a photograph. A second suspect, James Curtis Giles, was identified in a lineup. Despite an alibi and inconsistencies in physical appearance, in 1983 Giles was convicted of aggravated rape, and sentenced to 30 years in prison.

In 1984 the first rapist pled guilty to the crime and signed an affidavit that James Curtis Giles did not participate in the crime. That was not sufficient to exonerate Giles. The Innocence Project began its investigation of Giles's conviction in 2000. Because there were three rapists, the investigators at the Innocence Project began by identifying the DNA profiles of the rapists from the rape kit. They learned that none of the profiles matched that of James Curtis Giles. That by itself, however, was not enough to exonerate Giles. It appeared that the only way to free Giles would be to obtain the identities of the three males whose DNA profiles were in the semen admixture taken from the rape victim.

The investigation by the Innocence Project continued and led the investigators to a suspect by the name of James Earl Giles (unrelated to James Curtis Giles). They also had evidence that a man named Michael Brown was one of the rapists. Both Brown and James Earl Giles were deceased, so they could not get a DNA sample from the suspects. In their continued seven-year investigation, the investigators at the Innocence Project obtained DNA from family members of the two deceased suspects. Through an analysis of their DNA they were able to show the court that the family members were relatives of the actual perpetrators and that James Curtis Giles had been convicted on the basis of mistaken identity.[38]

According to Garrett's study "Judging Innocence," out of 200 exonerees, there were 59 cases (29.5 percent) in which a cold hit in a DNA data bank resulted in identification of the actual perpetrator. In 25 other cases the actual perpetrator was identified in other ways.[39]

Why should a defense attorney have to play the role of criminal investigator in order to find the real perpetrator of a crime for which his client has been falsely convicted when DNA has proved his client's innocence? Echoing the belief of a great majority of those involved in the criminal justice system, Garrett notes, "DNA testing provides the most accurate and powerful scientific proxy available to establish biological identity; it sets the 'gold standard' for other forms of forensic analysis."[40] If the gold standard of identity demonstrates beyond reasonable doubt that there is a mismatch between the biological evidence left at the crime scene and the DNA of the individual convicted of the crime, does our sense of justice require that we add to the gold standard yet a higher burden that involves finding the real perpetrator before innocence is established?

In a number of cases failure of a DNA match was insufficient in the eyes of the court to prove the innocence of a wrongfully convicted individual. Of the first 200 exonerees, 12 had been convicted at trial despite the fact that DNA testing excluded them.[41] An analogy can be made to defective theories in science. Once a theory has been established, it is often retained, despite newly found falsifying evidence, until a new theory replaces the old one. One can wonder whether a similar model holds for convicted felons proved innocent by DNA, namely, that prosecutors are unwilling to override a guilty verdict until they have found the real perpetrator. Fortunately, this is not the general rule in criminal justice. Many individuals who have been proved innocent by DNA have been released without knowledge or prosecution of the real perpetrator. Of the first 200 exonerees, 126 were released by DNA evidence while the perpetrator remained at large.

The Innocence Project compiled data on cases (both its own and cases managed by others) of individuals exonerated of wrongful conviction between 2004 and 2008. Of 232 cases, 100 (43 percent) resulted in the real perpetrator being identified. In some cases the identification of the real perpetrator led to the exoneration.[42]

In science a theory is assumed false (or tentative) unless proved true; in criminal law a theory of guilt is assumed false (presumed innocence) unless the triers of fact find that the weight of evidence supports guilt beyond a reasonable doubt. The difference is that there is no finality in science; it is constantly evolving and changing, whereas the finality principle in law makes it difficult to overturn a theory of guilt that has been corroborated by a jury. The finality principle was reaffirmed by Samuel Gross and colleagues in their study of exonerations that occurred between 1989 and 2003.[43]

An argument might be made that a universal DNA data bank could be helpful in exonerating wrongfully convicted incarcerees. But before we jump to this conclusion, we should draw attention to the presuppositions behind it, and ask whether there are less invasive ways of achieving justice. Defense attorneys investigating innocence claims may be helped by a universal database under conditions where the real perpetrator has not been convicted of a felony (and therefore does not already have a DNA profile in CODIS) or when prime suspects are not available for DNA testing. In those cases in which a serious felony was committed by a first offender, and where there is nondegraded DNA evidence at the crime scene, a universal database could provide the clue to the real perpetrator of a crime. But if we are judging how we can best achieve justice for those wrongfully convicted, the most important factors have nothing to do with universal DNA databases. They involve the availability of postconviction legal services and financial resources, access to postconviction DNA testing, preservation of DNA samples over the length of the sentence, timely processing and CODIS uploading of crime-scene DNA samples, and the recognition that a DNA mismatch should be sufficient to establish innocence in most circumstances, such as rape-murder cases involving a single individual.

Legal services for postconviction claims of innocence and DNA testing are, for the most part, a matter of philanthropy. As an example, between 2004 and 2008 the Innocence Project received a total of 12,614 new letters from prisoners and their families seeking help to prove wrongful conviction. They received 3,208 letters in June 2008 alone. A mere 1 to 2 percent of all persons sending letters were accepted as clients by the Innocence Project. A 2009 study by the National Academy of Sciences (NAS) "calls into question the scientific merit of virtually every commonly used forensic method, including analysis of fingerprints, hair, fibers, blood spatters, ballistics and arson."[44] The Innocence Project selects only cases where a claim of innocence can be resolved through DNA testing. One can only speculate how many wrongfully convicted individuals have remained in prison from the 98 percent whose cases were not accepted because of scarce resources or discarded or improperly stored biological evidence. We have evidence of the exoneration rate for the chosen cases. Within a period of five years the Innocence Project closed 233 cases, of which 98 were closed subsequent to DNA testing. Of those 98, 42.9 percent showed that the prisoner was excluded, and 41.8 percent showed that there was a match between the prisoner's DNA profile and that of the crime-scene DNA. Many cases were closed without DNA testing for a variety of reasons, including death of the client, lack of sustained client interest, or poor DNA crime-scene evidence.[45]

Once all the states meet the responsibilities recommended by the NAS study and provide full access to postconviction DNA testing, and the appropriate exculpatory role of DNA is accepted, the burden of the wrongfully convicted to find the real perpetrator should fade into the background, and the claims of the need for a universal database for exoneration should become moot.

The two forensic faces of DNA are its role as evidence in guilt and in innocence, but these faces are not symmetrical. The state commits extensive resources toward deploying DNA evidence for prosecuting crimes but rarely exercises responsibility in using DNA after conviction to test claims of innocence. Many prisoners remain incarcerated because the DNA evidence that could potentially exonerate them has been lost or purposely discarded. Although a number of states have passed legislation to ensure the preservation of evidence, about half fall short of such protections. As a result, the claims by incarcerees of wrongful convictions cannot be resolved by the use of DNA testing.

Even when the evidence is preserved, once a conviction has been finalized, the burden of proof for demonstrating wrongful conviction is raised by cultural, technical, and financial barriers, not the least of which is the disappearance of exculpatory evidence. Police and prosecutors are rewarded for prosecution and conviction but are shamed for wrongful incarceration. Postconviction applications of DNA have made clear that the often-expressed ideal that the criminal justice system is designed to minimize false convictions has not yet been realized.

Chapter 8

The Illusory Appeal of a Universal DNA Data Bank

Everybody, guilty or innocent, should expect their DNA to be on file for the absolutely rigorously restricted purpose of crime detection and prevention.

—Sir Stephen Sedley, Lord Justice of Appeal (Judge) for England and Wales[1]

We would be appalled, I hope, if the State mandated nonconsensual blood tests of the public at large for purposes of developing a comprehensive . . . DNA databank. The Fourth Amendment guaranty against unreasonable searches and seizures would mean little indeed if it did not protect citizens from such oppressive government behavior.

—Justice J. Utter, Washington State Supreme Court, *State v. Olivas*[2]

In January 2007 *Glamour* magazine ran an editorial titled "Let's Catch More Rapists Before They Strike Again." The editorial calls for the massive expansion of state DNA databases. The idea behind the expansion is based on a few known cases where a recidivist rapist, once arrested for a nonfelony violation, was not apprehended until he had committed multiple rapes. Had the individual's DNA been placed in a data bank after the first arrest, the argument goes, then the second and subsequent rapes might have been prevented. *Glamour* wrote:[3]

> Currently, most states get DNA samples only from convicted felons; victims' advocates and law enforcement officials also want to collect them from people guilty of misdemeanors, like minor assaults or burglary, and those *arrested* for crimes, even without a conviction. . . . It is absolutely necessary [to take DNA upon arrest], experts say, given that criminals often progress from less serious crimes to more violent ones.

The article goes on to state, "Preventing rape and saving lives should be a national priority—and a beefed-up DNA database is our best hope," and urges women readers to "check dnaresource.com to see what your state's DNA collection laws are, then urge your legislators to broaden those laws and push for more funding."

No one would disagree with the goal of preventing rape. But if one takes the stated goal of this article to its logical implications, then the next step after expanding the databases would be to universalize them. After all, if a rapist strikes and is not arrested, then the only way to prevent him from striking again might be to ensure that his DNA is in the database in the first place, and the only way to do that is to make it compulsory that everyone's DNA be in the database. Indeed, if everyone's DNA was in the database from birth, then theoretically every DNA sample collected from every rape would result in a hit.

Formal proposals for universal databases have been put forward by a handful of politicians and individual scholars. In 1999 Rudolph Giuliani, then mayor of New York City, proposed collecting DNA samples from all newborns for both medical and law-enforcement purposes.[4] A similar proposal was made in Michigan.[5] In 2006 Prime Minister Tony Blair called for a universal forensic DNA database to include every British citizen.[6] Proposed collection methods for such a database include linking law enforcement with state newborn screening programs, taking samples as part of child vaccination requirements for entering school, making provision of a DNA sample a requirement for obtaining a driver's or marriage license, and creating a national identification card that incorporates DNA information.

Certainly it would be a great benefit to public welfare and to the safety of women if the technology of DNA identification could reduce the incidence and increase prosecutions of sexual violence against women. But is an ever more inclusive database the best way to prevent rape? Do we need a database of everyone's DNA to address the problem of sexual assault? What will be gained and what will be lost by developing a universal DNA database, where collections of genetic profiles begin in infancy, and a national DNA database stores an identity profile of every resident and visitor to the country? What will be the impact on crime solving? What are the dangers? And how feasible is it to create such a database? In this chapter we consider each of these questions as we explore the pros and cons of a comprehensive, compulsory DNA database.

Crime Prevention, Efficiency, and Exonerations

The most common argument advanced in support of a universal DNA database is that such a database would prevent crimes, especially sexual assault. As stated in the *Glamour* editorial: "It's [DNA fingerprinting] especially helpful in case of rape, because while fingerprints can't be taken off a woman's body, fluids and other biological remnants can. But experts say we are underutilizing this breakthrough technology, leaving rapists on the street, free to attack again."[7]

A truly universal, nationwide database should result in a 100 percent match rate against the database. Any DNA collected from the scene of a rape would presumably link us to an individual. But some important caveats need to be considered in understanding the actual impact this might have on rape incidence. First, DNA is not relevant for most rape cases. According to the Department of Justice, most rapes committed against women are perpetrated by someone the victim knows.[8] According to the Centers for Disease Control, in 2000, 8 out of 10 rape or sexual assault victims reported that the offender was a friend, acquaintance, or relative.[9] Second, even after the name of the rapist is identified, there are usually problems in proving that the alleged sexual assault was not consensual, which cannot be resolved by the analysis of DNA. Third, only about one-third of rapes even get reported to police in the first place. A DNA database, no matter how inclusive, will do nothing to help solve unreported cases.

No one doubts that midnight rapists have been identified and successfully prosecuted largely because of DNA evidence. But several things must happen in order for this to occur. First, DNA evidence must be found at the scene of a crime. It has to be of sufficient quantity and quality to produce a full DNA profile that unambiguously matches a profile in the database. Often in cases of sexual assault DNA evidence is in the form of a mixture of more than one individual's DNA. This can complicate the analysis (see chapter 16). In addition, other factors in the case have to implicate and not exonerate the suspect. Therefore, a universal database will not place all members even of this smaller subset of rapists—all midnight rapists—behind bars.

Even if a universal database will not solve all rape cases per se, there are people who argue that it would still have a dramatic impact on crime rates generally. Advocates of a universal database for crime reduction base their arguments on two theories. First, they argue on the basis of a recidivist theory that with a universal database police will be able to catch perpetrators on their first offense, before they have a chance to commit a second crime.

There is considerable debate about recidivism rates and whether it is possible to predict that an individual who has committed one crime is likely to commit another. Complex social behavior cannot be generalized or easily reduced to probability statistics. Sex offenders are thought to have a particularly high rate of recidivism, and as a result, a number of policies targeting this category of offenders have been adopted at both state and federal levels, including community registration, notification, civil commitment, and compulsory treatment. The Combined DNA Index System (CODIS) database itself initially targeted this group. But despite significant attention to this category of offenders, results of studies on sex offender recidivism have been tentative and, at times, contradictory. Some studies have reported recidivism levels as high as 70 percent, whereas others have found levels as low as 4 percent, depending on the category of the offense.[10] Part of the difficulty is that "sexual offending" is a broad term and includes a wide variety of offenses, such as voyeurism and exhibition, as well as both violent and nonviolent sexual offenses. Recidivism rates among these different categories of offenders are unlikely to be similar. Also, recidivism can be defined broadly as any new offense or more narrowly as either any new sexual offense or a sexual offense of the same type, and it can be measured on the basis of rearrest, reconviction, or reincarceration. The range in recidivism estimates can be explained, in part, by these methodological disparities. A 2005 study by Kristen Zgoba and Lenore Simon found that 33 percent of sexual offenders in New Jersey followed over a four- to seven-year period after release recidivated, but fewer than 10 percent of sexual offenders were reconvicted for sexual offenses.[11]

Regardless of actual recidivism rates, it can be argued that every unsolved crime provides reinforcement for the recidivist to commit another criminal act. Career criminals are associated with minor burglaries, for which DNA is almost never sought because of the difficulty of collecting DNA known to have been left by the perpetrator, the expense incurred by the DNA analysis, the lower profile of the crime compared with crimes of violence, and the low yield from CODIS matches. With universal data banks the suspect yield and the prosecutions from minor burglaries might increase (if we assume that DNA can be collected from the crime scene and processed). If caught, convicted, and incarcerated, the offender would be prevented from committing another offense and possibly from graduating to more violent crimes.

In the Washington Supreme Court case *State v. Olivas* the court considered whether the state could obtain a DNA sample of a convicted felon when it was not needed by law enforcement. The Washington State Supreme Court concluded that the state could obtain a DNA sample under a "special-needs" provision based on the evidence of rates of recidivism of convicted felons. Rebecca Sasser Peterson interprets the court's ruling:

> Prisoners' knowledge that their DNA fingerprints are in a database, making them easily identifiable, could deter those same prisoners from re-offending after their release. Without further discussion, the court concluded that the government interest in deterring recidivist criminal activity through the use of an offender DNA database was a need "beyond normal law enforcement" and therefore qualified as a "special need."[12]

Catching an offender on the first act will not prevent the initial act from occurring. The recidivist theory, to the extent that it applies, only serves to prevent future criminal acts. A second theory in support of a universal database is that having everyone's DNA in the system would deter crime. According to David Kaye and Michael Smith, "In practice, settling for a DNA identification database restricted to convicts, or to convicts and arrestees, is sure to . . . further compromise public safety by halting far short of the deterrent and investigative capability that a population-wide database would afford."[13] If people know that they are likely to be caught and prosecuted, the argument goes, they will think twice about committing a crime. A career criminal might also be discouraged from acting in the face of a universal database because it would be exceedingly difficult for him to lie about his identity, especially if DNA databases were generalized to other countries. Ben Quarmby, former senior editor of the *Duke Law and Technology Review*, argues that a universal DNA database "would make it more difficult for the same individual to strike twice and would make it considerably more difficult for terrorists to assume false identities."[14]

Currently there is no empirical evidence to support the often-stated claim that DNA databases deter crime.[15] Also, the presumed deterrent effects of other criminal justice programs and surveillance technologies have often not panned out. For example, social science research on the death penalty, long justified as a way of deterring murder, indicates that the death penalty does not have an additional deterrent value beyond that of a lengthy prison term.[16] States that have imposed the death penalty over the last 20 years have not experienced reductions in homicide rates. Murder rates have been higher for death penalty states compared to non-death penalty states.[17] Similarly, more punitive prison sentencing policies appear not to deter crime. Prison populations have been growing steadily for a generation, but the crime rate today is about what it was in 1973 when the prison boom began. A 2005 study by the Sentencing Project showed that states with above-average increases in incarceration have below-average declines in crime rates. Moreover, the study found that states with harsher drug-sentencing schemes are correlated with more frequent drug use.[18]

Widespread surveillance technologies perhaps provide a more appropriate opportunity for comparison, but here, too, the promised benefits have not occurred. Video cameras, or closed-circuit television (CCTV), have been extensively deployed in public places and are becoming increasingly widespread in the United States. But studies have shown that there is little evidence to support the notion that cameras have either prevented or helped solve crimes. According to a study released by the Scottish Office Central Research Unit in 1999, although more than 70 percent of 3,000 people interviewed believed that CCTV would prevent crime and disorder and more than 80 percent thought that it would be effective in catching perpetrators, it turned out that widespread installation of CCTV cameras in the city of Glasgow "appeared to have little effect on the clear up rates for crimes and offences generally."[19] More recently, the U.K. Parliament reported:

> It has been difficult to quantify the benefits of CCTV in terms of its intended effect of preventing crime. We recommend that the Home Office undertake further research to evaluate the effectiveness of camera surveillance as a deterrent to crime before allocating funds or embarking on any major new initiative. The Home Office should ensure that any extension of

> the use of camera surveillance is justified by evidence of its effectiveness for its intended purpose, and that its function and operation are understood by the public.[20]

The deterrent effect of a universal database, if any, certainly would not apply to those individuals who commit a "crime of passion" or who are convinced that they can escape a crime without leaving their DNA at the scene. The possibility that criminals could act without leaving traces of their DNA is a real one; prisoners have in fact been overheard coaching one another on ways to plant biological evidence at crime scenes and how to avoid leaving their own DNA behind.[21]

Proponents of a universal database also argue that a universal DNA database will improve crime-investigation efficiency. There may be potentially valuable DNA evidence at a crime scene, but if there is no match in CODIS, the national forensic DNA database, police must resort to conventional old-fashioned policing to come up with suspects. In contrast, if they had DNA profiles for every resident in the country, crime-scene DNA would generate suspects rather efficiently, which Paul Monteleoni, 2006–2007 editor of the *New York University Law Review*, argues "represents a shortcut past a number of intrusive and error-prone investigative techniques."[22] Improving the efficiency of criminal justice has a certain appeal, especially during a period of strained public budgets. "Law and economics" advocates such as Judge Richard Posner believe that the concept of efficiency, or, as economists say, "Pareto optimality," is relevant in the operational aspects of law.[23]

It is reasonable to assume that the larger the DNA database, the greater the chances that there will be a cold hit. In fact, however, although a universal database would generate a higher number of suspects overall, the probability of getting a hit does not change as you increase the database with innocent people. It changes only when you increase the database with people who have already committed or who will likely commit a crime in the future. There is no way to predict who will commit future crimes. In addition, as is discussed in chapter 17, the limitation of the databases in detecting crime thus far appears to be driven not by the number of individual profiles in the database, but instead by the number of crime-scene profiles. In the United Kingdom, DNA is obtained in less than 1 percent of all recorded crimes.[24]

DNA evidence has been highly beneficial in proving that a person was wrongfully convicted. The Innocence Project of the Benjamin Cardozo Law

School has assumed responsibility for undertaking legal defense in over 250 exonerations. Most of these cases have hinged on the analysis or reanalysis of DNA evidence. But does society need a universal DNA data bank in order to protect the innocent from being falsely convicted? If DNA evidence left at a crime scene does not match that of the suspect, it would seem sufficient exculpatory evidence. In what way would a national DNA database contribute to this social good?

Columbia University sociology professor Amitai Etzioni is closely associated with a social theory called communitariansim. In Etzioni's words, "Communitarianism is a social philosophy that maintains that society should articulate what is good, and asserts that such articulations are both necessary and legitimate."[25] There are several grounds, according to Etzioni, on which large-scale DNA banking would be of such benefit to the common good that it would outweigh the losses in privacy, which he recognizes are not trivial. If nothing else, DNA data banks can be used to exonerate falsely convicted individuals.

There is one sense in which Etzioni appears to be correct. Even when falsely incarcerated individuals can prove that the DNA at the crime scene was not theirs, this information has not always been sufficient to obtain their freedom. That was the situation facing Darryl Hunt, a 19-year-old African American male who was convicted in North Carolina of the rape-murder of a white woman and sentenced to life imprisonment in 1984 (see chapter 7). Hunt's conviction rested on weak evidence, at best, in the form of a tentative eyewitness identification. Hunt testified that he did not know the victim, and there was no other evidence linking him to her or to the murder. In 1994 Hunt's lawyers requested DNA testing in the case. The DNA testing clearly excluded Hunt from the crime—his DNA did not match semen found on the victim's body at the crime scene. Despite the results, however, Hunt's appeals were rejected. Not until the evidentiary semen DNA profile was submitted to the state DNA database and resulted in a partial match that led to the investigation of Willard Brown, who eventually admitted to the crime, was Hunt finally exonerated in 2004 after 18 years of incarceration, and 10 years after he was first excluded by DNA.[26]

The Darryl Hunt case suggests that there are times when the only way to exonerate a falsely imprisoned person, even when his DNA profile does not match the evidence, is to find the real perpetrator. But it also points to deep-seated failures of our justice system. Many have attributed Hunt's wrongful conviction in the face of clear exculpatory DNA evidence to deep class and racial bias that continues to permeate the South and the American criminal justice system. Although a universal DNA database might help avoid these injustices, it will do little to address these fundamental concerns.

Fairness in Simplicity?

A universal database has an egalitarian appeal to it. Having everyone in the database might be fairer than having certain populations of individuals, particularly when those populations (e.g., those convicted or arrested) are skewed toward particular racial and socioeconomic classes. Inclusivity carries with it a certain simplicity in that it avoids having to grapple with questions of whose DNA should or should not be in the database. It would also obviate the need for a number of controversial law-enforcement practices, including familial searching, DNA dragnets, surreptitious DNA collection, and John Doe DNA indictments, that are often criticized on notions of fairness.

Some proponents of a universal database have indicated that on racial justice grounds they prefer a database that is all-inclusive rather than the one that the United States is fast approaching that contains DNA profiles from persons arrested. For example, Kaye and Smith support universal data banks, in part because they believe that "there can be no doubt that any database of DNA profiles will be dramatically skewed by race if the sampling and typing of DNA becomes a routine consequence of criminal conviction."[27] A universal (population-wide) DNA database would be "more effective and more fair" than any system based on conviction or arrest.[28] Limiting the national database to convicts and arrestees, they claim, "is sure to aggravate racial polarization in society, undermine the legitimacy of law and law enforcement, and further compromise public safety by halting far short of the deterrent and investigative capability that a population-wide database would afford."[29]

Racial disparities in our criminal justice system are well established.[30] They are systemic and occur at all levels—arrest, prosecution, conviction, and sentencing—but they have proven especially significant at the level of arrest, which is perhaps most prone to subjective judgment or bias on the part of police officers. For example, it is widely recognized that minorities are stopped more frequently than nonminorities in drug searches and traffic citations.[31] Those states where DNA data banks include the profiles of arrestees who have not been charged with or convicted of a crime inevitably will be weighted disproportionately toward minorities in CODIS. Pilar Ossorio and Troy Duster predicted an outcome for which there is already confirmation: "If a DNA database is primarily composed of those who have been touched by the criminal justice system and that system engages in practices that routinely and disproportionately target minority groups, there will be an obvious skew or bias in the databases and repositories."[32] Similarly, Michael Smith, a law professor at the University of Wisconsin and former chair of the working group of the National Commission on the Future of DNA Evidence, has argued: "If the [DNA] database has a wildly disproportionate representation of African American males, they will be the ones found to have committed previously unsolvable crimes. And when patterns of enforcement are racially skewed, the laws themselves are de-legitimized."[33] The racial justice disparities in DNA data banking are discussed in chapter 15.

England, which has the most permissive criteria of any nation for entering someone's DNA profile in the national DNA data bank, also has a very high percentage of its black population profiled. In 2006 it was estimated that one-third of young black males were profiled in the national DNA database, compared with one-eighth of young white males.[34] In an interview with BBC Radio in 2007, Sir Stephen Sedley, Lord Justice of Appeal (Judge) for England and Wales, stated that these "indefensible" anomalies in the system have resulted in a situation where

> if you happen to have been in the hands of the police, your DNA is permanently on record and if you haven't, it isn't. . . . It means that where there's ethnic profiling going on, a disproportionate number of members of ethnic minorities get onto the database. And it also means that a great many people who are walking the streets and whose DNA would show them to be guilty of crimes go free.[35]

For Sedley, a fairer system would be one where DNA is collected from every citizen and every visitor upon entry to the country. If every man, woman, and child has his or her DNA profile in the forensic DNA database, then at least one aspect of the criminal justice system would not be weighted disproportionately with respect to any racial or ethnic group.

But although a universal database might help level the playing field at the level of the database, it would do nothing to address the overall racial and ethnic bias in the criminal justice system: "Bias would still exist because law enforcement's lens does not focus equally on all types of crimes. Some kinds of activity, such as drug related street crime, receive far more attention from police than parallel kinds of crimes, such as cocaine sales at predominantly White college fraternities."[36] The makeup of the database will have no bearing on who is targeted for suspicion and arrest. Even if everyone is in the database, the majority of hits will continue to identify minorities so long as the types of crimes, neighborhoods, and populations monitored and investigated are racially driven. In fact, a universal database might give the impression that the system is fairer and could serve to mask the racial bias that is operating in the overall system.

There are, of course, other means to redress racism in criminal justice. For example, by restricting the collection of DNA to convicted felons, some of the racial disparities in the composition of the data bank will be diminished. Moreover, racial and ethnic disparities in the composition of the forensic DNA data bank are not the most onerous form of prejudice that is manifest in the criminal justice system. African Americans are more likely to be falsely convicted for rape than other population groups. With all African American DNA on the database, it is reasonable to assume that false convictions in that population group will actually increase because DNA fallibility follows racial lines. In cases where individuals have been falsely convicted because of mishandled DNA, they have disproportionately been people of color.

As previously noted, a universal database would obviate the need for several controversial law-enforcement practices that target specific groups of individuals and are criticized as unfair. Familial searching would be redundant if everyone were in the database. The reason for familial searching is to triangulate toward a precise match of a suspect starting with crime-scene DNA by getting a close match with a relative. That would be unnecessary if everyone's DNA profile were in the national database accessible to all prosecutors and officers of the law. Similarly, law enforcement would not have to stop hundreds or even thousands of individuals on the street in a community as a last-ditch effort to solve a crime. As Peterson writes, "A universal DNA database containing a DNA fingerprint from every citizen in the United States could be used to identify otherwise missed first-time offenders and to render unnecessary discriminatory dragnets."[37] Police also would not need to follow anyone around to pick up DNA off discarded items or create ruses to encourage individuals to leave their DNA behind. Their DNA would already be in the custody of the police. Not only would these controversial practices come to an end, but law enforcement also would not have to expend time and resources on seeking DNA samples from innocent people under each of these scenarios.

Although it is true that a universal database would render these practices unnecessary, it is important to recognize that what is controversial about these practices is that people are uncomfortable having their DNA collected and stored by law enforcement in the first place. In the case of DNA dragnets, for example, people are not comfortable voluntarily giving their DNA to law enforcement when they do not have any assurance that their DNA will not be misused or that it will be destroyed after they are excluded from the crime in question (see chapter 3). Simply creating "the dragnet to end all dragnets" is hardly a solution to this concern.

A universal database would also eliminate the need for "John Doe" DNA indictments. These indictments are sometimes sought by prosecutors as a way of extending statutes of limitations. Such statutes provide time limitations after which legal proceedings can no longer be brought against a person. The lengths of the periods vary from state to state and depend on the type of legal action and the type of crime. Statutes of limitations are designed to protect suspects against claims made after evidence has been lost or degraded, memories have faded, or witnesses have disappeared or are no longer alive. Most states have a statute of limitations that applies to all crimes except murder, and the time limit generally starts to run on the date on which the offense was committed. Once the statute has expired, a court lacks jurisdiction to try or punish a defendant.

Most rape cases fall under a statute of limitations. When law enforcement fails to indict someone on a rape case that is about to reach the critical year under a statute of limitations (SOL), some prosecutors have sought extensions to the SOL from state legislatures. In cases where DNA evidence is available, some prosecutors have filed a "John Doe" DNA indictment. The theory behind these indictments is that the perpetrator, whose identity is not known, is uniquely identified by his DNA, which serves as a surrogate for a physical description or a name. The DNA must be collected and tested within the statute of limitations, and the court must find probable cause that the individuals whose genetic profile is described in the indictment committed the charged offense. This will allow a prosecutor to bring a person whose DNA matches that of "John Doe" to trial without limit of time. Some courts have accepted John Doe indictments on DNA profiles alone; others have required additional information with the DNA, such as a limited visual description of the individual.

Proponents of John Doe indictments argue that using a DNA profile rather than a name or detailed physical description is a legitimate way to vindicate victims, prevent offenders from escaping conviction for their crimes, and possibly prevent future crimes because of the deterrence element.[38] Also, because DNA does not degrade easily when it is stored in dry environments, unlike memory, which does fade easily, the quality of DNA evidence over time does not circumvent the purpose of the SOL, which is to protect defendants against fading personal memory and testimony that lose their probative value over a period of years. As already noted, nonconsensuality is difficult to prove in many rape cases, and faded memory exacerbates this problem.

Critics of the John Doe DNA indictment counter that it evades an important constitutional guarantee for noncapital offenses.[39] The DNA, after all, may or may not represent the identity of the perpetrator. It is simply biological material of interest left at the crime scene and does not necessarily define the identity of the perpetrator of the crime. In addition, even under optimal conditions, errors can and have occurred in the collection, handling, and storage of DNA samples that can result in errors in identification (see chapter 16). Because DNA evidence can remain intact over many years while other evidence, such as memory, can fade, the relative weight of DNA evidence increases with respect to other evidentiary sources. If the DNA found at the crime scene is not that of the perpetrator, its mere presence in a courtroom can override fading memories that provide an alibi for the suspect. If there were a truly universal DNA database, then there would be no need to engage in John Doe DNA indictments, since presumably every DNA sample collected from every crime scene would result in a match with an individual in the first instance.

The Expense to Privacy and Autonomy

If the database included everyone in the country, then introducing a crime-scene DNA profile into the database might pick up the guilty individual, but at what expense to the erosion of privacy and to the vast majority of innocent persons who will have their DNA on file? As is discussed in chapter 14, DNA contains a vast amount of information about a person. In contrast to fingerprints, which can be used only for identification, DNA can be mined for endless amounts of information about a person, such as insights into familial connections, physical attributes, genetic mutations, ancestry, and disease predisposition. As science advances, the phenotypic information available from human DNA will necessarily grow. Genetic information could be used in discriminatory ways and may include information that the person whose DNA it is does not wish to know. And repeated claims that human behaviors such as aggression, substance addiction, criminal tendency, and sexual orientation can be explained by genetics render law enforcement's retention of DNA potentially prone to abuse. These privacy concerns come into play because of current law-enforcement practice, where the DNA samples are retained along with the generated DNA profiles (see chapters 1 and 14).

The taking of DNA is generally considered a "search" under the Fourth Amendment, in part because of the substantial and uniquely personalized nature of the information contained in the DNA sample. Although the courts have generally upheld DNA data banks of convicted offenders on the notion that such individuals have a "diminished expectation" of privacy, as balanced against society's need to promote law and order, it seems unlikely that an all-population data bank containing the DNA of millions of innocent individuals who have never been in contact with the law would withstand Fourth Amendment scrutiny. Moreover, as a matter of policy, the notion that innocent individuals should not have DNA retained permanently in a database is reasonable for a society that values freedom and individual privacy.

A mandatory, comprehensive DNA data bank renders DNA a tool for surveillance. Our DNA is shed in many places—on coffee cups, saliva, sweat, and hair follicles. The appearance of our unique DNA at a particular place is testimony to our being at that location unless someone planted the DNA there. Should ordinary citizens, for whom there is no suspicion of a crime, be the subjects or even the potential subjects of surveillance?

Suppose that a group of people meet to discuss strategies for a national protest against U.S. policy. Police enter the meeting space after the discussion is over and collect DNA to determine who came to the meeting so they can be placed under further surveillance. Individuals currently have no clearly established protective rights to their "abandoned DNA" (see chapter 6).

With a national DNA data bank, DNA can be used to track people's movements. Perhaps for the moment, DNA surveillance is a bit far fetched beyond its use at a crime scene because of the expense and time for analysis. But the technology is rapidly changing. DNA surveillance is recognized by legal scholars as a new technological means for privacy breaches of innocent people while serving as a new tool for criminal investigation and biometric profiling of crime-scene DNA:

> This power to reconstruct past events, however partially, will be invaluable to criminal investigators, but it must be recognized that it will diminish our spatial anonymity—the privacy of our movements by reducing our ability to enter bedrooms (or other locations that might prove embarrassing) without the risk of our presence there later being discovered.[40]

Currently, most citizens do not have fingerprints on file with the police. A population-wide database of fingerprints can be misused in violation of one's privacy and spatial anonymity. Police can collect fingerprint samples after people leave a gathering and determine the identity of those who attended without obtaining a warrant or taking photographs. With a universal DNA data bank, the opportunities for such breaches in privacy are even greater. Mark Rothstein and Sandra Carnahan frame the distinction between fingerprints and DNA as follows: "A fingerprint reveals only unique patterns of loops and whirls. In contrast, a DNA sample is the information-containing blueprint of human life, revealing one's genetic predispositions to disease, physical and mental characteristics, and a host of other private facts not evident to the public."[41] Fingerprints cannot be used to determine paternity, family relationships, or medical predispositions. A sperm donor who wishes to maintain his anonymity cannot be found through fingerprinting the child born from his sperm. If American society has been resistant to accept a mandatory national fingerprint database, why would it accede to a population-wide DNA database?

There are many other technologies that can be used to solve crimes, but we might not wish to adopt them all as such. One can imagine a mandatory requirement that each person wear a Global Positioning System (GPS) bracelet with an encoded DNA microchip—a type of permanent identity card. When police wish to keep an eye on someone, they can activate a program in the central computer (or obtain a court warrant to do this), and the GPS system will be turned on, allowing police to fully monitor the comings and goings of the individual. No doubt many crimes would be solved, and many might even be prevented.

If we have learned anything from history, once put into the hands of agents of government, surveillance technologies at times are likely to be misused. When we examine the social implications of any new technology with surveillance capabilities, in addition to examining its idealized use in criminal justice, we should also examine its use under less-than-ideal conditions, such as when government authorities operate covertly, circumventing legally mandated and constitutional limitations.

Some have proposed including a DNA profile as part of a national ID card. Such information could go hand in hand and be linked with a centralized, national DNA database and raises additional privacy concerns because of its potential to be accessed by third parties. Most people are unaware of how their Social Security numbers or other personal identifiers are used in centralized data banks. Think about getting your credit background check when you apply for a new credit card or a loan. We rarely get to see the information held by the credit-checking companies that provide this service for all major department stores and national chains. Not only is our privacy being compromised, but the personal information may be incorrect. Increasingly, supermarkets and chain pharmacies issue discount cards that provide company surveillance of consumer purchases. In addition, chain pharmacies sell information on prescribing habits of individual physicians to drug companies. A drug representative can show a physician a spreadsheet of all his or her prescribed drugs over the year, comparing generics with trade drugs. Presumably that information will allow the drug representative to change the prescribing behavior of the physician in the interest of his or her company. The collection of consumer information in centralized data banks will reach its apotheosis with the creation of a national personal identification system that could include genotypic and phenotypic information.

A hacker could potentially do significant damage to a system of everyone's DNA profile. In the United Kingdom, where debates about a potential universal database abound, even those who are generally supportive of such concepts have stressed the problem of potential security breaches. Tony McNulty, a former police minister in the United Kingdom, has stated: "I understand the debate around universalism. . . . How to maintain the security of a database with 4.5 million people on it is one thing—doing it for 60 million people is another."[42]

A universal database would also make it easier for a reasonably motivated criminal to frame someone for a crime. If everyone's DNA is in the system, simply placing a hair sample, cigarette butt, coffee cup, used condom, or other personal item could implicate someone other than the perpetrator or at least send investigators off in the wrong direction. The potential for DNA to be planted has been generally underacknowledged and is particularly problematic in light of the prevailing mischaracterization of DNA as infallible (see chapter 16). Jill Saward, a victims' rights activist, advocated for a universal DNA database as follows: "If we have everyone on the database, we don't have to worry about whether we think they might be guilty or not; everyone's there."[43]

Some supporters of a universal DNA criminal justice database recognize the possible risks, especially to suspicionless and innocent individuals. In response, they propose a range of safeguards. According to Michael Smith, a law professor at the University of Wisconsin, "The solution to the privacy problem is to destroy the biological samples in which individuals' DNA resides."[44]

For Smith, there is an inevitability of the universal database, and he prefers that we design it deliberately rather than falling into it along current patterns of piecemeal expansions: "If we do not design and build an inclusive DNA identification database, the databases we have will expand anyway—without necessary protections and without realizing the full public safety benefit of the technology."[45] Similarly, Richard Stacy, a former federal prosecutor, has remarked: "I sense that eventually a universal database will come; it seems as inevitable as the Internet. It may be that someday every child will have a DNA sample taken at birth and entered into a national database. If that happens, it will be up to us to make certain that it is done in a way that privacy is protected and abuse, from either the government or the private sector, cannot occur."[46]

Etzioni's solution for a national DNA database to serve the common good while protecting individual liberties is to institute an "independent public accountability board" that would ensure that civil liberties were protected from the misuse of biometric data. Because a universal DNA database will help ensure that a DNA profile of the perpetrator of a violent crime is correctly matched with the crime-scene DNA, he argues, it will help ensure that suspects are not wrongfully convicted, and if they were, the correct match will be made and the falsely convicted individual will be exonerated.

These proposals are well meaning. But if we take seriously the premise that DNA banking is contrary to our Fourth Amendment rights or is an affront to personal privacy and autonomy, then the logical solution is to limit the databases, not to expand them indefinitely. Similarly, we need not accept the notion that the expansion of the databases is inevitable. Although it may take significant political will and momentum to reverse current trends, there is no reason that we cannot restrict or roll back the databases if we can agree that innocent persons do not belong in DNA databases.

Feasibility

If we were to decide as a society that we wanted to move forward with creating a nationwide, universal database, how would we do it? Currently, more than 8 million persons have had their DNA collected, profiled, and stored in the CODIS database. The population of the United States exceeds 300 million. How feasible would it be to expand the database by a factor of 40?

The most commonly proposed idea for building a universal database is to piggyback on the infrastructure provided by newborn screening programs. All 50 states require that hospitals test newborns for certain genetic diseases, for example, phenylketonuria (PKU), an inherited metabolic condition that can easily be treated by diet if it is caught immediately after a child is born. Infants with PKU who are not treated will suffer severe mental retardation and organ damage. To test for PKU and other genetic conditions, hospital nurses extract a few drops of blood from the heels of newborns. The blood is preserved on paper called "Guthrie cards."

What happens to the blood samples today? Are they destroyed after the tests are completed? States handle the blood samples differently. In 2002 it was revealed that South Carolina was storing more than 300,000 blood samples of newborns.[47] The blood samples on the Guthrie cards could be used to create a national DNA data bank by profiling the samples of the newborns. Eventually the entire population of those born in the United States would have DNA profiles in a central data bank.

The sheer costs of such an expansion provide a considerable barrier, at least for the near future. At a highly conservative estimate of $50 a sample, it would cost the government $15 billion to test the population's DNA, and $200 million per year to test the newborns born each year in this country. If we look just at the cost of testing newborns, we would be spending $3 billion over the first 15 years of testing with no expected return to law enforcement (since children under the age of 15 are unlikely to commit serious crimes).

A universal database also seems unconscionable when one considers the persistent and ongoing state of crime laboratory backlogs. In 2002 the Bureau of Justice Statistics of the U.S. Department of Justice (DOJ) conducted an analysis of the 50 largest crime labs in the United States and concluded that for every DNA analysis request completed that year, nearly two requests remained outstanding at the end of the year.[48] In 2003 a special report released by the National Institutes of Justice estimated that more than 350,000 rape and homicide cases awaited DNA testing.[49] These backlogs are intensifying with each DNA expansion. In March 2009 the DOJ released an audit of the convicted-offender DNA backlog-reduction program that showed that despite the more than $1 billion that the government has poured into labs around the country, and although a federal backlog-reduction program had helped reduce backlogs to some extent, we still have a nationwide backlog of somewhere between 600,000 and 700,000 convicted-offender samples. Presumably that does not include arrestee samples or crime-scene samples, including rape kits. The report projects that the backlogs are likely to increase because of "recent legislative changes."[50]

In addition to newborn screening repositories, there are also an increasing number of "biobanks"—large collections of blood or tissue samples that often include genetic and other information derived from those samples and linked medical, family history, or lifestyle information. These biobanks can be publicly or privately maintained and are used for research purposes. In 1999 the National Bioethics Advisory Commission estimated that each year DNA from 20 million people in the United States is collected and stored in tissue collections ranging from fewer than 200 to 92 million samples.[51] Some countries have begun to develop national biobanks that could be accessed and used for research. The Icelandic government passed a law in 1998 that allowed for the creation of the world's first national biobank, but the project has stalled following significant public opposition over concerns about privacy and consent, as well as economic and technical difficulties. In the United Kingdom several proposals have been put forth for creating a national biobank, and at least one of those proposals explicitly called for linking the data bank directly with the forensic DNA database.[52] Other countries that have started to create national biobanks include Norway, Sweden, Estonia, and Canada.[53]

Advocates for national biobanks see the prospects of a universal database as stemming inevitably from health databases. Philip Reilly, a member of the National Commission on the Future of DNA Evidence, stated in 1999:

> If I was asked to prognosticate 50 years hence, . . . I think it is inevitable that we will move to universal DNA databanking, inevitable, because there will be such powerful, positive reasons to do so from the point of view of protecting the public health, avoiding disease and moving to a more preventive approach to medicine, that we'll get those for non–law enforcement reasons and then it will be irresistible to use that database.[54]

A universal forensic DNA database based on newborn screening programs would not be truly universal, since not everyone living in the United States was born in the United States. Approximately 12 percent of the U.S. population is foreign born. One possibility is that DNA testing would be added to a growing litany of paperwork and tests—including fingerprinting and retinal scans—that visitors are currently required to undergo upon entering the United States. Many have pointed out that implementing a DNA-collection program at the borders could be immensely complicated. Richard Thomas, information commissioner in the Home Office in the United Kingdom, has stated:

> I think we have to think very long and very hard before going down the road of a universal DNA database. . . . There are some risks involved. . . . And then there are the practicalities. It's not just the millions of people coming through Heathrow or Gatwick every day and how we actually *get* their DNA. We have to have addresses, we have to record our own population and UK visitors. How do we keep track of people to match the DNA to individuals? So there are some immense practical issues as well as the civil liberties and the data protection issues.[55]

Another proposal for creating a universal DNA database is to do so through a national ID card. Every working person in the United States, whether or not he or she is a citizen, is required to have a Social Security number. Children who have bank accounts also must have a Social Security number for tax purposes. The personalized Social Security number is as close as Americans get to possessing a national identity card. Efforts to make that number appear on all passports and driver's licenses have not succeeded. On the contrary, many states have turned away from requiring people to use their Social Security number on their driver's licenses.

After the terrorist events of 9/11 the idea of a universal identity card for every adult American has opened up a new policy debate on the extent to which "proof of identity" through a mandatory federally issued card will help curtail terrorism. While identity cards are being discussed as a national mandate, they are being implemented in select populations. Beginning in March 2007, the Department of Homeland Security required that all port workers, including longshoremen and truck drivers, apply for identification cards, which include a fingerprint and a digital photograph.[56] Although the driving force behind universal identification cards has been fear of foreign and/or domestic terrorism, other motives have begun to surface. Some of these include preventing crimes, especially for recidivistic offenders; identification of human remains; solving crimes; reducing racial and ethnic disparities in criminal data banks, where there have been disproportionately high numbers of minorities; preventing identify theft; and solving paternity disputes.

A future scenario in which every piece of DNA evidence lifted from the scene of a crime and run against the database results in a hit has some appeal. After all, a 100 percent hit rate surely would result in more crimes being solved. Although there may be disagreements about how many crimes would be solved by adopting a universal forensic DNA database, there is generally no disagreement that the database would likely generate more suspects, lead to more prosecutions, and be responsible for more convictions. It seems improbable but is possible that crime rates would also decrease because a potential rapist or burglar would know that he was almost certain to be caught. A universal database also has an egalitarian feel to it, even if it does not do much to address the real causes of racial disparities in our criminal justice system. At the level of the database, we would all trade some aspects of our privacy rights in exchange for some real or perceived improvements in safety. This diminished privacy would be distributed across the nation rather than burdening particular groups of individuals at the level of the database, even if certain groups of individuals continue to be the ones who are targeted for arrest and prosecution.

The allure of DNA data banks goes well beyond identifying people who may have been at a crime scene. The databases are much more valuable than fingerprints for surveillance purposes, establishing biological relations in immigration, and supporting genetic theories of criminal behavior. Therefore, the debates over a universal data bank have destabilized the balance between social welfare goals and personal privacy. Under the guise of crime solving, DNA affords government an unprecedented technology of mass surveillance, which comes into direct and irreconcilable conflict with the Fourth Amendment to the Constitution.

Collecting everyone's DNA for deposit in a searchable database would be the equivalent of a mass, permanent, and continuing dragnet. In *Davis v. Mississippi*[57] the Supreme Court ruled that mandatory dragnet fingerprint campaigns, where there were not warrants, probable cause, or individualized suspicion, would violate the Fourth Amendment. Others have argued that mandatory universal DNA databases would alter "how we perceive ourselves as a free and democratic society."[58]

For these reasons, universal databases should, and most likely will, be opposed. According to Jeffrey Rosen:

> A universal database that can be consulted for any crime, serious or trivial, is one that many citizens would resist. It opens us to a world in which, based on the seemingly infallible evidence of DNA, people can be framed or tracked, by their enemies or by the government, in ways that liberal societies have traditionally found unacceptable.[59]

Part II

Comparative Systems: Forensic DNA in Five Nations

Chapter 9

The United Kingdom: Paving the Way in Forensic DNA

It is not necessary to destroy the DNA profile if an individual is arrested and subsequently cleared of the offence, or a decision is made not to prosecute.

—Association of Chief Police Officers, United Kingdom[1]

It is arguable that the general retention of profiles from the un-convicted has not been shown to significantly enhance criminal intelligence or detection.

—Police liaison officer, Scottish DNA Database[2]

The United Kingdom has paved the way in forensic DNA technology. Its massive National DNA Database (NDNAD) is the oldest, largest per capita, and most inclusive national forensic DNA database in the world.[3] Founded in 1995, it currently contains DNA samples and profiles from nearly 4.5 million individuals, or approximately 7 percent of the U.K. population of 61 million. On file are DNA profiles drawn from people convicted of a wide range of crimes, including serious violent crimes and minor offenses, children as young as 10, and arrestees, many of whom have not been convicted of any crime.

It is curious that violent crime rates have been significantly lower in the United Kingdom than in the United States—including during the last decade when key policy decisions about data bank expansion were made—but England and Wales have, per capita, a far larger DNA data bank, indeed, the largest in the world. The United States has 20 times as many murders as the United Kingdom and 3.8 times the murder rate. The U.S. population is about 300 million. The number of murders committed annually in the United States from 2000 to 2005 averaged around 16,200, or 5.7 homicides per 100,000 people. In contrast, the United Kingdom, which has about 61 million people, experiences somewhere around 800 murders annually; from 2000 to 2006 the average murder rate was 1.5 per 100,000. If we list countries by their per capita murder rate, with the highest ranked being number 1, the United States ranks 24th and the United Kingdom ranks 46th.[4] If the United States had the same murder rate as the United Kingdom, it would be seeing about 4,590 murders annually rather than four times that many (a rate of 1.5 per 100,000 applied to 306 million people).

What can explain why England and Wales have, by percentage of population, the largest forensic DNA data bank in the world? There are a few possible explanations for why the United Kingdom was a leader in setting up DNA data banks. First, the persons who discovered DNA profiling were the British scientist Sir Alec Jeffreys and his colleagues at the University of Leicester. Therefore, it was a matter of national pride that the United Kingdom exploit this technology to its fullest capacity. Second, there has been a rising xenophobia within the British Isles responding to threats from Muslim and North African extremists. National identity cards and ubiquitous cameras in central London are part of a national response to a fear of criminal activities, especially terrorism. Third, the first DNA dragnet took place in the United Kingdom, which brought international attention to the role DNA can play in criminal investigation. Fourth, there is a tradition in Britain of professionalized police work that has encouraged the ready adoption of new technologies to solve crimes. This tradition is reflected in Britain's rich literary tradition of crime novels: for example, Sir Arthur Conan Doyle's famous novels of Sherlock Holmes, a fictional consulting detective who brought skills of forensic science and deductive reasoning to criminal investigation, and Agatha Christie's mystery novels, which created a mythic role for Scotland Yard in criminal investigations. Fifth, Britain does not have a bill of rights that protects personal privacy in the way in which it is designed to function in the United States and that could have provided some resistance against the United Kingdom's collection of personal data for criminal investigation.

Thus the initiation of and drive for DNA collection in Britain occurred within a climate of public receptivity (or at least a lack of strong opposition) and a general expansion of police powers in response to national security threats. When DNA profiling became available and was proved to work, there was a strong receptivity within the British government to integrate it into the criminal justice system. In 1993 a royal commission report held that "it is important that the police make the most effective use possible of the technical means at their disposal including forensic pathology, forensic science, fingerprinting, DNA profiling and electronic surveillance."[5] Within this context and without constitutional or other initial legal barriers, the United Kingdom emerged as the world's most aggressive nation in collecting and storing forensic DNA from its citizenry.

The World's First and Largest Forensic DNA Database

In 1995 the United Kingdom established its National DNA Database (NDNAD) as the world's first forensic DNA database. According to Robin Williams and Paul Johnson, "There is no singular and distinctive legislative instrument authorizing the collection and storage of DNA samples by the police in England and Wales or the retention and comparison of DNA profiles on the NDNAD."[6] Instead, a series of legislative proposals enacted by Parliament between 1994 and 2008 created and incrementally expanded police powers to obtain, store, and access biological samples and DNA profiles (see table 9.1).

In 1993 the United Kingdom's Royal Commission on Criminal Justice formally recommended the creation of a national forensic database. The legal foundations for the database were provided the following year through the passage by Parliament of the Criminal Justice and Public Order Act of 1994. This act, described by Paul Johnson and colleagues as "a direct legislative measure enabling both the establishment of the database and the facilitation of its immediate growth,"[7] gave police the power to take DNA samples without consent from anyone charged with any recordable offense. The law reclassified DNA swabs and hair samples from "intimate" to "nonintimate." This allowed rapid expansion of the number of DNA samples collected by the police because, under British law, nonintimate samples can be taken without a person's consent from anyone charged with a recordable offense. The law does not make a distinction whether the sample is relevant or irrelevant to the crime being investigated. The NDNAD went into operation the following year, with government funding by the Association of Chief Police

TABLE 9.1 Key Legislation Related to Criminal Justice, Evidence, and DNA Data Banking in the United Kingdom, 1984–2008

Date	Legislation	Description
1984	Police and Criminal Evidence (PACE) Act	Allowed police to ask doctors to obtain blood samples for DNA testing in investigation of serious crimes.
1994	Criminal Justice and Public Order Act	Enabled the NDNAD to be created; established police powers to take DNA samples without consent from anyone charged with any recordable offense.
1997	Criminal Evidence (Amendment) Act	Allowed DNA samples to be taken without consent from individuals who were still in prison, having been convicted for a sex, violence, or burglary offense before the NDNAD was initiated in 1995.
2001	Criminal Justice and Police Act	Allowed all DNA samples and fingerprints to be retained indefinitely, whether or not the person had been acquitted; allowed samples from volunteers to be retained indefinitely on condition of consent.
2003	Criminal Justice Act	Allowed law enforcement to take DNA profiles and fingerprints without consent from anyone arrested on suspicion of any recordable offense and to retain them indefinitely, regardless of whether the person was charged.
2005	Serious Organized Crime and Police Act	Enabled profiles from DNA samples taken from any deceased persons to be checked against the U.K. NDNAD for identification purposes, irrespective of whether there was any suspicion of their involvement in a crime.
2008	Counter-Terrorism Act	Extended police powers to allow DNA and fingerprints to be taken from persons subject to control orders, and to be collected during any authorized secret surveillance, retained indefinitely, and used "in the interests of national security."

Sources: Chief Constable Tony Lake, Lincolnshire Police, chairman of the National DNA Database Strategy Board, *National DNA Database Annual Report, 2006–2007*, September 2006, http://www.npia.police.uk/en/docs/NDNA_A4L_Section1-08.pdf (accessed May 26, 2010); United Kingdom, "Counter-Terrorism Act 2008," http://www.uk-legislation.hmso.gov.uk/acts/acts2008/pdf/ukpga_20080028_en.pdf (accessed May 26, 2010).

Officers in partnership with the Forensic Science Service (FSS). Initially, only DNA profiles collected in England and Wales were uploaded to the system. Subsequently, Northern Ireland and Scotland developed forensic DNA databases with their own criteria for inclusion. However, they too submit their profiles to the NDNAD.

Two databases make up the NDNAD. One consists of individuals whose DNA was collected for recordable offenses or who gave their DNA voluntarily subject to the authority of the Police and Criminal Evidence Act of 1984. The second database consists of profiles obtained from materials from unsolved crimes. There is also a Police Elimination Database that contains DNA profiles from many police investigators so that inadvertent contamination of crime-scene samples during collection and handling of the evidence may be detected. This database is maintained separately from the other databases in the NDNAD and is not subjected to routine searches. Unlike the Combined DNA Index System (CODIS, the U.S. federal database), the U.K. national database contains the names of individuals, their date of birth, gender and ethnic appearance (as defined by the police), and geographical locators along with the DNA profiles. As in the U.S. system, the biological samples collected as part of the United Kingdom's forensic DNA program are retained indefinitely by the laboratories.

DNA collection and retention were initially limited to individuals convicted of a recordable offense. The definition of a "recordable offense" itself has expanded over the years and currently includes many minor offenses, such as being drunk in a public place, begging, taking part in a prohibited public procession, and minor acts of criminal damage caused by children kicking footballs or throwing snowballs. Funding considerations further limited collection in practice to those convicted of violent and sexual offenses and domestic burglary. From 1996 to 2003 U.K. police powers to take and retain DNA were continually expanded. Amendments to the Criminal Evidence Act passed in 1997 allowed DNA to be collected from individuals in prison who had been convicted of crimes before the NDNAD went into effect. In 2000 Prime Minister Tony Blair launched the DNA Expansion Programme, which called for the DNA collection and profiling of "virtually the entire active criminal population"—an estimated 3 million people—by 2004. That goal was nearly achieved: by the end of 2004 the database contained 2.5 million profiles.

The national database has also been affected by changes in the laws that dictate the retention of samples. Before 2001 samples and profiles taken from individuals who were not charged or prosecuted or who were acquitted of charges had to be removed from the database and destroyed, but that changed through passage of the Criminal Justice and Police Act of 2001. This act gave police authority to retain profiles and biological samples indefinitely from persons not prosecuted or who were acquitted of a crime. In addition, the 2001 act allowed DNA samples provided voluntarily to police for the purpose of exclusion to be uploaded to the U.K. NDNAD on the condition that the individuals provided consent. Once consent is given, the decision becomes irrevocable. Consent also gives the police the right to use the sample for any purpose permitted under the law,[8] even if those uses were not in practice at the time consent was provided.

In 2003 the NDNAD was expanded once again, this time by allowing law enforcement to take DNA profiles and fingerprints without consent from anyone arrested on suspicion of any recordable offense. The law came into force on April 5, 2004.[9] Samples and profiles obtained from arrestees are retained indefinitely, regardless of whether the person is charged or convicted. This change in the law made England and Wales the first and only countries in the world to keep indefinitely DNA from persons who were not convicted of any crimes. Since then the United States has moved toward a similar policy of collecting DNA routinely from those arrested, although federal and many state laws enacted thus far either require expungement in cases where an individual is not charged with or convicted of a crime or allow individuals to request expungement of their information in such cases (see chapter 2).

Northern Ireland has a law on DNA collection that is more or less identical to the law governing England and Wales. It allows the police to obtain DNA samples without consent from anyone aged 10 or over who is arrested or detained in connection with a recordable offense. A leading nonprofit policy research and genetics watchdog group in the United Kingdom, GeneWatch UK, notes that "this law was introduced by the Secretary of State for Northern Ireland whilst the Assembly was suspended, in order to bring Northern Island's legislation into line with England and Wales."[10] Starting in October 2005, Northern Ireland began exporting DNA profiles to the NDNAD.

The situation is somewhat different in Scotland, which has proceeded far more cautiously than England and Wales. The information commissioner for Scotland believes that the indefinite retention of DNA profiles of individuals who are arrested but are not convicted of any offense is an intrusion into their private lives. In May 2006 the Scottish Parliament rejected the permanent retention of DNA taken from individuals who are not convicted of a crime. Under these circumstances the DNA profiles are deleted from the Scottish database and the U.K. NDNAD, and the biological samples from which the profiles are generated are destroyed.

Between 1995 and 2001 the NDNAD grew to over 1 million profiles. By 2006 it contained DNA profiles of more than 3 million individuals (about 5.2 percent of the population), including more than 18,000 volunteer samples and some 285,000 crime-scene samples of unsolved crimes.[11] As of September 2008 more than 4.6 million people—or 7.6 percent of the population, including 32,000 volunteers—had their DNA profiles retained in the database.[12] Approximately 1 million of these individuals had never been convicted of or charged with any crime.[13] In comparison, other European Union (EU) country databases contain samples representing less than 1 percent of their population. For example, in 2004 Austria, Belgium, Germany, and the Netherlands had forensic DNA databases that contained 0.7, 0.03, 0.3, and 0.02 percent of their populations, respectively.[14] The United States' database numbered approximately 4.8 million at this time, or 1.6 percent of its population, less than one-quarter of the per capita size of the United Kingdom's database.[15] The gender of subject samples in the U.K. NDNAD as of January 2009 was 79 percent males, 20 percent females, and 1 percent unknown. All data held in the national DNA database are overseen by a tripartite board consisting of the Home Office (the lead government department for police, immigration, passports, drug policy, and counterterrorism in the United Kingdom), the Association of Chief Police Officers, and the Association of Chief Police Authorities. Recently the board has expanded to include two representatives of the Human Genetics Commission (a government advisory body) and the chair of the new DNA ethics group, an advisory nondepartmental public body (NDPB), established to provide independent advice on ethical issues to Home Office ministers and the strategy board. The data held in the NDNAD are owned by the police force that submitted the DNA to the database.

Other Applications of the NDNAD

Other applications of the U.K. NDNAD, which were introduced through the Serious Organized Crime and Police Act of 2005, include the use of the database to check profiles of casualties from catastrophic events for identification purposes and to investigate illegal immigrants who claim to have family in the United Kingdom. Lord Triesman, the prime minister's special envoy for returns, was quoted as saying, "Scientific and technical identification of nationality [will be] an important tool" in determining the status of illegal immigrants and their possible return to their countries of origin.[16]

The NDNAD and its accompanying stored biological samples have been accessed on many occasions for research. A freedom-of-information (FOI) request filed by GeneWatch UK in 2006 revealed that the NDNAD had been used in a series of research projects. These included "operations requests" by police to search for named individuals or individuals having a specified ethnicity or last name that corresponded with a particular ethnicity (e.g., those "having typical Muslim names"). The biological samples have also been used in research on the Y chromosome that seeks to predict ancestry or ethnicity from DNA. The FOI request also revealed that a commercial company, LGC, kept a "minidatabase" of information sent to it by the police, including individuals' demographic details, alongside their DNA profiles and samples.[17]

For any serious crimes, police can obtain DNA by asking, by employing a ruse, or, if that does not work, by recovering something that a suspect discards. A distinction is made between acquiring a DNA sample and loading it into the national database. The official position has been that covertly obtained DNA samples may be compared with crime-scene evidence but cannot be searched against the database because they have not been obtained under the Police and Criminal Evidence Act (PACE). Uploading DNA to the U.K. database is limited to profiles overtly acquired from suspects investigated under PACE or samples from volunteers who wish to be eliminated as a suspect in an investigation.

There are some restrictions on the use of information contained in the NDNAD. U.K. law limits the use of the NDNAD to investigating crimes or identifying remains—as in war, fire, or natural disasters—where there are unidentified victims. As written into the Police and Criminal Evidence Act

1984 (as amended), "Samples and profiles may only be used for purposes related to preventing and detecting crimes, investigating an offence, conducting a prosecution, or identifying a deceased person or a body part."[18] Analysis of samples to provide information on genetic disorders is generally unlawful, although exceptions can be made in cases where such information might aid a criminal investigation. It is also expressly prohibited to access profiles held in the NDNAD to assist in determining the paternity of a child in civil cases.[19]

Familial DNA Searches and Low-Copy DNA Testing

In addition to paving the way in DNA data-bank expansion, the United Kingdom has also spearheaded the way in the development of highly controversial DNA techniques, including familial searching and low-copy-number (LCN) DNA analysis. The United Kingdom was the first country to use its database to generate crime suspects from incomplete matches of crime-scene evidence profiles with profiles in the data bank. These so-called familial searches bring criminal investigators to the "suspicionless" family members of an individual whose DNA is a close match to the profile found in the crime-scene evidence.[20] The United Kingdom was also the first country in the world to convict someone following identification through a familial search result and is currently the only country in the world that uses this technique routinely in high-profile investigations. By the beginning of 2008, the United Kingdom analyzed 148 cases using familial searching techniques; only 15 of them had been resolved with 9 convictions.[21]

Familial searching is highly controversial. Although proponents of the technique claim that it can save time and money in high-priority investigations, critics point out that trawling the database for possible relatives radically alters the nature and intent of the database system and effectively places an entire class of innocent people—those who happen to be relatives of convicted offenders—under lifelong genetic surveillance by way of their relation to individuals in the database. (For a detailed discussion and analysis of familial searching, see chapter 4.)

The United Kingdom has also championed LCN DNA testing, a technique that seeks to generate a DNA profile from a minuscule amount of DNA, such as trace DNA left behind when a person touches an object. Standard DNA analysis is employed for DNA samples that contain as little as 1 ng of DNA, or as few as 160 human cells, or the size of a tiny blood spot down to 250 picograms ([pg], where 1 pg is one trillionth of a gram) of DNA or about 40 human cells. LCN, by contrast, is used on samples of less than one-tenth this size (100 pg) or 16 cells, or about 1,000 times smaller than a grain of salt.[22] This technique has been highly controversial among forensic scientists, and many have questioned its accuracy and reliability. Because the technique relies on such small amounts of DNA—such as DNA transferred to a murder weapon or left behind from a fingerprint—the analysis is highly vulnerable to contamination and other sources of error. Mixtures are even more difficult, if not impossible, to ascertain and separate out than in standard DNA testing. Interpretation of LCN analysis is highly subject to "allelic dropout" (where some alleles do not appear in the analysis because the signal is so low) and increased "stutter" ("stutters" are usually small peaks in the output of the DNA analyzer that are artifacts from DNA amplification and form as a result of halted polymerase activity from the polymerase chain reaction [PCR]).[23]

The scientific controversy around LCN rose to public attention through the trial of Sean Hoey, a 38-year-old electrician who was arrested and charged with Real IRA's bombing in the city of Omagh in Northern Ireland. The 1998 bombing in a crowded market area killed 29 people and injured more than 200. Hoey's trial, which lasted 56 days, hinged on LCN DNA evidence. Judge Reg Weir strongly rejected the evidence, raising significant concerns about the validity of the technique, as well as the careless handling of the DNA:

> It is not my function to criticize the seemingly thoughtless and slapdash approach of police and officers to the collection, storage and transmission of what must obviously have been potential exhibits in a possible future criminal trial, but it is difficult to avoid some expression of surprise that . . . such items were so widely and routinely handled with cavalier disregard for their integrity. . . . I find that the DNA evidence . . . cannot satisfy me either beyond a reasonable doubt or to any other acceptable standard.[24]

Furthermore, the judge cast doubt specifically on the LCN DNA technique, expressing concern over the range of opinion in the scientific community about its reliability and its lack of adoption by most other countries.

The discrediting of LCN in the Omagh bombing case prompted the U.K. police to impose a brief moratorium on the use of LCN DNA testing that was lifted following a three-week review of pending prosecutions involving the technique.[25] The U.K. government also commissioned an expert review of LCN analysis that was spearheaded by Professor Brian Caddy of Strathclyde University. Caddy's review concluded that LCN typing was "robust" and "fit for purpose" while offering a number of recommendations for improving and standardizing the methodology.[26]

This review has not quelled the discomfort and significant doubts that many forensic scientists continue to have about the reliability and use of LCN DNA. Allan Jamieson at the Forensic Institute in Glasgow, who testified in the Omagh trial, has pointed out most fundamentally that since one is starting with DNA from which there was no visible stain, it is impossible to know how it got there, and whether it was relevant to the crime in question.[27] Bruce Budowle, former senior scientist of the FBI's Forensic Science Laboratory, has questioned how the British commission could have come to its conclusion, given that the technique by its nature is not reproducible, and has pointed out the myriad problems that can arise through LCN analysis, including the high potential for error in process and in interpretation, the lack of standard protocols for collecting and handling such low-level DNA samples, and the inability to know the source type of the DNA (e.g., hair, blood, semen).[28] Budowle recommends that the technique's use be limited to developing investigative leads and to the identification of human remains, but that it not be presented as evidence in court. Furthermore, he advocates that the limitations of the technique be fully disclosed and stated up front.[29] Citing the lack of consensus in interpretation and the availability of viable alternative approaches, Jason Gilder and colleagues agree with Budowle in response to Caddy's conclusions on LCN: "Superficial characterizations such as 'robust' and 'fit for purpose' are a denial of the serious scientific questions that remain about the reliability and validity of LCN testing."[30]

Despite the ongoing controversy about LCN, the United Kingdom has used the technique in more than 20,000 cases and remains the only country to use it routinely.[31] Sweden and Australia have also allowed LCN to be presented as evidence in a few high-profile cases. In the United States the New York Medical Examiner's Office is the only lab to date that admits to using the technique in some cases. The FBI does not allow the technique to be used in criminal investigations and uses it only in missing-persons investigations where trace DNA samples are known to have come from a single source.

Public Opposition to the NDNAD

One of the United Kingdom's most astute scholars on the use of DNA in criminal justice wrote about the seemingly inexorable acceptance of DNA profiling: "The blinding by science of criminal justice professionals and the public (ultimately juries) has met little resistance and widespread acceptance, and indeed judicial encouragement for the NDNAD, which may ultimately see the creation of a national, comprehensive and compulsory DNA database."[32] Indeed, the development and expansion of the NDNAD benefited from broad public support, up to a point. The move to collect DNA upon arrest, and Scotland's refusal to go along with the policy, invited significant scrutiny within the United Kingdom and around the world. Public confidence in the system was also rattled by the government's aggressive approach to familial searching, LCN DNA analysis, and the retention of DNA from young children and volunteers.

The permanent retention of biological samples is one issue that was controversial within the United Kingdom before the start-up of the arrestee testing program in 2004. The U.K. Human Genetics Commission, a government advisory body that is made up of experts in genetics, ethics, law, and consumer affairs, concluded in a 2002 report that the reasons given for retaining samples were "not compelling."[33] Since then, the Home Office has recognized that retaining samples is "one of the most sensitive issues to the wider public."[34]

The DNA profiles of those who gave DNA samples voluntarily to police for the purpose of excluding themselves as a suspect may also be loaded into the NDNAD with their consent. However, once consent is given, the decision becomes irrevocable. Thus the consent form gives the police the right to use the sample for any purpose they deem important without updating the consent form. The home secretary published a report in May 2009 that sought a change in this policy:

> In giving their consent to the sample, the volunteer is also asked whether they wish to give their consent for their profile to remain on the NDNAD. If such consent is given, the volunteer is not then able to subsequently require that the sample and profile are destroyed. We are proposing that a volunteer who gives their samples for elimination purposes are not placed on the NDNAD. Whilst consent will continue to be required for the taking

> of the sample, consent will not be sought for the sample or fingerprints to be retained on a national database and subject to future speculative searches.[35]

The expansion of the NDNAD to anyone arrested for any recordable offense in 2003 sparked considerable public debate within the United Kingdom and around the world. GeneWatch UK questioned whether more aggressive policies on inclusion of DNA samples have been equitably distributed across ethnic and racial groups.[36] If police are more likely to stop, detain, and arrest people of color, then their rate of inclusion in the data bank will be disproportionately higher than their population or their contributions to the crime rate.

By 2006 more than one-third of black males in the United Kingdom were profiled in the NDNAD, compared with 6 percent of adult white males[37] and about 13 percent of Asian men.[38] Three out of four black males between the ages of 15 and 34 had profiles in the DNA database.[39] These figures, once brought to the attention of the Black Police Association, prompted the association to call for an investigation of the racial disparities of the NDNAD. A report of the Select Committee on Home Affairs of the Parliament focused on the reasons behind young black people's overrepresentation in the criminal justice system. The committee found that

> Black people constitute 2.7 percent of the population aged 10–17, but represent 8.5 percent of those of that age group arrested in England and Wales. As a group, they [blacks] are more likely to be stopped and searched by police, less likely to be given unconditional bail and more likely to be remanded in custody than white young offenders.[40]

Baroness Scotland is quoted in the report in expressing her concerns about the racial disparities of the database as follows:

> It means that young black people who have committed no crime are far more likely to be on the database than young white people. It also means that young white criminals who have never been arrested are more likely to get away with crimes because they are not on the database. It is hard to see how either outcome can be justified on grounds of equity or of public confidence in the criminal justice system.[41]

The results of the parliamentary investigation prompted further prodding of the NDNAD, and while the report was being aired in the media, it also became apparent that DNA from 108 children under the age of 10 had found its way into the NDNAD system, even though the warehousing of DNA profiles from children younger than 10 is outside the scope of the law. The stories further reported that the NDNAD contained DNA profiles from 46 people over the age of 90 and 883,888 children between the ages of 10 and 17, and that at least 50,000 of these juveniles had not been charged with or convicted of any crime.[42] Nick Clegg, a Liberal Democrat Home Affairs Committee spokesman who requested the information from the NDNAD, stated in response:

> The Government's onward march towards a surveillance state has now become a headlong rush. As an increasing number of young children well under the age of criminal responsibility appear on the database, it is clear the Government sees no limits to its invasion of our privacy. Worse still, by harvesting the data of many people who are not even charged with an offence, let alone convicted, the fundamental principle that we are innocent until proven guilty is further undermined. Why should anyone be on this database if they are innocent of any wrongdoing?[43]

Another subject of debate in the United Kingdom has been the actual efficacy of the database. According to U.K. authorities, in 2006 the chance of a new crime-scene profile loaded into the NDNAD immediately matching the profile of an individual already in the database was 45 percent. However, during that same period crime-scene DNA profiles were loaded into the database for less than 1 percent of all the recorded crimes committed annually.[44] Helen Wallace of GeneWatch UK has pointed out that the efficacy of the database is limited not by the total number of profiles in the system, but instead by the number of crime-scene profiles.[45] Wallace has further elucidated the differences among reported "matches," "crime detections," and "convictions." It turns out that only about half of DNA matches made against the database lead to a DNA detection (where the crime is considered "cleared up" either through an arrest or other resolution). In addition, not all DNA detections lead to convictions; sometimes DNA is found at the scene of a crime but turns out to be irrelevant to the crime in question, and at other times sufficient evidence cannot be brought against the individual in question. The Home Office has estimated that in the United

Kingdom approximately 50 percent of detections lead to convictions, and some 25 percent lead to a custodial sentence. Thus fewer than one-quarter of DNA matches lead to convictions of any type, and fewer than one-eighth of matches lead to convictions of offenses serious enough to merit incarceration. Therefore, the high number of DNA matches reported by the Home Office is potentially misleading with respect to the benefit of the expanded NDNAD (see chapter 17).[46]

In September 2007 the Nuffield Council on Bioethics, an independent think tank based in London, released a report, *The Forensic Use of Bioinformation: Ethical Issues*, that provides a comprehensive analysis of the United Kingdom's policies and procedures regarding forensic DNA. The report embraced a rights-based approach in balancing the need to protect the public from criminal activities while also protecting individual liberty, autonomy, and privacy. The council also framed its recommendations on the principle of "proportionality," which is based on the idea that any interference with legally enforceable human rights must be justified by the state with evidence to support that such interferences are proportionate to the need to fight crime. The report states that the United Kingdom has the lowest threshold for holding DNA profiles of any EU country and that its percentage of the population with a banked DNA profile is higher than that of any other EU country.[47]

The council recommended that

> the law in England, Wales and Northern Ireland should be brought into line with that in Scotland. Fingerprints, DNA profiles and subject biological samples should be retained indefinitely only for those convicted of a recordable offence. At present, the retention of profiles and samples can be justified as proportionate only for those who have been convicted. In all other cases, samples should be destroyed and the resulting profiles deleted from the National DNA Database.[48]

The council further recommended that volunteers who provide their DNA to the database for exclusion purposes be able to have their DNA withdrawn from the system at any time and without having to provide a reason. The council questioned the necessity of retaining biological samples and called for the government to convene an independent commission to examine the full impacts of retention.[49] The report pointed out that other than England and Wales, no European jurisdiction systematically retains the profiles or samples of individuals who have not been convicted of a crime. DNA samples are destroyed immediately in some EU countries, including Germany and Belgium. In Switzerland they must be destroyed within three months of successful profiling.[50] The report further recommended that familial searching be used only in cases where it is specifically justified, and that there be a presumption in favor of removing DNA taken from children from the database.

The DNA-collection system in the United Kingdom has been brought to the attention of the courts. In one case the British courts ruled that a sample had to be removed from the U.K. NDNAD. This case involved a teacher who was accused of assault but then, for lack of sufficient evidence, had her charges dropped. In this case the woman's DNA sample was taken after the decision was made not to prosecute. The British court ruled that the woman had the right to have her fingerprints and DNA destroyed because they were taken after the decision not to prosecute. Had her DNA been taken before the decision was made not to prosecute, the court held that the samples would have been considered lawful, and they would have been retained even though her charges were dropped.[51]

The practice of retaining DNA samples from arrestees who were acquitted was challenged in August 2004 before the European Court of Human Rights by petitioners on the grounds that the United Kingdom did not have a right to retain fingerprints and DNA samples once a person was cleared of criminal charges, under articles 8 and 14 of the European Convention for the Protection of Human Rights and Fundamental Freedoms (the "convention"). Those articles state:

> *Article 8—Right to respect for private and family life: Everyone has the right to respect for his private and family life, his home and his correspondence.*
> There shall be no interference by a public authority with the exercise of this right except such as is in accordance with the law and is necessary in a democratic society in the interests of national security, public safety or the economic well-being of the country, for the prevention of disorder or crime, for the protection of health or morals, or for the protection of the rights and freedoms of others.
>
> *Article 14—Prohibition of discrimination*
> The enjoyment of the rights and freedoms set forth in this Convention shall be secured without discrimination on any ground such as sex, race, colour,

> language, religion, political or other opinion, national or social origin, association with a national minority, property, birth or other status.

The plaintiffs in this case were two individuals who requested that their DNA be destroyed and their records removed from the database. One was a juvenile (Mr. S.) who was charged with attempted robbery at the age of 11 but was then acquitted. The other was an adult, Michael Marper, who was arrested and charged with harassment of his partner, but the charges were subsequently dropped. The challenge, *S. and Marper v. The United Kingdom (Marper)*, was initially heard in the Divisional Court and rejected. Then it was brought to the Court of Appeal, which dismissed the plaintiffs' claim, ruling that although the retention of fingerprints and DNA samples was in breach of provisions of Article 8(1), it was proportionate and justified under Article 8(2) and thus did not violate Article 14. It was then brought to the House of Lords in 2004 and dismissed. Lord Steyn of the House of Lords argued that interference in privacy by taking one's DNA was minimal compared with the benefits the database had to society. That benefit was in crime detection and reduction, in which the public had a clear interest. Lord Steyn opined that there would be no adverse impacts on the individuals from whom the samples were taken unless they were implicated in a future crime.

Marper was appealed to the European Court of Human Rights and was heard in the Grand Chamber in Strasbourg on February 27, 2008. On December 4, 2008, the European Court of Human Rights ruled that the blanket and indefinite retention of DNA profiles and samples pursuant to U.K. law from individuals accused but not convicted of certain crimes violated Article 8 of the convention. The court found that it was not necessary to reach the question whether the law was also in violation of Article 14. Specifically, the court found that the retention of S. and Marper's DNA breached human rights law, stating:

> In conclusion, the Court finds that the blanket and indiscriminate nature of the powers of retention of the fingerprints, cellular samples, and DNA profiles of persons suspected but not convicted of offences as applied in the case of the present applicants, fails to strike a fair balance between the competing public and private interests and that the respondent State has overstepped any acceptable margin of appreciation in this regard. Accordingly, the retention at issue

constitutes a disproportionate interference with the applicants' right to respect for private life and cannot be regarded as necessary in a democratic society.[52]

The *Marper* court grounded its conclusion in the notion that DNA contains highly private information about an individual. In comparing DNA with fingerprints, the *Marper* court provided that "the Court has distinguished in the past between the retention of fingerprints and the retention of cellular samples and DNA profiles in view of the stronger potential for future use of the personal information contained in the latter" and that "because of the information they contain, the retention of cellular samples and DNA profiles has a more important impact on private life than the retention of fingerprints."[53] The *Marper* court also affirmed that "an individual's concern about the possible future use of private information retained by the authorities is legitimate and relevant to a determination of the issue of whether there has been an interference"[54] with the individual's right to privacy under the convention, and, furthermore, that "the mere retention and storing of personal data by public authorities, however obtained, are to be regarded as having direct impact on the private-life interest of an individual concerned, irrespective of whether subsequent use is made of the data."[55]

The court recognized the distinction between biological samples and DNA profiles and stated that although the personal information contained in the profiles is more limited, "DNA profiles could be, and indeed had in some cases been, used for familial searching with a view to identifying a possible genetic relationship between individuals"; furthermore, "the DNA profiles' capacity to provide a means of identifying genetic relationships between individuals . . . is in itself sufficient to conclude that their retention interferes with the right to the private life of the individuals concerned."[56] The court also found that "the retention of the unconvicted persons' data may be especially harmful in the case of minors such as the first applicant, given their special situation and the importance of their development and integration in society."[57] Finally, the court based its conclusion within the context of other member-state laws and practices, stating that "the United Kingdom is the only member State expressly to permit the systematic and indefinite retention of DNA profiles and cellular samples of persons who have been acquitted or in respect of whom criminal proceedings have been discontinued."[58]

The U.K. government's dramatic loss in *Marper* is perhaps a clear sign that it has overstepped its authority in taking and storing DNA from innocent persons. In response to the *Marper* decision, the Labour government adopted a new retention framework as part of the Crime and Security Act of 2010. Under the new law, DNA profiles collected from adults arrested but not convicted must be destroyed after 6 years. DNA profiles of juveniles arrested but not convicted may be retained for 3–6 years, depending on the severity of the offense. Most notably, under the new framework all biological samples must be destroyed once the DNA profiles are obtained, and these samples may not be retained for more than 6 months. At the time of this writing, the new framework still had not been brought into force as a result of continued debate over the retention of arrestee DNA profiles.

The U.K. government adopted its highly aggressive policies on DNA collection incrementally and—except in Scotland—without vigorous public debate. The development and expansion of the U.K. NDNAD received fairly broad public support until concerns about the racial composition of the database and the collection of DNA from young children began to surface. The dramatic shift in policy to include anyone arrested for a recordable offense, whether ultimately convicted or not, also spurred significant attention and, ultimately, significant opposition. Familial searching and the premature use of LCN DNA and its inability to resolve the Omagh bombing case raised additional alarm bells concerning the United Kingdom's policies and practices on the collection and use of DNA. Organizations such as GeneWatch UK and the Nuffield Council on Bioethics helped bring the social and ethical concerns associated with DNA data banking into the public sphere.

British criminal justice authorities overseeing the U.K. NDNAD were, from a relative standpoint, transparent in providing information about the ethnic and racial composition of the individuals whose DNA was acquired and profiled. In contrast, no public information about the composition of the U.S. national database is available. Public awareness in the United Kingdom has just started to contribute to a shift in DNA policy, although the extent to which this might result in a long-term suspension of database expansion is unclear. Despite the outcome in *Marper* and changes instituted by the home secretary, considerable controversy remains over the length of time samples and profiles are retained, the protocols for the collection and use of DNA, and the retention of volunteer elimination samples.

Chapter 10

Japan's Forensic DNA Data Bank: A Call for Reform

The DNA Profile Information Database System, currently being operated by the National Police Agency, must be established and operated in accordance with laws, not by regulations, so that the right to privacy and the right to control personal information are not infringed. Hence, National Public Safety Commission Regulation No. 15 should be abolished.

—Japan Federation of
Bar Associations[1]

DNA testing was first used as a tool in criminal investigations in Japan in a 1992 rape trial. For the first decade crime laboratories in Japan relied on DNA analysis using two different loci: MCT 118 and HLADQ-a. In August 2003 the National Police Agency (NPA) introduced the "STR-9 locus multiplex kit." Currently the NPA uses an analysis of 15 loci in generating a profile.[2]

The taking of DNA is considered a "search" under Article 35 of Japan's Constitution, which guarantees:

> The right of all persons to be secure in their homes, papers and effects against entries, searches and seizures shall not be impaired except upon warrant issued for adequate cause and particularly describing the place to be searched and things to be seized. . . . Each search or seizure shall be made upon separate warrant issued by a competent judicial officer.

As in the United States, the Japanese courts have adopted a balancing test for weighing privacy interests against the needs of law enforcement in cases where these interests collide.[3]

Japan's Criminal Procedures Act requires that the police obtain a warrant before taking DNA samples, at least in principle.[4] However, as in the debate in the United States, there is some question whether the privacy invasion associated with DNA is limited to the physical intrusion associated with the taking of DNA, or whether it also relates to an invasion of informational privacy. Some commentators have argued that the taking of DNA does not constitute a "search" in cases where it is taken without physical intrusion.[5]

Japan's DNA Database

In December 2004 the NPA began operating a national database of DNA profiles generated from crime-scene samples. The following summer Japan's National Public Safety Commission issued an administrative rule, Regulation no. 15, announcing the creation of a national DNA database of suspects for use in criminal investigations. The stated purpose of the rule was "to enact necessary regulations in order to systematically create, administer and operate Suspect DNA Type profiles and to contribute to criminal investigations."[6]

Before 2005 some of Japan's 47 prefectures were collecting and storing DNA from some convicted offenders or suspects, but there was no national system or set of guidelines in place to share DNA profiles across jurisdictions. The 2005 regulation created a centralized system, operated by the NPA, for collecting and sharing suspect DNA profiles and comparing them with stored crime-scene profiles.

Japan's database is operated and managed by the NPA's Criminal Bureau. Responsibility for the profiles and accompanying information, including prevention of leakage, disappearance, or destruction of the data, rests with the director for criminal identification. In addition to the DNA profiles, the database contains the name, sex, and date of birth of the suspect, the date of arrest, the name of the offense, the type of sample collected (blood or saliva), and "other useful information."[7]

According to Dr. Kazumasa Sekiguchi, assistant director of the Criminal Identification Division at the time at which the database was established,

DNA is taken from suspects only in cases where it is necessary for the investigation of the crime.[8] So although the database can be used to profile and register DNA from an individual for any criminal act, as a matter of current practice, only those suspects in cases where DNA was collected at the scene of a crime have their DNA collected and uploaded to the system. This marks a significant difference from the practices of the United States and Britain, where the determination whether DNA is collected is based on crime classification (e.g., felony, misdemeanor, sex offenses).

Suspect DNA profiles are generally retained regardless of whether an individual is convicted of a crime. The regulations require expungement in cases where the suspect has died, or where the director determines that there is no longer a need to retain the said profiles. Although the 2005 regulations do not require the police to routinely destroy biological samples collected from suspects after a profile has been obtained, the Japanese police assert that they do so[9] in accordance with a set of internal guidelines established by the NPA.[10]

The regulations also allow for the expungement of crime-scene profiles where cases have been adjudicated and closed, or where it is determined that the profiles are no longer needed. There is no regulatory requirement that the crime-scene samples be retained, and defendants appear to have had difficulty obtaining access to samples for retesting. According to Hikaru Tokunaga, professor of law at Konan University in Osaka, defendants are almost never given an opportunity to reanalyze a sample.[11] In addition, in five cases where the entire portion of a DNA sample was exhausted by initial testing, the Japanese courts ruled that the evidence was admissible, even though there was no possibility for retesting.[12]

Access to the database is restricted to the director for criminal investigation. Although there are no known uses of the database to date for research purposes, the 2005 regulations do not explicitly prohibit access to the database for these purposes.[13]

There is no restriction on the age of individuals in the DNA database, although Japanese law stipulates that a person less than 14 years of age is not punishable for a crime. The Japanese database is not currently used to conduct familial searches.[14] The 2005 rule appears to limit DNA analysis to the 15 specified loci.

Japan's DNA database is extremely small in comparison with those in the United States and the United Kingdom. As of June 2007 Japan's DNA database contained 9,964 DNA profiles from suspects, or less than 0.008 percent of the population (compared with more than 2.7 percent of the U.S. population and approximately 7 percent of the U.K. population).[15] The number of suspect files approximately doubled from the prior year. Of the nearly 10,000 profiles, 1,313 were from rape suspects; 1,908, theft; 969, robbery; 795, homicide; 873, sexual assault; and 577, illicit drug use. There were 6,750 crime-scene samples on file at this time. About one-third (2,375) of these were from theft crimes; 996 from rape; 577, sexual assault; 525, robbery; 568, property destruction; 366, trespassing; and 184, homicide.

Questions About the Legality of the Database and a Call for Reform

Absent specific legislative authority, the regulations that established Japan's DNA database have been met with fierce criticism by several legal authorities in Japan.[16] In December 2007 the Japan Federation of Bar Associations (JFBA), consisting of 52 local bar associations and over 24,000 member attorneys, released a biting opinion, referring to the regulations as illegal and unconstitutional and calling for their immediate repeal and the development and enactment of legislation in this area.[17]

Japan's NPA appears to claim its authority to retain DNA profiles by likening them to fingerprints both in function and in quality.[18] The JFBA points out, however, that the fingerprint database that was authorized by legislation—specifically, Article 218, Section 2, of the Code of Criminal Procedure—was clearly intended to be limited to fingerprints. Furthermore, the JFBA points out that DNA is fundamentally different from fingerprints in that far more information about an individual may be revealed by DNA analysis. Finally, the difference between fingerprints and DNA is well recognized in the law, since the taking of DNA from a suspect requires a court order, whereas the taking of fingerprints does not.[19]

The JFBA opinion outlined specific recommendations for legislation in the areas of registration (who should be included in the database), retention, use, and expungement on the basis that DNA information is of "utmost individual privacy." In the area of registration, or inclusion, the JFBA suggests that DNA collection from suspects be limited to cases where there is a specific need to investigate a specific crime and that those cases be further limited to those crimes that constitute a felony against life and physical integrity. In considering the U.S. and U.K. models for DNA collection, the JFBA found that taking

DNA without individualized suspicion of a specific crime is contrary to Article 35 of Japan's Constitution. Furthermore, the JFBA concluded that obtaining a DNA profile in cases where the DNA profile is not necessary for investigating the crime in question (in other words, collecting DNA merely for future forensic investigations) violates the due-process protections guaranteed by Article 31 of the constitution.[20] The JFBA also recommended that DNA collected voluntarily from individuals not be included in the database absent secondary consent from those individuals, and that juveniles generally be barred from inclusion in the database, especially those under the age of 14.

In the areas of retention, use, and expungement the JFBA recommended strict access limitations to the database and that any illegal use of the database constitute an imprisonable offense. The opinion suggests that retention of DNA profiles from suspects be limited to a period of 5 or 10 years, and that the database be used only when there is a need to investigate a specific case. It also recommends that expungement be obligatory in cases where a defendant is acquitted or a person is cleared of suspicion, and that persons wrongly registered in the database have a right to request expungement of their information.

Japan does not currently have an accreditation system of experts and laboratories. The JFBA opinion recommended increased monitoring and improved quality-assurance measures for operation of the database and the use of DNA by law enforcement. Some legal scholars have suggested the development of an accreditation system.[21] Box 10.1 describes the first case in Japan (known as the "Ashikaga case") of a person whose sentence was overturned on the basis of postconviction DNA testing.

BOX 10.1 The Ashikaga Case

In May 1990 a 4-year-old girl was found dead on the bank of the Watarase River in Ashikaga, Japan. The corpse was naked, saliva was found on her body, and semen was found on her underwear. A suspect by the name of Toshikazu Sugaya was tailed on the basis of a neighbor's comment that "his behavior was suspicious." The police collected bathroom tissues with semen from his garbage. The National Research Institute of Police Science performed an MCT 118 DNA profile analysis and found that the semen on the tissues "matched."[a] Sugaya was arrested and indicted in December

1991. Using the test results as leverage, investigators extracted a confession from Sugaya. During his trial before the Utsunomiya District Court he repeatedly denied the accusation and stated that he had been forced to make a false confession as a result of the DNA testing. He was convicted of murder in July 1993 and sentenced to life imprisonment.

Sugaya's defense attorneys appealed his case. They claimed that the DNA tests were inadmissible for two reasons. First, they argued that the DNA analysis method was not scientifically reliable. In particular, they argued that most of the sperm stain was consumed when the police analyzed it to determine whether it was human sperm. In addition, they argued that the remaining part of the sample was not analyzed for a full 15 months after the victim's body was found. Second, the defense argued that the defendant's DNA was illegally obtained from the garbage without a warrant.

The case went to the Supreme Court. On July 17, 2000, the court ruled that the DNA evidence was admissible and upheld the decision of the lower court. The court was silent on the issue of inadvertent DNA collection, but the Appeals Court held that there is no illegality in collecting DNA from garbage.

Sugaya's lawyers continued to question the reliability of the DNA tests that were carried out in 1991 and repeatedly requested, and were denied, additional DNA testing. They requested a retrial in December 2002, submitting evidence that a DNA analysis of Sugaya's hair did not match the results of the analysis that was conducted in 1991, as well as expert testimony from two forensic scientists who pointed out that a fine white foam had been found in the air passage of the victim in an autopsy, and it was therefore rational to assume that the cause of death was drowning rather than strangulation. The Utsunomiya District Court rejected the retrial request in February 2007.[b]

On June 4, 2009, Sugaya was released from prison following a reexamination of the DNA evidence in the case that was ordered by the Tokyo High Court. The DNA testing revealed that the DNA of the crime-scene samples did not in fact match Sugaya's DNA. Sugaya was released after spending more than 17 years in prison, and the prosecutor's office requested that the court open a retrial, stating that the results were "likely to

(*continued*)

serve as clear evidence to absolve [Sugaya]."[c] A formal apology was issued to Sugaya by the Tochigi prefectural police chief, and on June 23, 2009, the Tokyo High Court ruled to retry the case.[d] On March 26, 2010, Sugaya was acquitted of all charges. He urged the court to ascertain the causes that led to his wrongful conviction.[e]

The "Ashikaga case," as it is referred to, is the first case in Japan to be overturned as a result of postconviction DNA testing. It has called into question the judicial decisions and interrogation tactics of Japanese investigators and has prompted significant public attention to the need for establishing a right of convicted individuals to postconviction DNA testing, particularly in cases where those individuals were convicted on the basis of outdated DNA testing methods or other unreliable forensic techniques.

[a] *Hosokai* (Lawyers' Association), Admissibility of Expert Evidence based on MCT 118 DNA Analysis-*Ashikaga* Case, Adjudicated on July 17, 2000 (2003), summarized and translated by E. Omura, January 2008.
[b] "Court Rejects Retrial Request from Man Convicted of Killing Girl Through DNA," GaijinPot.com, February 13, 2007, http://gaijinpot.com/search/index/lang/en?q=court+rejects+retrial+request (accessed May 18, 2010).
[c] See Japanese Law Blog, "DNA and the Criminal Justice System," June 6, 2009, http://japaneselaw.blogspot.com/search?q=sugaya (accessed April 15, 2010).
[d] "Retrial of 1990 Murder Certain to Acquit Sugaya," *Japan Today*, July 2, 2009, http://www.japantoday.com/category/crime/view/retrial-of-1990-murder-certain-to-start-to-acquit-sugaya (accessed April 15, 2010).
[e] Asahi Shimbun, "Sugaya's Ordeal Finally Ends in Acquittal," Asahi.com, March 27, 2010, http://www.asahi.com/english/TKY201003210138.html (accessed April 15, 2010).
Source: Authors.

Future Expansion?

There appears to be an interest on the part of at least some Japanese law-enforcement representatives in expanding the scope of the database along the lines of the British model.[22] The Japanese government points to security threats, the need for improved efficiency in the criminal justice system, and increasing crime rates as justifications for the existence and expansion of the database.[23] Foreigners are commonly cited as the cause of alleged increases in violent crime[24] compared with 20 to 30 years ago, despite at least some evidence to the contrary.[25]

In 2004 an article published in *Nature* indicated that the National Research Institute of Police Science was planning to develop a four-year, $1.4-million project to develop ethnic profiles of all the crime-scene samples stored in its crime-scene database. The plan was to generate data on ethnicity, blood type, metabolic enzymes, hair and skin pigment proteins, mitochondrial DNA, and signs of asymptomatic viral infections that can be used to distinguish ethnicities.[26] All this information would be included along with the DNA profile information for all the crime-scene samples. The head of Amnesty International's Tokyo office described the project as a "sign of moral panic." Koichi Hamai, a criminologist at Ryukoku University in Kyoto, expressed concern that the project would perpetuate a tendency in Japan to scapegoat foreigners for Japan's social and economic problems.[27] It is unclear whether this plan has since been initiated.[28]

Japan's policy to collect DNA from individual suspects only in cases where it is useful for the investigation at hand—where DNA evidence is available at the scene of the crime for which the individual is suspected—has resulted in a database that has a far more limited reach than those in the United States and the United Kingdom. Nonetheless, the establishment of the database by means of an administrative rule under the NPA rather than through a legislative process has brought with it intense scrutiny. For example, the JFBA has questioned the constitutionality of Japan's database and has lobbied for reform of the current criteria used by police to acquire and retain forensic DNA profiles. The JFBA's call for reforms that respect the privacy of citizens who are not charged with a crime or who have not been convicted of a serious offense echoes ongoing efforts in the United States and the United Kingdom to limit or roll back the databases to exclude innocent people. In Japan there is also uncertainty whether phenotypic information will be contained in the forensic DNA database beyond the STR profile numbers used exclusively for identification. Whatever limits and restrictions there are on collecting and storing forensic DNA in Japan, they are either self-imposed by the NPA or they arise from the requirement for warrants, which police must obtain when they want to check crime-scene DNA with the DNA of a suspect. Japan's transition to a mixed-jury system for serious crimes in 2009 means that its system of forensic DNA will have to be transparent to a broader sector of society if it is to gain the confidence of the public as the Japanese society decides whether and how to build a legal foundation that balances collecting DNA and incorporating it into its judicial proceedings, and protecting the privacy of its citizens.

Chapter 11

Australia: A Quest for Uniformity in DNA Data Banking

Privacy and respect for human dignity need not be abandoned when balancing civil liberties with community safety. In many ways, privacy principles will enhance the integrity and legitimacy of DNA profiling by limiting collection to the minimum necessary to achieve the legitimate aims of law enforcement agencies, requiring its use to be in accordance with these aims, demanding secure storage of DNA material, and requiring its destruction or de-identification when the information is no longer needed. . . . Transparency and accountability reassure the community that what is sacrificed for greater safety and security is done so legitimately.

—Office of the Victorian Privacy Commissioner,
August 8, 2002[1]

Soon after DNA evidence was first introduced into the criminal justice system as an investigatory tool, the Australian government began a process to develop a comprehensive legislative scheme for collecting DNA for its use by law enforcement. In 1990 the Standing Committee of Attorneys-General established the Model Criminal Code Officers Committee, which was tasked with developing a Model Forensic Procedures Bill. After 10 years the committee produced a final draft of a model bill that gave authorization and specified procedures for the collection and use of DNA (as well as fingerprints, photographs, impressions, and other forensic materials) by law enforcement.

The Australian government recognized early on that it had limited capacity to influence state and territory laws on DNA collection because of the nature and structure of Australia's commonwealth and criminal justice systems. At the same time, it was clear that national harmonization in forensic procedures would be essential both for the sharing of DNA and for ensuring the protection of individual rights across state and territorial jurisdictions.

> It is important to note that the development of the Model Bill, particularly its provision concerning the exchange and matching of DNA samples between jurisdictions, was predicated on the expectation that it would be uniformly enacted by all States and Territories as well as the Commonwealth Government. Unless the Model Bill is adopted uniformly, the arrangements for the DNA system as a whole would allow an agency in one State to obtain information collected in another jurisdiction in circumstances that would not be allowed in its own State. This would be a diminution of the rights of the citizens of that State as established under that State's laws.[2]

Therefore, the model bill was designed with the intention that it would legalize and regulate forensic DNA testing for law-enforcement purposes in a manner that could be adapted uniformly by all states and territories, as well as the commonwealth government.

The final model bill was agreed to in 2000. It provided the government with the power to request or require DNA from three groups of individuals (serious offenders, suspects, and volunteers), all subject to certain procedural rules and privacy protections. "Serious offenders" are those convicted of crimes that carry a penalty of five or more years in prison.[3] Suspects could be asked to provide DNA if the individual was suspected of an indictable offense with a penalty of two or more years in prison. An important and noteworthy feature of the model bill was that it differentiated between *intimate* and *nonintimate* forensic procedures. The significance of this distinction in terms of the legislation relates primarily to the ordering of forensic procedures in cases where an individual does not consent; intimate procedures could not be performed on suspects or offenders without their consent unless a magistrate's order was obtained. DNA collection from individuals, whether by way of blood drawing or buccal swabbing, was considered an intimate procedure.

The model bill also provided for the establishment of Australia's national DNA database and a scheme (similar to the U.S. Combined DNA Index System [CODIS]) for the sharing of DNA profiles among jurisdictions. It was recommended that the DNA database have several separate indexes with guidelines that determined which indexes could be compared with one another. In addition to a serious-offenders index and a crime-scene index, the database was to have a separate suspect index and two indexes of volunteers' DNA, one for those who had provided DNA for "limited purposes" and another for those who had provided it for "unlimited purposes." DNA collected for "limited purposes" could be used only for a specific criminal justice purpose (i.e., for investigating a specific crime) and could not be compared with DNA profiles from other indexes.

Adoption of the Model Bill: Inconsistencies in Police Powers

A version of the model bill was ultimately adopted by the commonwealth, as well as all the states and territories, as planned. However, there were a few key differences in each case. The commonwealth's legislation—the Crimes (Forensic Procedures) Act 1998 and subsequent amendments in 2000, 2001, and 2006—was closest to the model bill. Under Australian commonwealth law a senior constable, or a person authorized by a senior constable, of the Australian Federal Police (AFP) may ask a suspect (other than a child or an incapable person) to consent to yield a biological sample. A suspect is defined in the national legislation as a person whom a constable suspects or has reasonable grounds of suspecting has committed an indictable offense, is charged with an indictable offense, or has been asked to appear in court in relationship to an indictable offense. The commonwealth defines an "indictable offense" as an offense punishable by imprisonment for a period exceeding 12 months. This language gives the police broader powers to request DNA profiles from suspects than did the model bill, which limited such requests to those who were convicted of offenses with a penalty of two or more years in prison. As in the model bill, DNA testing is considered an intimate forensic procedure.

Consistent with the model bill, serious incarcerated offenders could be required by court order to give a DNA sample in cases where consent was not granted. But while the model bill was written to be applied retroactively, the commonwealth's legislation limited this provision to convicted offenders in prison or on parole; those with past convictions but no longer in the system could not be ordered to provide a sample. A serious offender is a person either under sentence or who has been sentenced for an offense punishable by a maximum penalty of five or more years of imprisonment. Destruction of DNA is required where proceedings are not instituted against the suspect within 12 months or are discontinued, the suspect is acquitted, or the conviction is overturned.[4] Improperly obtained samples or samples that should have been destroyed but were not are inadmissible.

More significant departures from the model bill occurred in laws enacted by the states and territories. For example, the laws differed largely over the question whether buccal swabs should be classified as "intimate procedures." Tasmania, the Australian Capital Territory (ACT), and the Northern Territory classify buccal swabs as "nonintimate procedures." Victoria's legislation was consistent with the model bill in classifying buccal swabs as "intimate," and South Australia went further in categorizing them as both "intimate" and "intrusive." New South Wales determined that buccal swabs were neither intimate nor nonintimate and placed them in a separate category entirely. Classifying buccal swabs as "nonintimate" generally provided law enforcement with substantially more power in DNA collection by enabling police to take samples from suspects or offenders without having to obtain a court order. For example, the Tasmanian Forensic Procedures Act 2000 allows a senior police officer to order a nonintimate procedure on a charged person who is aged 15 or over and who is in custody, as well as a suspect or charged person not in custody if the officer is satisfied that there are reasonable grounds to believe that the procedure may produce evidence.[5] A similar provision was enacted in New South Wales and Queensland.[6] In contrast, Victorian legislation did not provide police with any power to compel bodily samples from either suspects or convicted persons; samples could be obtained only with consent or by order of a magistrate, or a children's court in the case of a juvenile.[7]

Amendments have been considered among states and territories for destroying samples if a person is not charged within a fixed period of time (e.g., 6 months) from the time of taking of the sample or if a person is acquitted of the charges.[8] New South Wales and Queensland adopted a provision similar to that of the commonwealth legislation; DNA samples must be destroyed if a person is acquitted or a conviction is quashed, or where proceedings against the individual are not initiated within 12 months.[9]

Database Operations

In or around August 2002 the commonwealth government started approaching serious offenders convicted of murder, rape, and sexual offenses against children to collect DNA samples. Routine DNA collection was initiated in some states and territories before this; Victoria began collecting DNA from individuals convicted of serious sex offenses in 1996. A year later Victoria expanded collection to those found guilty of drug trafficking, arson causing death, and aggravated burglary.[10] Over time sample collection expanded throughout the states and territories to include "high-volume" crimes, such as breaking and entering.

DNA dragnets have also been conducted in Australia. The first took place in Wee Waa, New South Wales, in 2000 in response to the rape of a 91-year-old woman. Nearly the entire male population of the village (aged 18 to 45) was tested, totaling over 400 samples. Stephen Boney confessed to the crime shortly after his DNA was collected, but before it was processed. Following resolution of the case, all the samples collected from volunteers were destroyed, as required by law.[11] Although the dragnet did help resolve the crime, some have pointed out that the individual ultimately convicted was well known to the police, had a history of aggravated rape, and had moved to the neighborhood just before the crime. Therefore, they argue that the individual should have been caught using conventional policing techniques and without having to subject many innocent individuals to testing.[12] In addition, the dragnet resulted in significant upset to the community. Those who refused to provide a sample or who expressed concerns were singled out and stigmatized. As a result, the Green Party of New South Wales proposed an amendment to the Crimes (Forensic Procedures) Act 2000 that would have required the police to seek the approval of a magistrate before carrying out a dragnet.[13]

By early 2003 Australia had established its National Criminal Investigation DNA Database (NCIDD) and its Disaster Victim Identification Database. Both national databases are operated by CrimTrac, an executive agency and a prescribed agency of the commonwealth government. CrimTrac enters into agreements with each state and territory—the participating jurisdictions. These agreements determine the procedures of interjurisdictional access to and sharing of DNA profiles. Amendments to the commonwealth legislation adopted in 2006 stipulated the legal status of the NCIDD as being an "amalgam of state and territory databases."[14]

Although the NCIDD went live in 2001, it did not become fully operational until 2007. The delay was primarily due to the need for each state and territory to have passed its own legislation and reach an agreement with the commonwealth government.[15] Inconsistencies between the federal legislation and each of the state laws created further obstacles to implementation.

In contrast to the United States, which uses 13 loci for DNA profiles, the Australian forensic units analyze DNA at 9 loci. Police typically obtain buccal swabs, hair and root samples, or blood samples. By June 30, 2007, the NCIDD had a total of 300,624 profiles, of which 111,404 were classified as suspects, 52,963 were serious offenders, 31,446 were offenders, 15,868 were from volunteers for unlimited purposes, and 63,327 were from crime scenes.

Privacy and Other Public Concerns

The creation of the national DNA database, along with several other developments in biotechnology and the life sciences, raised considerable question among the public whether Australia's regulatory agencies were equipped to deal with the broad social and ethical impacts of genetic technologies.[16] In response, the Australian government initiated the most extensive undertaking worldwide to examine the uses of genetic information.[17] Conducted jointly by the Australian Law Reform Commission (ALRC) and the Australian Health Ethics Committee (AHEC) of the National Health and Medical Research Council (NHMRC), the two-year inquiry culminated in a final report, *Essentially Yours: The Protection of Human Genetic Information in Australia*, that contained 144 recommendations for how to best protect genetic privacy, guard against discrimination, and ensure the highest ethical standards in research and practice.[18] Among the recommendations were several that called for significant changes to the DNA-collection and retention provisions in the Crimes Act. These can be summarized as follows:

1. *DNA collection from individuals*: The report recommended removing the consent provisions in relation to suspects and serious offenders so that a court order is required before the conduct of any forensic procedure. In addition, it recommended strengthening protections for victims of crime by

specifying that a known victim must be treated as a volunteer and that steps must be taken to ensure that the victim's DNA is separated from that of a crime-scene sample and not stored in the crime-scene index of the database system or compared against the crime-scene index. The report also called for the development and publication of formal guidelines for carrying out DNA dragnets.

2. *Destruction of DNA samples and profiles*: The report called for the destruction of DNA samples collected from suspects or offenders after profiles have been generated. In addition, it recommended that forensic material taken from a suspect, and any information obtained from its analysis, be destroyed "as soon as practicable"—and no later than 12 months following collection—after the person has been eliminated from suspicion, or police investigators have decided not to proceed with a prosecution against that person. It further recommended that "the destruction of forensic material, and information obtained from it" be defined in terms of "physical destruction of samples and permanent and irreversible de-identification of profiles." Finally, the report called for the assignment of ultimate responsibility for managing the destruction of forensic material, as well as the establishment of formal policies and procedures to enable persons to obtain confirmation that such destruction has occurred.
3. *System oversight*: The report recommended the creation of a Human Genetics Commission (modeled after that of the United Kingdom) to provide advice to the government about the social, legal, and ethical implications of human genetics. The commission was to be created to encourage public engagement and was to operate in an open and transparent manner. In regard specifically to forensic DNA, the report recommended that all DNA testing be performed by accredited laboratories; that CrimTrac's board of management be expanded to include independent members; and that requirements be instituted for periodic public auditing and annual reporting to Parliament about the number and category of samples obtained, the authority under which they were obtained, and compliance with destruction requirements and dates.[19] It also called for nationally consistent rules protecting genetic samples and information that ensured that disclosure occur only with the consent of the sampled individual or pursuant to a court order.

The final report of the inquiry was tabled in the Australian Parliament in May 2003, and as a result a few of the recommendations have found their way into other sources of legislation but have not been adopted in their original form. For example, instead of a Human Genetics Commission, the government established a Human Genetics Advisory Committee that is made up of health and genetics experts and lacks community representation.

In contrast to the sentiments of the recommendations laid out in *Essentially Yours*, the Australian government has taken the position, similar to that of the U.S. Department of Justice, that the DNA profiles used in forensic investigations are not "personal information" and therefore fall outside the legal bounds of the Commonwealth's Privacy Act 1988. As a consequence, the Australian government has no obligation to remove DNA profiles, for example, where convictions are overturned. CrimTrac's goal is to make the database as comprehensive as possible, while the jurisdiction contributing the information holds the personal data. There are variations among jurisdictions in how they handle personal information.[20]

According to the 2006–2007 annual report of CrimTrac,

> The [NCIDD] database does not contain personal information as defined in the Commonwealth *Privacy Act 1988*. The profile associated with an index is set out in legislation—for example, crime scene, offender or suspect. Only the forensic laboratory in the police agency that supplied the identifier can discern names and circumstances associated with the profiles.[21]

The DNA profiles can be automatically removed from NCIDD if a predetermined sunset date is specified in the system. If the profile is so removed, no trace of the profile remains other than a security audit log. According to Crimtrac, the DNA profiles held in NCIDD do not reveal personal details such as physical identity, age, ethnicity, race, appearance, or medical conditions. Users of the NCIDD are not able to ascertain from the DNA any medical conditions of an individual.

Research by Richard Hindmarsh has shown that the failure of the Australian government to respond to public concerns about forensic DNA databanking and genetic privacy, combined with a general technocratic policy approach to decision making about genetics issues, has resulted in increasing public distrust in the operation of forensic DNA databases in Australia.[22] Failure to enact the recommendations stemming from the public inquiry and enduring coordination problems between state and federal laws have fueled this distrust. Furthermore, the discovery in 2006 that police in South

Australia had been illegally retaining and using DNA samples that were supposed to have been destroyed pointed further to the need for greater oversight of the system.[23]

Australia has had significant problems with backlogs in DNA testing. A 2006 New South Wales ombudsman's report showed that demands for testing in New South Wales increased nearly ten-fold from 2000 to 2004, creating a backlog of 7,000 cases. The same report identified 13 cases where errors had occurred.[24]

Overall, the Australian system has built in more privacy protections than that of the United States. Forcible collection of DNA can occur only from those convicted and only in cases where a court order has been obtained. It is possible that these greater protections can be attributed to Australia having developed its system after the United Kingdom and the United States had theirs up and running. During deliberation of the Crimes (Forensic Procedures) Act 2000 the advocacy group Justice Action was active in highlighting for the government the pitfalls of the U.S. and U.K. systems, along with the limitations of DNA evidence. For example, the group applauded the drafters of the bill for placing more responsibility on police and magistrates to justify the ordering of forensic tests, since this would "reduce the number of spurious or trivial tests performed under the Act and help avoid the US experience of rampant overtesting and massive test backlogs which led to the US Justice Department's 1998 'Inquiry Into the Future of DNA Evidence.'"[25]

Because of its strong federalist system, consisting of independent state and territorial laws, the Australian government recognized the importance of a harmonized network of forensic DNA data banks. The challenge for the commonwealth government was to protect individual rights while ensuring that all regional and national police authorities could share and check DNA profiles. Despite the commonwealth's best efforts at uniform criteria, there remain differences in the way states and territories exchange, retain, and manage forensic DNA information. Each jurisdiction legislates its own criteria for matching forensic DNA profiles with those maintained by other jurisdictions (see table 11.1). Each jurisdiction establishes its own restrictions regarding the retention period for certain samples or DNA profiles and the matching that is permissible for their data banks.

CrimTrac was set up in 2000 as an agency of the Australian government to provide solutions to the states and territories for sharing forensic DNA

TABLE 11.1 Australian DNA Data-Bank Chronology

1989: Desmond Applebee is the first individual in Australia to be convicted on the basis of DNA evidence. The federal government and several states and territories initiate a process to develop regulatory standards for DNA collection and use by law enforcement.

1990: The federal government establishes the Model Criminal Code Officers Committee, tasked with developing a Model Forensic Procedures Bill.

1992: The National Institute of Forensic Science commences operations. Among its roles are the development of national standards of quality control and accreditation of forensic laboratories throughout Australia.

1996: Victoria begins collecting DNA from persons found guilty of a serious sexual offense.

1997: Police services endorse the establishment of a national criminal DNA database; forensic agencies from each jurisdiction adopt a commercially available multiplex polymerase chain reaction (PCR) system for routine use in their labs; Victoria enacts legislation regulating the use of its database and expands collection to include those convicted of drug trafficking, arson causing death, and aggravated burglary.

1998: Australian forensic laboratories agree to a common national standard for obtaining DNA profiles.

1998: The Australian Federal Government commits $50 million to establish CrimTrac, with a national DNA database as a central element.

1999: Victorian police obtain the first cold hit from the state DNA database when the DNA profile of convicted thief Wallid Haggag is matched to blood found in a car used in a burglary.

2000: The first DNA dragnet in Australia is carried out in Wee Waa, New South Wales. Samples are "voluntarily" collected from approximately 500 men. Stephen Boney confesses to the crime shortly after his DNA is collected and before it is processed.

2000: The Model Forensic Procedures Bill and the Proposed National DNA Database are introduced in the federal legislature; the CrimTrac agency is formed to facilitate sharing within and between jurisdictions.

2001: CrimTrac launches the National Criminal Investigation DNA Database (NCIDD) to allow the nine Australian jurisdictions to match DNA profiles. The jurisdictions begin to prepare legislation and ministerial arrangements to allow their participation in NCIDD. The Australian attorney-general and the minister for health and aged care initiate an inquiry conducted jointly by the Australian Law Reform Commission (ALRC) and the Australian Health Ethics Committee (AHEC) of the National Health and Medical Research Council (NHMRC) on how to best protect genetic privacy and guard against discrimination in health, employment, and law-enforcement contexts.

2003: The final report of the joint inquiry initiated in 2001, *Essentially Yours: The Protection of Human Genetic Information in Australia*, which includes 144 recommendations pertaining to human genetic databases, is tabled in the Australian Parliament.

2005: NCIDD DNA profile matching commences between Queensland and Western Australia and between Queensland and the Northern Territory.

2006: A North Queensland member of Parliament calls for DNA samples to be taken from all Australians and entered into a national database. A South Australian judge rules that police regularly broke laws controlling the state's DNA database and illegally retained DNA from suspects cleared of crimes. NCIDD DNA profile matching commences between Western Australia and the Northern Territory and between Queensland and the commonwealth.

(*continued*)

TABLE 11.1 *(Continued)*

2007: The commonwealth reaches agreement with and incorporates Western Australia, New South Wales, and the Northern Territory into the NCIDD system. Seven jurisdictions sign a single ministerial arrangement to share DNA data through CrimTrac, removing the last obstacle to a truly national DNA system. Victoria and New South Wales have committed to signing when their legislation allows.
2008: New South Wales initiates Forensic DNA database matching agreements with Saskatchewan and Queensland.
2009: Forensic DNA database matching agreements are worked out between Victoria and the Northern Territory and between New South Wales and Queensland. The NCIDD becomes a national network with agreements for matching DNA profiles among all states and territories.

Source: CrimTrac, http://www.crimtrac.gov.au/systems_projects/ (accessed May 27, 2010).

information across law-enforcement jurisdictions. Each jurisdiction enters into a memorandum of understanding (MOU) with each other jurisdiction on the conditions (called "matching tables" criteria) under which DNA samples can be matched. Interjurisdictional index matching across all jurisdictions and with the commonwealth's DNA database was not finalized until 2009 when all the MOUs were signed. Protecting individuals from the unlawful retention of DNA samples is a high priority of the Australian system.

Chapter 12

Germany: From Eugenics to Forensics

> It is not just the risks of false prosecution that make genetic fingerprints one of the most dangerous instruments in the arsenal of the investigative authorities. Their use also reverses the principle of presumption of innocence. Innocent bystanders who unwittingly find themselves at the scene of a crime could suddenly find themselves prosecuted by police due to traces of DNA they left behind.
>
> —Martin Kreickenbaum[1]

Germany was initially resistant to establishing a DNA database because of public concerns over personal privacy and data protection. Because of the eugenics movements during German fascism in the mid-twentieth century, genetic information was considered very sensitive in postwar German society. However, in the aftermath of a few highly publicized cases of sexual abuse and child murder in Germany during 1996–1997, the German minister of justice asked for an inquiry into the possibility of creating a forensic DNA database. A spokesperson for the Ministry of the Interior issued a public statement that Germany could no longer afford to get along without a DNA database that would help police solve crimes. In 1997 Germany passed the Statute on Identification Through DNA Testing and amended its Code of Criminal Procedure to allow for the creation of such a database.

Among the high-profile cases that initiated a change in Germany's stance on DNA collection was the murder in 1996 of an 11-year-old girl, Christina Nytsch, who had disappeared on her way home from an indoor swimming pool. Her body was found five days later in the woods near her home in Strücklingen, a small village of about 3,500 people in Lower Saxony. Within weeks of the murder the police began collecting saliva samples from volunteers in what became the most massive DNA dragnet in the world. In the end, 16,400 men between the ages of 18 and 30 provided DNA samples. The girl's killer, Ronny Riken, was arrested in 1998 after his DNA was found to match DNA collected from the scene of the crime. He confessed to the murder, as well as to the rape of another 11-year-old.[2]

Legal Foundations for Taking DNA

Before the creation of the DNA database DNA had been used in Germany as an investigative tool for individual cases. The German Parliament and the courts began framing policies and decisions on collecting DNA in the early 1990s. The German Supreme Court (Bundesgerichtshof or BGH) rules on civil and criminal matters, whereas the Federal Constitutional Court (Bundesverfassungsgericht or BVerfG) is responsible for constitutional matters such as privacy. In August 1990 the BGH ruled that DNA analysis of crime-scene evidence is admissible in court so long as it involves noncoding regions of the human genome.[3] In August 1992 the BGH decided the question of whether DNA analysis alone could be the exclusive reason for a guilty verdict. The court ruled that criminal proceedings could not rely exclusively on DNA but must take into consideration all available evidence and testimonies.[4] The court also accepted the "multiplication rule" for the calculation of allele frequencies and the probability of a random match (see chapter 1). However, it raised the question of the selection of an appropriate population sample for a suspect of a particular ethnic or racial group.[5]

The BVerfG took up the question whether the police can request DNA from individuals for whom there is no a priori suspicion. In August 1996 the court ruled that in serious criminal cases a judge may order the taking of blood samples from persons for DNA analysis even when there is no individualized suspicion.[6] This opened the door for mass screenings. The German courts have ruled that under certain circumstances DNA dragnets involving a large number of potential suspects do not interfere with a person's constitutional rights.[7] Mass screenings for DNA must be voluntary and require a court order. Also, all biological samples and profiles must be destroyed immediately after the investigation.[8]

An issue debated by the German Parliament was whether DNA taken for one purpose (to investigate a specific crime) could be used for another purpose (to determine a match for another crime). In December 1996 the German Parliament agreed on a law that extended Section 81 of the German Code of Criminal Procedure (Strafprozeßordnung or StPO) to allow DNA collected during a criminal investigation to be used for purposes other than what the courts had originally approved. According to the new law, blood samples and other bodily tissues may be taken from a suspect under court order and may be used in the criminal investigation in which he or she has been accused, as well as other relevant, pending ("einem anderen anhängigen") criminal proceedings. Thus, if the suspect is charged with a murder, his DNA can be used to determine whether he was at the crime scene in a similar unsolved murder where crime-scene DNA evidence is preserved. At the same time, the law sought to protect individual privacy by requiring a defendant's samples to be destroyed when the investigation is over.[9] To further protect against any possibility of DNA profiling, the law also made it clear that DNA analysis could not be conducted for any purpose other than to determine the source of a DNA sample. It is thus expressly prohibited to troll the DNA in search of unusual inherited genetic diseases, mutations, or rare single-nucleotide polymorphisms (SNPs). Furthermore, the law required that DNA analysis could be carried out only upon written order by a judge and by an impartial expert named by the judge.[10]

In 1997 the German Parliament enacted § 2, DNA-Identitätsfeststellungsgesetz (the Statute on Identification Through DNA Testing), and amended § 81g of the German Code of Criminal Procedure, which authorized the creation of a national DNA database. Under the new law DNA could be compelled from individuals who had been convicted of serious crimes, but only by way of a court order. In addition, DNA profiles could be stored in the database only where there was probable cause to assume that the person would be involved in a similar crime in the future.[11]

Some Germans have expressed concern that retention of DNA profiles from convicted individuals represents an infringement of their constitutional rights. The convicted felon is stigmatized on the basis of a prognosis about his or her future behavior. It is also the case under German law that for an individual convicted of a crime, there is no clear limit of time on the storage of data. However, all database records are subjected to a case review after 10 years,[12] at which time they may be deleted if a determination is made that they are no longer required.[13] By a 1999 amendment to the Statute on Identification Through DNA Testing, state prosecutors are allowed to access the Federal Central Register of Criminal Offenders to seek a match of genetic profiles of convicts currently serving their prison terms with the profiles of unsolved crimes.

The BVerfG ruled on three cases in January 2001 where plaintiffs challenged the constitutionality of mandatory DNA testing and the retention of DNA profiles in the database on the grounds that they violated privacy and the right of self-determination over personal information. The high court held that the creation of a DNA database for use in facilitating future investigations into major crimes was constitutional. The court found that there was a compelling public interest in the state's development of DNA databases to help criminal investigators solve crimes that justified the intrusion into privacy associated with the retention of DNA profiles. However, the court also accepted the statutes' requirement that storage of the DNA test result must be based on a finding of probable cause that the person would be involved in a similar crime in the future, and it clarified that the justification for database inclusion must be provided on an individual basis and must be dependent on a careful prognosis of future risk. A court must consider how each case meets the criteria for inclusion and cannot use the argument that it is always possible that the individual will commit another crime.[14]

Whose DNA Can Be Taken?

In general, police can upload the DNA profile of a person to the national database if he or she is suspected or arrested for severe crimes, a criminal offense of substantial importance, or sexual assault punishable by more than one year in prison. The police cannot carry out a DNA analysis on any suspect exclusively on the basis of their own interpretation of the criteria. A judge must decide whether a person's DNA can be taken and analyzed.

To avoid any conflict of interest, a judge also decides which expert can do DNA analysis. Initially, as noted above, this was required for all DNA testing, including the analysis of crime-scene samples. However, this resulted in a significant backlog of court decisions. In 2005 the law was amended to exclude crime-scene samples from authorization, as well as suspects who have provided informed consent.[15] For others, the expert authorized to conduct DNA testing cannot come from the same institutional unit (police division) that is responsible for carrying out the investigation of the suspect. The expert must guarantee that neither he nor his unit or institution is engaged in any legal analysis associated with the case and that the results are not open to unauthorized individuals. In addition, the identity, date of birth, and address of the donor of the DNA are blinded to the person doing the DNA analysis.

Section 81e of the StPO permits the investigation of DNA only for well-defined purposes, such as determination of parentage, ascertaining the identity of the DNA source in criminal investigations, or determining the gender of the source of the sample. The police cannot troll the DNA for disease alleles, genetic dispositions, or so-called personality or behavior genes. The only phenotyping permitted under German law is gender determination.[16]

> In context with forensic applications in Germany it was and still is necessary to explain that DNA analysis as used in forensics has nothing to do with genetic diagnostics. It has to be cleared up that it only has the potential to prove whether a biological sample in a crime case can originate from a suspected individual or whether a falsely suspected person can be excluded respectively. People have to be convinced that the methods as well as the loci just are useful for identification purposes and that it has not the potential to explore the privacy of an individual.[17]

The Germans have also established a public representative for data privacy protection (Datenschutzbeauftragter). This person provides independent oversight of the privacy of the data.

Two types of institutions in Germany carry out most of the forensic DNA analysis: police laboratories and university institutes of legal medicine. Police laboratories may come under state or federal authority. Private DNA laboratories are also capable, available, and sometimes called on to carry out DNA analysis. The criminal office in each state is called Landeskriminalamt (LKA). The federal office of criminal justice is known as the Bundeskriminalamt (BKA). Among the 25 institutes of legal medicine at the medical faculties of German universities are some that can provide the police with DNA analysis.

The state laboratories began running forensic DNA analyses in 1987 after the heads of the forensic science institutes established a working group to set standards for forensic casework. By 1989 the group started to undertake DNA analysis in casework by applying restriction fragment length polymorphism (RFLP) techniques. In 1992 the federal laboratories started introducing the method of short tandem repeats (STRs) and fluorescently labeled primers that became the standard of Applied Biosystem's DNA sequencer. Like many countries, Germany was slow to set up an accrediting system for its DNA laboratories. Forensic DNA laboratories operated for over 10 years without an accreditation system.[18]

Destruction of Biological Samples

One of the most notable aspects of Germany's laws governing DNA collection is a requirement that the DNA samples collected from suspects or convicted individuals be destroyed immediately after profiling. The use of DNA for any purpose other than to obtain the information needed to create a DNA profile is illegal. In addition, DNA profiles must be removed from the database in cases where a suspect is acquitted or the proceedings are discontinued for other reasons without a verdict.[19] All database records are subjected to a case review after 10 years to determine whether they should be retained in the database. These provisions, along with those cited earlier in regard to DNA collection, represent perhaps the most stringent privacy protection measures enacted to date for forensic DNA databases.[20] Belgium, Switzerland, the Netherlands, and Norway have similarly strict protections in regard to reference sample destruction.

By 2008 Germany had a forensic DNA database that contained profiles of 534,782 persons out of a population of 82 million people.[21] The law on forensic DNA and national DNA data banks grew out of a series of court decisions and parliamentary enactments. Today DNA profiles may be taken from persons suspected of or arrested for serious crimes or sexual assault punishable by more than one year in prison.

A number of safeguards are built into Germany's DNA data-banking system that are not seen in other systems, including those of the United States and the United Kingdom. Most notably, all DNA samples obtained from volunteers, suspects, and convicted individuals must be destroyed after successful profiling. DNA profiles are also destroyed automatically when they are no longer required for a case or where a suspect is acquitted, proceedings against the individual are not initiated, or a conviction is overturned unless the suspects are considered at risk for committing future crimes. DNA analysis may be carried out only by written order of a judge, who also designates the expert who will carry out the analysis. In addition, the samples must be analyzed anonymously (without names of the suspect or victim) in a laboratory that is not connected to the investigating agency. Forensic science institutes associated with law enforcement are presumed to be independent.

Chapter 13

Italy: A Data Bank in Search of a Law

> Only a centralized databank, involving all Police forces and outside academic experts, formalized in a structure such as the Ministry of Justice and with a Scientific Committee of very high profile (whoever inserts data must have proper qualifications and ascertainable credits), might be able to offer the required guarantees to avoid the usual problems of an "Italian style" management of the DNA data collected. It is necessary to watch and check the quality and quantity of the data to ensure that it would always be possible, when the judiciary think the moment has come—to erase one's own data.
>
> —Giuseppe Novelli, professor of medical genetics, Tor Vergata University[1]

Unlike in most European countries and the United States, police in Italy initiated operation of DNA data banks without proper legislative authorization. DNA was collected and stored by police and the Carabinieri Special Corps well before legislation worked its way through the Italian Senate in 2008 and the House in 2009. It is not clear how many databases were in existence in public or private hands during this period. This de facto database operation was criticized both within Italy and internationally. According to a 2007 report on European Union (EU) forensic DNA data banks:

> Although there is currently no act that allows for the creation of a forensic DNA database, the Italian police do collect DNA samples. . . . As there is no

statutory regulation regarding retention of DNA profiles, the current practice of the Italian police to store the DNA profiles that are derived from unidentified crime scene stains and from convicted offenders is probably illegal.[2]

A number of factors set the stage for Italy's legislative effort in this area. Most notably, the country had a strong scientific tradition in molecular biology and genetics, capable of supporting forensic DNA science. In addition, Italy was seeking to raise its preparedness to meet potential terrorist attacks, and its police authorities looked on DNA data banks as a critical tool for their antiterrorism campaign, as well as to fight criminal activity in general. Finally, the Italian state also had an abiding interest in the protection of individual privacy through its Special Commission for the Rights to Privacy, also known as the Italian Data Protection Authority (or, for short, the Privacy Authority).[3]

The Privacy Authority

Absent a formal law, Italy's Privacy Authority issued nonbinding opinions on DNA collection for identification and related subjects. Stefano Rodotà, former president of the Privacy Authority, stated in April 2004:

> These databanks must be reserved to particularly relevant purposes and must concern extremely limited categories of individuals, collecting the data most important for identification, specifying the relationship between the collected data and the genetic material from which they have been extracted. In no case must the aforesaid databanks in any way resemble a mass screening or a utilization, even if only partially, discriminatory.[4]

In March 1997 the authority adopted a provision for protecting genetic data, including data and samples stored in biobanks.[5] In past years the authority has given recommendations to Parliament and the government about the organization and management of DNA data and samples for use by law enforcement.[6] In October 2007 it issued an opinion specific to the issue of the creation of a national DNA data bank.[7] This opinion states: "It is urgent to regulate systematically this particularly delicate subject, [where] . . . the necessary points of reference for the legislators are lacking."[8] The text outlines existing policy gaps associated with the collection of biological material, the analysis of samples, the typology of profiles, the registration and security of data, and the storage of samples without the consent of the person concerned.

The Pisanu Law of 2005

On July 22, 2005, the Italian cabinet (a collective body formed by the ministers in the government and the prime minister) approved a law decree, "Urgent Measures Against International Terrorism." The decree was proposed by Giuseppe Pisanu, minister of the interior and a member of the leading right-wing parliamentary majority party, Forza Italia. The Pisanu decree entered the Parliament on July 27 and became official law on July 30. The law was approved as an antiterrorism act in response to terrorism threats in Italy following the attacks in Great Britain.[9] Under the decree DNA samples could be taken from a detained person who refused to give samples voluntarily after prior authorization was given by a prosecuting magistrate. Amendments were introduced stating that hair or saliva samples would be the source of the DNA profiles.

The law provides for "new regulations on personal identification" and is currently the only formal legal reference for collecting genetic data to solve crimes. The text of the Pisanu Law contains provisions on the collection of biological material, the analysis of samples, the typology of profiles, the registration of and access to data, storage of samples without consent, and security.

The Pisanu Law allows the collection of biological material for the purpose of forensic procedures: DNA can be obtained voluntarily or by force, or without the knowledge of the person concerned when it is collected from the scene of a crime.[10] When the Pisanu Law was approved, the Privacy Authority notified the Parliament that special attention should be given to the DNA collections already in existence and created by local police departments and laboratories. For instance, should these collections be transferred to a central location, and if so, how? What will be the end use of the aforementioned archives and data banks? While similarly motivated by the threat of terrorism, this law is much less comprehensive than the one approved by the Italian Parliament following the Prüm Treaty (discussed next).

The Treaty of Prüm

One of the factors that appears to have accelerated the Italian government's legislative agenda on DNA data banks was the European Union's announcement in June 2007 that the elements contained in the Treaty of Prüm (better known as Schengen III) "now become part of EU law and shall be enforced in all countries belonging to the Community." The treaty is considered the most rigorous document for international cooperation regarding security information. Initially signed on May 27, 2005, by seven member states of the European Union (Belgium, Germany, Spain, France, Luxembourg, the Netherlands, and Austria), the treaty seeks to strengthen transborder cooperation in an effort to fight against terrorism, criminal activity, and illegal immigration.[11] The treaty envisages the collection of personal information, including both digital fingerprints and DNA profiles. Italy was among the second tier of signatory nations, which currently number 17.[12]

As a member of the EU, Italy is required to pass a law implementing the treaty that is consistent with the criteria established by the EU. Each of the EU member states has three years (starting in 2007) to harmonize its national law with the common rules.

In 2006 Giuliano Amato, then Italian home secretary, sent a letter of intent to give his assent to the Treaty of Prüm.[13] Italy's assent to the treaty became operative in June 2009 upon approval of its law governing DNA data banks.

Emerging Databases

Two investigating departments in Italy have DNA profiles of felons stored in their own computerized databases: (1) the Carabinieri Special Corps, known as the Reparto Investigativo Speciale (RIS), which has four branches located in different cities, each with self-managed data banks of biological samples and genetic profiles, and (2) the Scientific Police (forensic investigators).[14] The protocols by which biological samples are obtained and stored by the two police corps are not the same, and standards in both cases appear far from those prescribed by the Privacy Authority.

The rules for postconviction DNA testing in cases where new evidence becomes available are described in the section of the Code of Criminal Procedure titled "Sentences Subject to Revision." Newly found genetic profiles, mistakes in attribution, or new elements in the evaluation of samples previously collected may be considered new evidence. The convicted individual or the attorney general in the Court of Appeal may ask for a reopening of the trial on the basis of this evidence.[15] The Court of Appeal judge may accept the request or turn it down.

DNA can be also acquired as evidence during court proceedings. But as Andrea Monti, a legal expert in data banking, notes: "Judges accept DNA as evidence in trials, but it is not necessarily probative for conviction or absolution."[16]

At present Italy is faced with rising xenophobia due to the politics of such parties as Lega Nord and Alleanza Nazionale. It is difficult to know whether DNA of non-European persons is held in national data banks for investigations. On occasion the media report that buccal swabs are taken from immigrants landing illegally on Italian soil or from individuals suspected of terrorism. There are no official data about this practice, but only newspaper reports about boats arriving in Sicily and other Mediterranean ports from Africa.[17]

A case of DNA data banking reported in the news tells of an Albanian citizen arrested in May 2006 at Gargazzone, a town near Merano in Trentino, for car theft. His DNA was found in a stolen car. The police were able to trace the DNA to him because his DNA was on file in an RIS data bank. In fact, years earlier his DNA had been analyzed together with 400 other biological samples during a previous rape investigation that had not identified any suspect.[18] From the deposition of a Carabinieri officer questioned under oath during the investigation, it became known that this data bank contained about 15,000 profiles, all filed illegally.

Although no official data have been released, an official at the Carabinieri has stated that all the police databases combined contain fewer than 100,000 DNA profiles, with fewer than 50,000 held by the Carabinieri.[19] Unofficial sources, mainly investigative journalistic reports, indicate that the RIS data bank contains 2,200 genetic profiles derived from the analysis of biological samples of individuals investigated in past cases. An additional 11,700 samples and profiles were taken from persons identified in the course of prosecutions but not under investigation, and another 5,100 were extracted from samples found at the scenes of crimes, so their sources are unknown. The total of 19,000 profiles seems to be the contents of the data bank run on the RIS software.

Italy shares national information with international police forces (Interpol). Robert Noble, Interpol general secretary, stated in July 2007 that a network was activated to allow the sharing of DNA profiles among the G8 countries (the United States, Canada, France, Germany, Russia, Japan, Italy, and Great Britain).[20] It includes some 65,000 to 79,000 profiles obtained from crime scenes. The following year Noble encouraged the G8 countries to support national law-enforcement efforts in using and further developing Interpol's global databases to combat organized crime.[21]

The Newly Passed Bill

The bill to join the Prüm Treaty (contained in the "security package" suggested by the present home secretary, Roberto Maroni) calls for the formation of a national DNA data bank.[22] It was introduced in the Senate in July 2008 and was approved by it in December 2008. While the bill was under consideration in the legislature, the Italian Ministry of Justice simultaneously initiated an administrative process to fulfill Italy's obligations to the Prüm Treaty. The proposed legislation would establish a national DNA data bank within the Department of Public Security in the Home Office, as well as a Central Laboratory of this data bank in the Department of Penitentiary Administration (DAP) in the Ministry of Justice. The bill was examined and approved by a commission from the Ministry of Justice and one from the Ministry of Foreign Affairs.[23] Passage of the bill by the Italian Parliament on June 30, 2009 as Public Law No. 85 formally ratified the Prüm Treaty and provided the legal foundations for the creation of Italy's national DNA data bank.

To set up the new DNA data bank, the legislation requests an appropriation of 11 million euros during the first year and an additional 15 million in the following three years. "By progressively substituting the fingerprint system this databank will make it possible to identify people," declared Home Secretary Maroni. "It will become a definite reference point."[24] Since passage, however, the law has continued to be met with criticism.[25] As of May 2010 Italy's national DNA data bank was still not operational.

Prior to the passage of the Italian Parliament's Law No. 85, Roberto Lattanzi, a member of the Italian Data Protection Authority, said in regard to the oversight of DNA data banking, "There is still what I would call a 'fluid' situation."[26] The Privacy Authority, an administrative institution with considerable autonomy and composed of four members (two selected from the Senate and two from the House), has provided guidelines pertaining to the conservation of biological samples, the privacy of human medical data, and the databasing of forensic profiles derived from biological samples.

Under the new law anyone who commits any crime, excluding financial crimes, punishable by three years or more in prison will have his or her DNA entered into a forensic database. Arrestees and juveniles who have not been convicted of a crime but who are charged with a crime punishable by three or more years in prison would also be required to provide DNA samples. The law contains provisions for forcing a suspect to yield a saliva sample under a magistrate's order. According to an official of the Carabinieri, undocumented immigrants are profiled and placed in the database only if they have committed a qualifying offense.

The law states nothing about a properly managed chain of custody or how a mismanaged chain of custody will impact the admissibility of the samples as evidence in a judicial proceeding. Law enforcement officers can access DNA profiles from the database without prior authorization from the prosecutor responsible for the investigation or the judge assigned to the case. The law does not clearly identify who is responsible or accountable for the destruction of samples or profiles.

According to Professor Amedeo Santosuosso of the University of Pavia Law School, one of the most controversial provisions in the law is the one that deals with expungement of DNA samples and profiles in situations where a case is dismissed. If a biological sample and DNA profile are acquired under criminal investigation, the police may hold on to the biological sample unless the suspect is released and a petition is made to have the sample destroyed and the profile expunged.[27] Similarly, a person convicted of a crime and subsequently exonerated must petition to have his or her DNA profile removed from the database. In all other cases, biological samples are retained for up to 20 years and the profiles for up to 40 years.

Currently police can obtain DNA from a suspect surreptitiously by following someone and picking up his or her discarded cigarette butt or cup. Police can also collect saliva when the law allows them to do so for establishing personal identification (in the absence of other identification) or if a person voluntarily gives a saliva sample. Italian police have also undertaken DNA dragnets to solve crimes, most notably in the murder of Maria Fronthaler in a village near Dobbiaco on April 1, 2002.[28]

The political Left has raised the possibility that DNA samples and tests could be used for purposes other than those officially stated. Francesco Pizzetti, current director of the Authority for the Right to Privacy (Guarante per la Protezione dei dati Personali), wrote about the ways in which DNA collection can violate the sphere of personal freedom. Pizzetti stressed in particular that

> it is perfectly possible to build a DNA databank coherent with the formula of the Treaty of Prüm if the security measures and the criteria set by the Authority for the Right to Privacy are observed and the scheme once agreed upon by the Ministry of Justice and the Home Office could be something that would not provoke alarm or reactions on the part of the citizens or on our part.

Pizzetti mused: "What is this database for? What information is to be put into it? There needs to be clarity about how this data is collected so that people's dignity is always respected. . . . Above all clarity is needed on which people can be forced to give samples and in which cases."[29]

According to Luciano Garofano, an RIS colonel at Parma, one of the four Italian cities where these corps are located, the DNA data bank is to be considered an indispensable instrument for fighting crime. "The political world pays more and more attention to such themes and the times are ripe," Garofano stated.[30]

Under the new system, the central laboratory of the data bank at the DAP in the Ministry of Justice will have the task of storing biological evidence and samples from known individuals, while the database will contain computerized profiles. Only authorized personnel will be granted access to the database. Abuses are punishable by imprisonment from one to three years. In cases where someone is proved innocent, according to the law, all acquired DNA profiles must be destroyed. Otherwise profiles are retained for a maximum of 40 years and biological samples for a maximum of 20. This is guaranteed by the Italian Data Protection Authority and the National Committee for Biosecurity and Biotechnologies, under the control of the Presidency of the Council of Ministers.

Law No. 85 adapts Italian regulations to the Prüm European Treaty.[31] However, the treaty states that only digital genetic profiles and not biological samples can be kept for identification purposes. The new law goes beyond this standard in authorizing retention of the biological samples. As stated by Francesco Pizzetti, "If we intend to collect data of both types we need articulated and complex rules. If we imagine a bank with only one type of data, we think it is difficult to accept that biological samples could be kept for all purposes."[32]

Civil Liberties Concerns

The Privacy Authority has found that such an "organized complex of data," even if not connected on the World Wide Web to other police services nor interactive with other archives, "makes a databank," and "its particular nature draws our attention to the adoption by the RIS of further measures and precautions for the protection of the profiles and samples stored by that special branch of Carabinieri." A call for clear rules on how the database will be used has been voiced by Francesco Pizzetti.[33]

In July 2007 the authority sent the RIS a series of recommendations on data management to be adopted within six months, but to date this has not occurred. The initiative of the authority was probably triggered by an inspection revealing that there were no clear rules on who had access to the database for consultation. There is a need to adopt a series of controls to ensure the integrity of the data, the traceability of biological samples, privacy constraints, and measures granting access to the data banks (after identification). Also needed are guidelines on how long the biological samples and the genetic profiles derived from them should be kept in storage. As of the fall of 2009 there were no clear procedures for requesting the removal of DNA either for investigation or for research (patients are not always informed of the fact that samples of their tissues are stored for research well beyond the period covering their hospitalization). According to provisions of the legislation, the data are to be expunged in case a person is exonerated of any charges. Within the European Union, Italy has the longest criminal trials, and therefore, even with this provision, DNA samples and profiles might remain in the data bank longer than in many other countries.[34]

According to Francesco Pizzetti, "A law regulating such matters is absolutely necessary, but such a law may be considered unreliable without a national debate on the implications of a database of this kind and the concept of the forcible taking of a DNA sample." He argued that the information we get from the media often exaggerates the public safety benefits provided by a national DNA data bank.[35]

The need for security has always been emphasized by politicians continuously looking for approval, and the media have influenced public opinion about the importance of and need for a DNA data bank. Giuseppe Novelli, professor of medical genetics at Rome, agrees, stating: "I wouldn't be against it myself, provided my DNA was stored in a secure place with the maximum guarantee, not as happens today, when it may be stored in different places without a guarantee of security, and, above all, without my knowing where it is kept."[36]

The international standards that have been established for the operation of DNA data banks, according to Amedeo Santosuosso, law professor at the University of Pavia, would not allow for "Italian-style" management of DNA data banks. As Andrea Monti, a legal expert in biocomputer science, has stated, "Even considering the amount of money involved in this project, I think that the risk of an Italian style [management] may still be possible. Rules are not enough; if the system gets positive results it will be thanks to the dedication of some *rara avis* who is at the helm in the public administration."[37]

The international nongovernmental organization European Digital Rights (EDRI), founded in June 2002 to defend civil rights in the information society, outlined the lack of safeguards in the Italian forensic DNA databank legislation. According to the EDRI, there are no security measures against unauthorized access to the forensic DNA data; there is no properly established "chain of custody" in the handling of biological evidence; law-enforcement officers can access the national database without prior authorization from a prosecutor or judge; and no one is clearly identified to order the destruction of biological samples and forensic DNA profiles. The EDRI also faults the new law for excluding perpetrators of white-collar crimes from having their DNA profiled and placed on the national database.[38]

Italian criminal justice authorities initiated the use of forensic DNA profiling and established DNA databases well in advance of legislative authority and controls. With its legislated mandate to protect the privacy of genetic data, Italy's Privacy Authority issued opinions on civil liberties protections regarding the collection and retention of DNA profiles by the police. The legal foundations for the creation of the Italian national DNA database, which grew out of the Prüm Convention, were established in civil law by the passage of Law No. 85 in June 2009. Many civil liberties issues remain contested, including the adequacy of existing safeguards to prevent unauthorized access to the database and to track the chain of custody of crime-scene biological evidence, the inclusion of DNA profiles from arrestees and juveniles in the database, and the placement of burden on exonerated or acquitted individuals to petition to have their DNA reports expunged.

Part III

Critical Perspectives: Balancing Personal Liberty, Social Equity, and Security

Chapter 14

Privacy and Genetic Surveillance

The privacy and dignity of our citizens is being whittled away by sometimes imperceptible steps. Taken individually, each step may be of little consequence. But when viewed as a whole, there begins to emerge a society quite unlike any we have seen—a society in which government may intrude into the secret regions of man's life at will.

—Justice William O. Douglas, *Osborn v. United States*[1]

It is difficult to imagine information more personal or more private than a person's genetic makeup.

—Senator Edward Kennedy[2]

Personal privacy is highly valued in most modern democratic societies, although there is a broad spectrum in the ways in which privacy is interpreted and protected among different countries. While generally thought of as simply the "right to be left alone," the concept of privacy is highly complex, involving a number of overlapping personal interests. These include having control over our personal information and decision making and an ability to exclude others from our personal things and places. In the United States privacy protection has a long and complex history. Our government was founded on the principle that privacy—as well as related notions of dignity, due process, and liberty—serves as a check on the abuse of state power. These values are found not only in the Fourth Amendment's prohibition of unreasonable searches and seizures but also in the Fifth Amendment's guarantees of due process, equal protection, and freedom from self-incrimination, in the Sixth Amendment's right to counsel, and in the Eighth Amendment's ban on cruel and unusual punishment. At the same time, the Constitution does not explicitly mention the word "privacy," and it is routinely debated whether a general right of privacy is guaranteed by the Constitution.

Today "privacy law" in the United States does not consist of a single statute but instead is a complex array of protections that are dispersed among multiple sources, including constitutions, statutes, regulations, and common law. Statutory privacy protections evolved in direct response to technological development. Anita Allen notes, "The word privacy scarcely existed in the law before 1890, when new technologies contributed to an explosion of interest in privacy among intellectuals and lawyers." Specifically, developments in printing and photography sparked considerable concern that "privacy would be lost in a world of unchecked curiosity, gossip, and publicity."[3] These concerns prompted two lawyers, Samuel Warren and Louis Brandeis, to publish a highly influential *Harvard Law Review* article that outlined their concept for a new "right to privacy" and served as a pillar for virtually all U.S. law and policy in the realm of privacy that was to come.[4] Over the next few decades notions of "privacy" and a "right to privacy" became fixtures of the legal apparatus as a number of state courts—many of them drawing directly from Warren and Brandeis's work—began to recognize a common-law privacy right and state legislators began to pass privacy-protection legislation. By comparison, it was not until 1965 that the U.S. Supreme Court expressly recognized a constitutional right of privacy in the landmark decision *Griswold v. Connecticut*.[5]

The fragmentation and ambiguities of privacy law have rendered the right to privacy somewhat vulnerable to the ebb and flow of historical conditions. Perceived threats to our civil order, in particular, have resulted in weakening privacy, as well as other civil liberties protections. Until recently we were able to assume that listening devices would not be secretly planted in our homes or on our telephone lines without a court warrant. That expectation has shifted after 9/11 through the enactment of the 2002 Homeland Security Act. That law gives the executive branch powers to execute warrantless wiretaps that invade the privacy of people whom it judges are a high security risk. Under the Terrorist Surveillance Program the U.S. National Security Agency is authorized by executive order to monitor phone calls and other communications involving a party believed to be outside the United States, even if the recipient of the call is on American soil. The Protect America Act of 2007 (signed into law on August 5, 2007) amended the Foreign Intelligence Surveillance Act of 1978 to give congressional standing to warrantless federal wiretapping.

Just as the notion of "privacy" is rich in ambiguity, so too do notions of "genetic privacy" shift in different contexts. Although it is well established in law and policy that people's expectation of privacy regarding their genome is not a frivolous concern, the standards for the way in which genetic privacy is treated vary widely from one context to the next. Broad public acknowledgment of the sensitivity, vastness, and potential misuses of genetic information ensured Congress's ultimate passage of the Genetic Information and Nondiscrimination Act (GINA) in 2008. This act provides certain baseline protections to help ensure that employers and health insurance companies cannot request, have access to, and make decisions on the basis of information in a person's genetic code. Similar and in some cases more comprehensive protections have been passed by most state legislatures.

Although the notion that an employer or insurance company should not have access to our genetic information has been codified as law, law enforcement's authority to collect DNA has ballooned over the last decade. Most recently people merely arrested by federal authorities and in select states, whether ultimately convicted or not, have been having their DNA routinely collected and permanently retained. Also, as discussed in chapter 6, law-enforcement officials are increasingly collecting DNA from individuals surreptitiously, the presumption being that if a person has "abandoned" his or her DNA, they have the right to collect it, analyze it, and perhaps use it as evidence against that person, all without a search warrant or the person's knowledge or consent. Protection of DNA information in the criminal justice context appears to be operating under a different set of principles than that of health or employment. Can we retain an expectation of privacy in our DNA in health and employment contexts and lose that expectation when our DNA becomes an object of interest in the criminal justice community? If we are moving toward a double standard of privacy, one for medicine and another for forensics, we should know why, whether, and under what circumstances it is justifiable.

This chapter considers these and other questions about our privacy interests in our DNA. We begin with a discussion of what is so private about our DNA, including an analysis of the often-heard debate between civil libertarians and law enforcement over whether DNA is different from a fingerprint. We then discuss the role of the Fourth Amendment in protecting our privacy in our DNA. We trace the direction of the law in this area and identify some of the important questions that are likely to be addressed in the coming decade. Finally, we conclude with a discussion of the particular hurdles for privacy advocates in protecting our genetic information in the law-enforcement context.

Genetic Privacy

The history and meaning of "genetic privacy" are in many ways parallel to those of "privacy" more generally. Just as the quest for privacy arose out of technological innovation at the turn of the nineteenth century, so too did concerns about "genetic privacy" stem from scientific and technological development—in this case the rise of molecular biology and computer science. A proliferation of DNA data banking and an increasing ability to extract information from DNA encouraged a shift in medical and behavioral research to focus more on genetic factors. Tissue repositories, such as newborns' blood spots, that had been initially established back in the 1960s became gold mines of genetic information as genetic techniques evolved during the last decade of the twentieth century. The completion of the draft human genome sequence in 2000 brought questions of genetic privacy front and center as biotechnology companies sought to stake claims on genetic resources through the patenting of human genes and further proliferation of privately held DNA data banks.

What is our privacy interest in our DNA? Privacy experts generally distinguish four privacy concerns pertaining to DNA. First, there is *physical privacy* or *bodily privacy*. This comes into play at the point of DNA collection, whether it occurs for purposes of genetic testing, medical research, or criminal investigation. DNA can be collected by taking blood, either by a pinprick of the finger or a venal draw. The collection of blood is generally viewed as a highly intrusive process in the medical or research arena, where genetic information generally cannot be collected without the informed consent of the individual providing the sample. In the law-enforcement context DNA collection can occur voluntarily, forcibly, or surreptitiously. Surreptitious collection does not trigger physical privacy per se, since the DNA is collected off objects that are no longer on the person. When DNA was first introduced into forensic evidence around 1986, blood samples were the main source. More recently, in the law-enforcement context, DNA has been collected through use of a buccal swab. This method of collection is generally considered less intrusive than methods that involve the taking of blood.

Second, genetic privacy refers to *informational privacy*. Information contained within our genome is considered highly sensitive because it can reveal a vast amount of information about us. The organization JUSTICE has described genetic information as "the most intimate medical data an individual may possess."[6] Genetic testing is currently available for over 1,700 diseases and abnormalities, with about 1,400 available in clinical settings, and this number continues to increase every year (see figure 14.1).[7]

Some genetic polymorphisms correlate directly with disease states, providing information about an individual's current health status. But others are *predictive*—they correlate with a statistical predisposition to disease of familial disease patterns, providing information about the possibility that an individual may develop a disease over the course of his or her lifetime (see box 14.1). The fact that DNA contains information about an individual's future health risks is the primary reason why some health law experts believe that DNA information is uniquely sensitive. George Annas has analogized DNA information to "a probabilistic, coded 'future diary.' . . . As the code is broken, DNA reveals information about an individual's probable risks of suffering from specific medical conditions in the future."[8]

In addition to providing medical information, genetic tests may reveal environmental and drug sensitivities as the field of pharmacogenetics advances. DNA sequence information may also contain information about behavioral traits, such as a propensity to violence or substance addiction, criminal tendencies, or sexual orientation.

Of course, our genetic status does not determine our medical future or our behavioral traits. That this information is probabilistic, not deterministic, makes the prospects that it might be released to third parties particularly dangerous. Concerns that this information could be used to discriminate against individuals seeking insurance and employment or to stigmatize individuals and families abound. There is some basis to this concern: in the 1970s several insurance companies and employers discriminated against individuals who were sickle-cell carriers, even though they would never develop the disease.[9] More recently a lawsuit filed by the U.S. Equal Employment Opportunity Commission (EEOC) against the Burlington Northern Santa Fe Railway Company revealed that the company was conducting

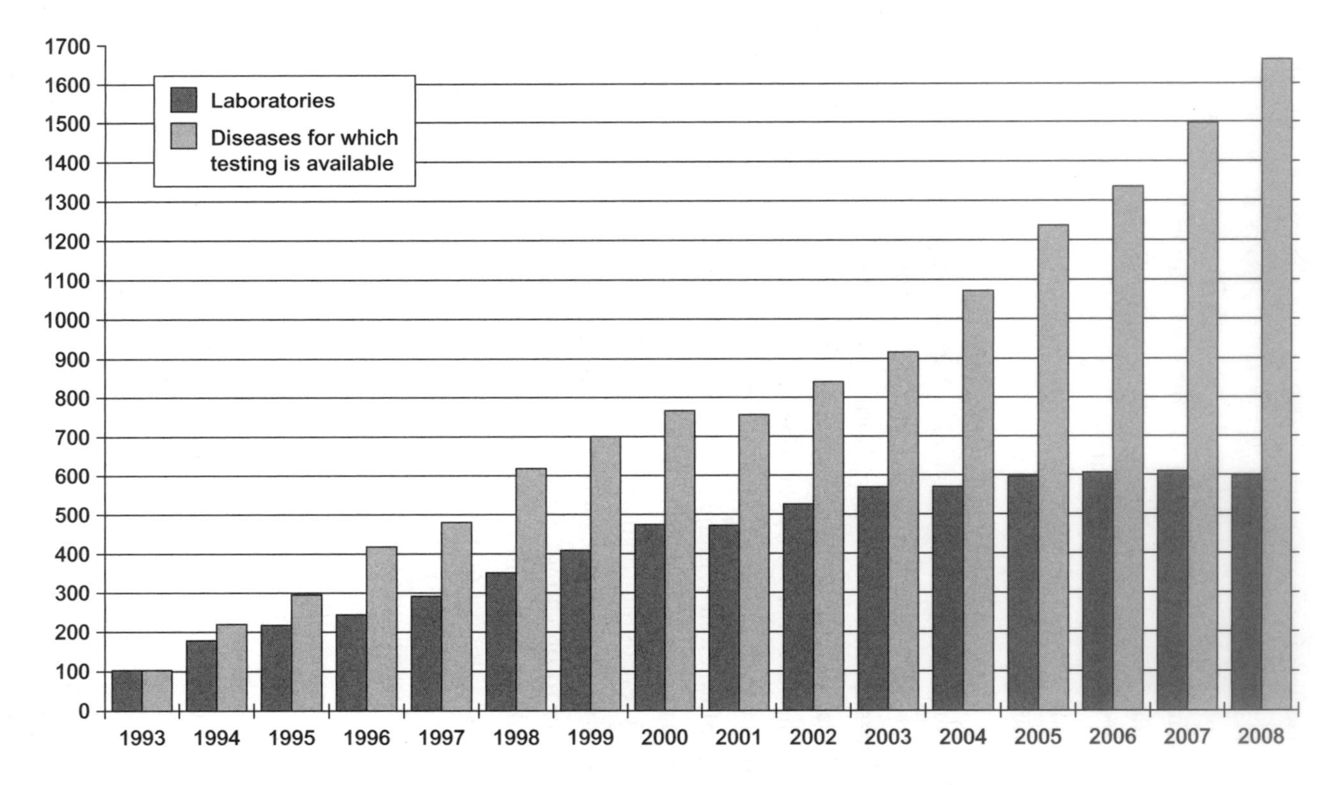

FIGURE 14.1. Growth of genetic testing, 1993–2008. *Source*: GeneTests database (2008), www.genetests.org. Copyright University of Washington, Seattle.

BOX 14.1 James Watson's Genetic Privacy

At age 79 Nobel laureate James Watson donated his DNA for sequencing and eventual public access to the 3 billion base pairs that made up his genetic code. The codiscoverer of the chemical and physical structure of DNA had one restriction. Watson expressed his expectation of privacy regarding one segment of his genome, which he believed might reveal whether he has a predisposition to Alzheimer's disease. One of his grandmothers contracted Alzheimer's, and Watson figures that his chances are about one in four. The relevant sequence, Watson stated, should be kept secret from others and from himself.

Source: D. R. Nyholt, Chang-En Yu, and P. M. Visscher, "On Jim Watson's *APOE* Status: Genetic Information Is Hard to Hide" [letter], *European Journal of Human Genetics* 17 (2009): 147–149.

genetic testing of its employees who had filed claims for work-related injuries based on carpal tunnel syndrome without their knowledge or consent. The company was asking those employees to provide a blood sample that was then tested for a rare genetic condition that had been associated with an increased risk of developing carpal tunnel syndrome. At least one worker was threatened with possible termination for failing to submit to a blood test. The case was settled quickly, with the company agreeing to all of the terms sought by the EEOC.[10]

What is perhaps unique about genetic information is that it both links us to and distinguishes us from all other human beings. As we discussed in chapter 1, our genetic code is thought to be unique—no two individuals, save perhaps identical twins, have the same DNA. At the same time, our genetic code is shared; it is part of our common heritage, passed on from one generation to the next. In this sense our precise genetic code is unique to each of us, but much of the information it reveals is likely to be shared by our relatives or entire groups of people. For example, if a woman undergoes genetic testing for BRCA1 and BRCA2, the two genes that have been associated with hereditary forms of breast and ovarian cancer, the results of this test have implications not only for that individual woman but for all her close relatives.

This leads us to the third notion of privacy that is relevant to our DNA—*familial* or *relational privacy*. Because DNA is inherited, it can be examined to infer whether two individuals are related. In the case of parental linkages DNA testing is quite unambiguous, since 50 percent of our DNA was contributed by our biological father, and 50 percent by our biological mother. Sibling relationships can also be inferred, but beyond this, DNA testing is far more limited. Revealing an unsuspected personal biological relationship can have serious consequences for individuals or their families.

Similarly, DNA can reveal information about one's ethnic origins. The Global Genome Project has introduced broad population groupings corresponding to a common set of genetic alleles called haplotypes. The extent to which our DNA contains these genetic loci gives population geneticists and anthropologists a clue to our geographical ancestry. Currently this information is of personal interest to some individuals who want to know what percentage of their inherited genome came from populations in Europe, Africa, Asia, North America, or Australia. However, this information can have significance beyond those interested in family genealogy. For example, American Indian heritage can sometimes be used in establishing tribal land rights. It can also be connected with social stigmas. When phenotype cannot reveal one's ancestry, some would look to genotype.

Finally, DNA can provide information about whether an individual was physically present at a certain location. A person's DNA found on a bedsheet at the scene of a crime is prima facie evidence that the person was at the location. In other words, DNA has implications for *spatial* or *locational privacy*. Traditionally, privacy has focused more on questions of who you are and what you are doing, and less on where you have been, where you presently are, or where you are going. What we do in our personal spaces, as well as whom we visit and where, should fall within our sphere of privacy. Increasing uses of radio-frequency ID tags (RFIDs), automated toll-payment systems, Global Positioning System (GPS) hardware and software, and surveillance cameras have directly challenged this notion of privacy. As an example, in 1996 the Federal Communications Commission (FCC) required that all cell-phone manufacturers equip their units with location technology. As a result, it is now possible to position a cell-phone caller in a geographical location.

DNA Privacy in the Context of Law Enforcement

How do the notions of genetic privacy just discussed play out in the context of forensic DNA techniques and practices? What specific privacy concerns come into play when law enforcement collects, analyzes, and retains our DNA? And what protections are currently in place?

First, there is the bodily intrusion associated with DNA *collection.* As mentioned earlier, when DNA was first introduced into the criminal justice system, DNA was collected by drawing a blood sample. Recently law enforcement agencies have switched over to collection by way of a buccal swab. The physical intrusion associated with swabbing the inside of an individual's mouth versus drawing blood is clearly lower. Nonetheless, the inside of one's mouth is still an intimate space, a body cavity that is not generally made accessible to others. Certainly it would be wrong to say that no privacy interest at all is raised in circumstances where an individual is forced to open his or her mouth so that a police officer can obtain a saliva sample. Aakash Desai, a graduate student at the University of California, Berkeley, who was forced to give a DNA sample after being arrested at a demonstration on campus protesting custodial layoffs and furloughs, as well as tuition hikes, described the experience as follows:

> I felt violated when the government took my DNA. I felt like *I* was being burglarized and not the other way around. I feel like the government now owns my genes. This hurts my sense of self, and makes me feel sick to my stomach. Each swab was like being coerced into giving up part of my being. The government had already taken my possessions and clothing, now they were taking the building blocks of my own body. It seems like some ownership of myself has been lost and my privacy violated.[11]

As discussed in chapter 2, when DNA data banks were first created in the United States, DNA collection was limited to individuals who had been convicted of serious, violent crimes. Otherwise, law enforcement could only either collect DNA "voluntarily" or require a DNA sample from individuals in cases where they had a warrant supported by probable cause. Over the last 15 years this standard has slipped dramatically. Today the overwhelming majority of states are collecting DNA from all felons, as well as those convicted of some misdemeanors, and 14 states have approved legislation to take DNA from individuals merely upon arrest. At the same time, a number of "suspect databases" have been created as a result of DNA dragnets, where thousands of individuals have had their DNA collected without probable cause, without any privacy protections for the subjects, and with little, if any, oversight. Similarly, as is discussed in chapter 6, there has also been a trend toward collecting DNA surreptitiously from individuals, without their knowledge or consent.

After DNA is collected, it is analyzed for information. The *analysis* of DNA raises additional privacy concerns. When the FBI published its *Legislative Guidelines for DNA Databases* in 1991, the agency was explicit that the DNA records held by the Combined DNA Index System (CODIS) should relate only to the identification of individuals, and that no records should be collected on physical characteristics, traits, or predisposition for disease. Currently all police departments that collect DNA samples create a forensic profile, containing information related to 13 loci. More recently some forensic laboratories have run Y-chromosomal short tandem repeats (Y-STR) typing on stored samples to help determine whether the source of that sample is likely to be a relative of the source of a crime-scene sample.

Law enforcement and other DNA database-expansion advocates have repeatedly characterized DNA profiles as nothing more than a "DNA fingerprint." UCLA law professor Jennifer Mnookin has observed that early promoters of the use of DNA testing in the criminal justice system initially referred to DNA typing as "DNA fingerprinting" in an effort to enhance its appeal and to encourage judges and lawmakers to view DNA as nothing more than a more rigorous and precise technique than traditional fingerprints.[12] For example, Assistant Attorney General William Moschella wrote in a letter to Senator Orrin Hatch, "By design, the information the system retains in the databased DNA profiles is the equivalent of a 'genetic fingerprint' that uniquely identifies an individual but does not disclose other facts about him."[13] DNA databasing advocates have underscored the DNA fingerprinting analogy with claims that the DNA profile used in the system consists merely of markers that correspond to repeating segments of so-called junk DNA. For example, in promoting the passage of the DNA Fingerprint Act of 2005, Senator Jon Kyl stated, "The sample of DNA that is kept in NDIS is what is called 'junk DNA.'"[14] However, the "DNA fingerprinting" analogy is fundamentally flawed, most notably because DNA samples, which under current laboratory practices are permanently stored alongside the generated profiles, have the potential to reveal far more than a fingerprint. But the analogy is even mistaken as applied to the DNA profiles (see box 14.2). Although it is true that none of the CODIS loci have been found to date to be predictive of any physical or disease traits, this does not mean that such a correlation will not be found in the future. In addition, it is not necessary for any of the markers to correlate directly with any stigmatizing information for there to be a concern; the specter of discrimination and stigma could arise where one or more short tandem repeats (STRs) are found to correlate with another genetic marker whose function is known, so that the presence of the seemingly innocuous STR serves as a "flag" for that genetic predisposition or trait. A finding of this nature has already occurred; a study in England from 2000 found that one of the markers used in DNA identification is closely related to the gene that codes for insulin, which itself relates to diabetes.[15]

BOX 14.2 DNA as a "Fingerprint": A Flawed Analogy

Perhaps the most common refrain provided by law enforcement in defense of the data banking of DNA is that DNA is no different from a fingerprint. The fingerprint analogy is deeply flawed. First, fingerprints are two-dimensional representations of the physical attributes of our fingertips. They are useful only as a form of identification. Fingerprints cannot be analyzed to determine whether two individuals are related. They cannot tell you your likelihood of developing Alzheimer's disease or breast cancer or whether you are a carrier for cystic fibrosis. Nor can they be read to determine which version of the monoamine oxidase A (MAOA) gene you have. There is no exponentially growing list of conditions that can be read from a fingerprint, or even significant research in this area.

Law-enforcement advocates underscore the fingerprint analogy by stating that the DNA profiles that are generated in DNA testing are merely "junk DNA"—that is, the 13 loci used in forensic analysis do not code for any phenotypic characteristics. But of course, they are focusing only on the DNA profile and quite blatantly ignoring the most significant privacy concern associated with DNA data banking, that of the DNA samples,

(*continued*)

which are stored indefinitely by forensic laboratories and have the potential to reveal almost unlimited information about ourselves.

Beyond this, even the reference to so-called junk DNA is misleading with regard to the DNA profile. Even noncoding regions of the DNA transmit more information than a standard fingerprint. Recent scientific studies have demonstrated that those regions are not devoid of biological function, as was once thought. And although they may never be found to have highly sensitive direct coding functions, they may very well be found to *correlate* with things we may care about and deem private. Since the completion of the Human Genome Project, no serious scientist refers to noncoding regions of DNA any longer as "junk." In addition, simply by way of basic principles of inheritance, DNA profiles can be used to signal parent-child and other close family relationships. An examination of additional so-called junk DNA markers can provide more definitive analysis. Law enforcement's recent ventures into familial searching are an indication that even criminal investigators recognize that DNA can provide far more information than a fingerprint (see chapter 4).

Source: Authors.

The analysis of crime-scene samples offers another example of how the privacy of DNA can be overlooked in the law-enforcement context. As discussed in chapter 5, there are currently no laws that we are aware of that prevent law enforcement from running any genetic tests of interest on DNA collected from the scene of a crime. Mining those samples for medical or other information that might be used to narrow a pool of suspects is certainly a possibility.

Finally, privacy concerns arise in association with the *uses* of the stored DNA samples and generated profiles. Those in the database are subjected to ongoing, repeated, suspicionless searches; every week the 7 million DNA profiles in CODIS are automatically searched against crime-scene profiles. In a sense, one's inclusion in the database makes him or her an automatic suspect during his or her lifetime for any future crime. And as we have discussed in chapter 4, recent uses of forensic DNA databases to search for "partial matches" that may implicate a family member of an individual in the database make the privacy intrusion of repeated searches apply not only to the individuals in the database but also to all of their close relatives. There is also a real but neglected risk of being falsely accused of a crime that comes from being in the database. We discuss the potential sources of contamination and error in detail in chapter 16.

Perhaps the most significant privacy concerns arise about the long-term storage of the original DNA samples. Although the information contained in the DNA profile is necessarily limited, every forensic laboratory in the country currently holds on to the biological samples from which the DNA profile was generated, and, as discussed earlier, these can be mined for an increasing amount of highly personal information. Thus criminal justice agencies around the country have all the medically relevant information on a profiled individual that can be gleaned from one's DNA. So long as the samples are retained, there exists the possibility that they could be disclosed to or accessed by third parties or used in ways that result in the disclosure of highly confidential information or for malicious or oppressive purposes. And because genetic information pertains not only to the individual whose DNA sample is stored but to others who share in that person's bloodline, potential threats to genetic privacy posed by the samples' long-term storage extend well beyond the millions of individuals who are currently in the system. Jean McEwen has warned: "The unique composition of forensic DNA banks . . . will make those repositories a nearly irresistible source of samples for *behavioural* genetics research or testing . . . without the informed consent of those from whom the samples were taken."[16]

Federal policy limits disclosure of DNA information "for law enforcement identification purposes," but of course there is nothing about "identification purposes" that prevents law enforcement from mining stored samples for information that is restricted to noncoding regions of the DNA. State laws provide little, if any, additional protection against unrelated use of DNA obtained for investigative purposes. Four states are silent on this issue, and most others parrot the federal standard. According to Mark Rothstein and Sandra Carnahan, "Many state statutes allow access to the samples for undefined law enforcement purposes and humanitarian identification purposes, or authorize the use of samples for assisting medical research."[17] The FBI has also conceded that most states do not have protections against the dissemination of DNA samples.[18] Most states limit uses of the database to those that are "authorized," but it appears to be left to the discretion of law enforcement to determine what is an "authorized use." Only eight states expressly prohibit the use of a DNA database to obtain information about human physical traits, predisposition to disease, or medical or genetic disorders, and Alabama explicitly authorizes the use of DNA information for medical research. Some states explicitly prohibit the use of DNA samples and profiles for purposes other than law enforcement, but this, of course, does not appear to prevent law enforcement from mining the DNA for additional information. Other states are either vague or have laws that permit the restricted use of their DNA forensic data banks. About 15 states have statutes that expressly permit the use of data banks for medical research and humanitarian needs.[19] Twenty-four states allow DNA samples collected for law-enforcement identification to be used for a variety of other non-law-enforcement purposes.[20] Massachusetts's law, for example, contains an open-ended authorization for any disclosure that is, or may be, required as a condition of federal funding and allows the disclosure of information, including personally identifiable information, for "advancing other humanitarian purposes." Two states prohibit the use of DNA samples in medical research.[21] Thirty-four states have statutory language authorizing the use of databases for statistical purposes related to improving law enforcement, such as developing better probability figures on false matches.

There is no national policy governing the retention of forensic DNA samples, and state laws vary on policies for destroying DNA samples and removing the profiles from the database in cases where convictions are overturned or charges are dismissed. Where sample and profile removal are possible, the burden is most often placed on the individual to have his or her records removed, rather than on the state. Wisconsin is the only state that requires that biological samples be destroyed after typing, but it appears that to date, no such destruction has occurred.

Some are dubious that forensic DNA data banks will ever be used by medical or behavioral scientists.[22] David Kaye states unequivocally that behavioral genetics researchers who desire to use the samples in CODIS (and in state and local databases feeding CODIS) are excluded from doing so: "Behavioral genetics researchers who come knocking on the doors of state or federal administrators for the DNA of convicted offenders will find them locked, and the key cannot be located within the disclosure and usage provisions of the current database laws."[23] Their research would not qualify under the federal DNA database statute, which states that CODIS must be used for law-enforcement identification purposes only.[24] This would include developing population statistics and validation criteria where all identifiers are removed, but not behavioral genetics or medical research. Thus, even if scientists had an interest in using the database, according to Kaye, in most cases they could not get access. Currently, no state explicitly allows DNA samples from the forensic database to be used in research on predisposition for disease or in behavioral research.[25] The Alabama law states that the DNA population statistical database, which shall include "individually identifiable information," can be used "to assist in other humanitarian endeavors including, but not limited to, educational research or medical research or development."[26] Davina Bressler implies that since the Alabama law prohibits research with DNA samples from the database of individually identified records, therefore, any such anonymized research would not be of any value for medical or behavioral science. In addition, some of those states that allow the DNA samples to be used for "other humanitarian purposes" refer more specifically to identifying human bodily remains or missing individuals, especially lost or kidnapped children. Therefore, these allowances were most likely not intended to open the database to broad medical or behavioral research.

To date, there is no evidence that any state has thus far permitted its forensic DNA data banks to be used for medical or behavioral research. In her analysis of the 50 state statutes Bressler concludes:

> A more careful reading of the statutes reveals that only one state allows for medical research with records (and they must be anonymized) and that no state allows for medical research with samples. The repeated assertions that many states allow researchers to use DNA samples from convicted offenders to be used in medical research—with or without consent—are incorrect.[27]

Although these arguments are noteworthy, it is also the case that current statutory restrictions are at best unclear, untested, and open to interpretation. Perhaps more important, statutes can be repealed or amended. If there is legitimate knowledge to be gained, will current restraints hold back the scientific pressures for access to public data? For example, if genetic screening of stored DNA samples could reveal whether individuals were "genetically predisposed" to criminal offending, would it be allowed? Consider, for example, a subset of felons convicted of pedophilia. If the database for that group of felons contains information about convictions, medical history, family associations, and profiles of crime victims and also has the pedophile's complete genome from a biological sample, there almost certainly would be medical and/or behavioral geneticists who would want to pursue the question whether pedophilia has a genetic basis. Although pedophiles appear to be highly recidivistic—having almost a compulsion suggesting a biological mechanism—that is a huge step from a genetic mechanism. There is also the prospect that a genetic-based crime-control strategy could ultimately include mandatory genetic screening from birth to identify individuals predisposed to certain so-called undesirable behaviors. Those who support such access see scientific merit in research that links genes to impulsiveness, aggressiveness, pedophilia, or novelty seeking and believe that ethics review boards can adequately oversee such uses of databases.[28]

In England, where, as is the case in the United States, the government contracts out some of its DNA analysis, concerns have been raised about private companies sharing DNA samples collected for forensic purposes with researchers. In 2006 the *London Observer* reported that a "private firm has secretly been keeping the genetic samples and personal details of hundreds of thousands of arrested people." The British Home Office has given permission for 20 research studies using DNA samples from the National DNA Database (NDNAD). The goal of some of the studies is to determine whether it is possible to predict a suspect's ethnic background or skin color from his or her DNA.[29]

A report issued by the public-interest group GeneWatch UK noted that British researchers who have access to the national DNA forensic database

> do not have to seek consent from participants or the approval of independent ethics committees to carry out their research. They have only to seek permission from the NDNAD Board. . . . Some of the research could be highly controversial, for example research on ethnicity and race . . . or research on "genes for criminality."[30]

GeneWatch UK has gone on record proposing that any research on the NDNAD should be reviewed by an independent ethics board to ensure that the research is morally and socially acceptable and that consent should be received in advance of the research by those whose DNA profile or biological source is part of the study.[31]

It is clear that the only sure way to prevent any misuses of the stored samples is to destroy the samples themselves. No one proposes destroying the crime-scene samples, since these might be the only evidence from the scene of the crime and the accuracy of the analysis could be contested. However, individual samples collected from known offenders or arrestees are another matter. A number of advocacy organizations, including the American Civil Liberties Union (ACLU), and ethics advisory committees have proposed destroying the known-offender biological samples after DNA typing is completed as a way of preventing possible misuse. Law enforcement has argued that the samples need to be retained for reasons of quality assurance (most significantly for retesting a sample in case of a mix-up) or to rerun the samples for purposes of upgrading the system (for example, for profiling a larger number of genetic markers). The U.K. Human Genetics Commission's 2009 report found these arguments unpersuasive:

> We cannot see any need for long-term retention of subject samples, for the following reasons: if the identity and whereabouts of the subject are known, it will be possible, and not disproportionately expensive or difficult, to obtain a new sample for analysis; conversely, if the subject's whereabouts are not known, having a DNA sample is unlikely to assist in locating them; and finally, the argument that there may be a future need to upgrade the profiles by analyzing more loci is unconvincing.[32]

Fourth Amendment Protection: Trends in Case Law on DNA Data Banking

The Fourth Amendment to the U.S. Constitution provides that "the right of the people to be secure in their persons, houses, papers, and effects, against unreasonable searches and seizures, shall not be violated, and no Warrants shall issue, but upon probable cause." For nearly 200 years courts limited the scope of this amendment by holding that it protected only against physical intrusions—for example, the entry into a house or the seizure and examination of private papers. Thus in 1928 the Supreme Court held that government wiretapping of a telephone call did not constitute a "search" or a "seizure" and was therefore not even subject to any Fourth Amendment scrutiny.[33]

This changed in 1967 when that same Court decided *Katz v. United States.* Charles Katz was convicted in California of illegal gambling. The crucial evidence against him was a series of recordings that the FBI had made, without a warrant, of calls Katz had made from a public pay phone booth in Los Angeles. Katz challenged his conviction, arguing that the recordings could not be used against him, and the case made its way up to the Supreme Court. By a vote of 8 to 1 the Court overruled its prior cases and ruled in favor of Katz, holding that the government had violated Katz's Fourth Amendment rights by secretly recording his private phone calls without a warrant.[34]

Katz extended the reach of the Fourth Amendment beyond physical intrusions. Writing the majority opinion, Justice Potter Stewart stated that "the Fourth Amendment protects people, not places. What a person knowingly exposes to the public, when in his own home or office, is not a subject of Fourth Amendment protection. . . . But what he seeks to preserve as private, even in an area accessible to the public, may be constitutionally protected."[35] The Court refused to continue to take the "narrow view" that the "search and seizure" language in the Constitution was meant only to protect against physical penetration of private, delineated space or objects and ruled that the Fourth Amendment protects people's concept of personal privacy: "The fact that the electronic device employed to achieve that end did not happen to penetrate the wall of the booth can have no Constitutional significance."[36]

The *Katz* standard provides the controlling test for determining whether government action constitutes a Fourth Amendment search: first, that a person have exhibited an actual (subjective) expectation of privacy; and second, that the expectation be one that society would recognize as reasonable. Government actions that intrude into these "reasonable expectations of privacy" are searches and therefore constitutional only if the government can show that they are reasonable. The Fourth Amendment also prohibits unreasonable seizures, whether those seizures be of property or of the person (or his or her bodily tissue).

How does the concept of privacy as articulated in *Katz* apply to DNA data banking and forensic DNA applications? Because the Fourth Amendment protects against "unreasonable" searches and seizures, courts must assess not only whether a search has occurred but also the reasonableness of that search. Is the taking of DNA a "search"? Is it a "seizure"? If so, is it reasonable under any circumstances? Is a warrant required? Do we have an "expectation of privacy" in our DNA? How is that expectation balanced against the public benefits of DNA collection for purposes of criminal investigation?

The Taking of DNA Constitutes a Search and a Seizure

A law-enforcement officer's sticking a needle into a person's arm for a blood sample against his or her will without evidence of suspicion is generally regarded as a violation of his or her privacy. The limits of taking blood samples forcibly were the issue in the U.S. Supreme Court case *Schmerber v. California* (see box 14.3), where the majority stated, "The interests in human dignity and privacy which the Fourth Amendment protects, forbids any such intrusions [blood samples] on the mere chance that desired evidence might be obtained."[37] Although the Court in *Schmerber* found that a blood test presents only a minimal intrusion, it nonetheless held that in order for the police to take a biological sample from an arrestee, they must either have a warrant or have probable cause to think that the sample will yield evidence of a crime *and* that exigent circumstances exist that make it impracticable to procure a warrant. The Court has found that similar collections of bodily fluids or tissues for analysis—such as urine tests,[38] breath tests,[39] and fingernail scrapings—are similarly privacy intrusions that constitute a "search."[40]

BOX 14.3 *Schmerber v. California*

Armando Schmerber was taken to the hospital after he had been involved in a traffic accident. The hospital performed blood tests involuntarily, which police used to determine whether Schmerber had been driving while intoxicated. He argued that the police had invaded his privacy by obtaining his blood and testing its alcohol level. The court ruled that the defendant exhibited physical features that gave police probable cause to believe that he was intoxicated and that police did not have time to obtain a warrant. Police argued they were under exigent circumstances that justified their warrantless search. The court agreed: "We today hold that the Constitution does not forbid the State minor intrusions into an individual's body under stringently limited conditions, [which] in no way indicates that it permits more substantial intrusions, or intrusions under other circumstances."[a]

[a] *Schmerber v. California*, 384 U.S. 757 (1966).
Source: Authors.

Following the analysis in *Schmerber*, lower courts have consistently held that the taking and analysis of DNA constitutes a "search." Most courts have focused, as in *Schmerber*, on the *physical intrusion* associated with the collection of DNA by way of either a blood draw or a buccal swab. Considerably less attention has been paid to the *informational* privacy aspect of DNA collection and analysis and the notion that the DNA profiles are subjected to ongoing, repeated searches, although some courts have acknowledged this aspect as well.[41]

Generally speaking, the Fourth Amendment requires that searches be supported by a warrant, issued by a magistrate only upon a factual showing that the search will likely uncover evidence of a crime. Over the years the courts have carved out a number of exceptions to this warrant requirement (for example, allowing police to stop and frisk people whom they have reason to think are armed and dangerous, and to search cars when they have probable cause to think the car contains contraband or evidence of a crime). Generally, the more intrusive the government action, the higher its burden is to justify that action. Thus a brief traffic stop requires less justification than does a search, and some searches—such as those requiring dangerous surgery—are per se unreasonable. But even for the most minor intrusion, the law is clear that searches and seizures for general law-enforcement purposes must be supported by some level of individualized suspicion.

When DNA was first introduced into forensic evidence around 1986, blood samples were the main source. As forensic DNA testing has shifted from blood draws to buccal swabs, its perceived level of intrusiveness has dropped precipitously. Nevertheless, courts have generally found that forcible collection of DNA by use of a buccal swab constitutes a "search" requiring a warrant or at least probable cause. Furthermore, because the later analysis of the sample reveals information about a person's genetic makeup—information that we as a society consider private—that analysis is itself a search.

Evolving Case Law on DNA Data Banks

Although there is little question that DNA collection and analysis constitutes a "search," all twelve circuits have nonetheless upheld mandatory DNA-collection statutes that apply to convicted offenders. In so doing, the courts have judged the reasonableness of the search by balancing the interests of the government against the privacy of the individuals involved. In weighing these interests, some courts have found that the government's interest in maintaining a DNA database of convicted offenders is one of "special needs, beyond the normal needs for law enforcement." These cases follow the reasoning in *Skinner v. Railway Labor Executives' Association*, where the U.S. Supreme Court upheld the collection of blood, breath, and urine samples from employees in safety-sensitive positions for purposes of randomized drug testing, without a warrant, probable cause, or individualized suspicion. Although the Court recognized that these tests were "searches" that invaded the defendants' reasonable expectation of privacy, it nonetheless upheld the testing program on the notion that it was not for general law-enforcement purposes but, instead, to investigate railroad accidents and to prevent injuries.

The Second, Seventh, and Tenth Circuits have each relied on the special-needs exception in upholding state DNA databasing of convicted offenders. Rothstein and Carnahan have questioned this line of reasoning as applied to database statutes, since it is hard to understand the purposes of DNA collection in these contexts—identification, investigation, and prosecution of criminals—as anything other than law enforcement.[42] After all, the mission statement of CODIS explicitly states that it was created for law enforcement purposes, and many of the state statutes describe the database as a "powerful law enforcement tool." Although the courts that have applied the special-needs exception have, in some cases, acknowledged that DNA testing of inmates is ultimately for a law-enforcement goal, they have concluded that this falls within the special-needs analysis because "it is not undertaken for the investigation of a specific crime."[43]

In recognizing the inherent tension in applying the special-needs exception to a law-enforcement database, some courts have avoided this line of reasoning and have instead upheld convicted-offender databases on the notion that persons under supervision of the criminal justice system following conviction have a "reduced expectation of privacy." For example, in *United States v. Kincade* a majority of the Court explicitly ruled that the special-needs exception could not justify DNA-collection schemes. Nonetheless, the court found that "parolees and other conditional releasees are not entitled to the full panoply of rights and protections possessed by the general public."[44] Just as the police can enter and search the house of a person on parole without a warrant or even any reason to think the resident has done anything wrong, they can intrude upon his genetic privacy and bodily integrity.

Courts that have followed this general balancing scheme in upholding DNA data banks have tended to analogize DNA testing to "fingerprinting," focusing narrowly on the use of DNA for "identification," and have largely ignored potential uses or misuses of stored biological samples. In *Rise v. Oregon*, for example, the Ninth Circuit found that once a person is convicted, one's identity is a matter of state interest, and the offender "has lost any legitimate expectation of privacy in the identifying information derived from blood sampling."[45]

The Question of Innocent Persons: Arrestees, Suspects, and Family Members

Although it appears that the question whether DNA can be collected and permanently stored from convicted felons is thoroughly decided, the issue whether the reaches of DNA data banks that extend to innocent persons can withstand constitutional muster is another matter. The court rulings that have upheld DNA data banking of convicted offenders do not provide law enforcement a blanket justification to collect and use DNA without limits. As a matter of policy, for a society that values freedom and individual privacy, the notion that innocent individuals should not have DNA taken without their knowledge or consent or retained permanently in a database, seems consistent with other protections against law enforcement's unfettered acquisition of personal information. But states and the federal government are giving law-enforcement agencies authority to override personal privacy as DNA data banks are expanded to arrestees, DNA is collected surreptitiously in the course of investigations, and DNA databases are mined for partial matches.

In following the reasoning of the courts in upholding DNA statutes of convicted felons, it is hard to see how the routine, forcible collection of DNA from arrestees—who are innocent under the law—could be tolerated. To date, of the four courts that have considered this issue, three have determined that the routine collection of DNA from arrestees is unconstitutional. In 2006 Minnesota's Court of Appeals held that taking DNA from juveniles and adults who have had a probable-cause determination on a charged offense but who have not been convicted violates state and federal constitutional prohibitions against unreasonable searches and seizures. Similarly, in 2009 the federal District Court of Western Pennsylvania, consistent with the long-standing recognition that arrestees enjoy the presumption of innocence and give up only those rights whose infringement is necessary to ensure jail security and safety, struck down the federal law allowing the testing of arrestees. The court found that "requiring a charged defendant to submit a DNA sample for analysis and inclusion in CODIS without independent suspicion or a warrant unreasonably intrudes on such defendant's expectation of privacy and is invalid under the Fourth Amendment."[46] In contrast, the Virginia Supreme Court upheld DNA collection from suspects alleged to be violent felons, finding that because DNA can be used to identify a person, taking DNA from an arrestee "is no different in character than acquiring fingerprints upon arrest."[47]

As is discussed in chapter 6, within certain limits, U.S. courts have upheld methods of law-enforcement authorities in obtaining nonconsensual DNA samples without warrants. Sometimes police use a ruse to obtain a suspect's DNA. Other times police investigators shadow an individual until they can retrieve a discarded object, including the individual's sputum on the sidewalk, with the desired DNA sample. If DNA privacy is to make any sense, then we have to distinguish between the abandoned object and the information that can be deciphered by a technical expert in the DNA left on the discarded object. We should not lose our expectation of privacy for our DNA, even though we have given up the object carrying it.

To date, no court has considered whether the initial collection of DNA or the subsequent use of familial searching violates the privacy interests of convicts' family members, who have not forfeited their privacy rights. In the case of familial searching the identification and analysis of partial DNA matches broadens the scope of a DNA database, subjecting family members to genetic surveillance. This means that they are more likely to be suspected of a crime they did not commit. Whether the privacy interests of family members in these situations could rise to the level of constitutional protection seems uncertain, at best, particularly since—as in the case of surreptitious sampling—there is no initial physical intrusion associated with the search of the family member's DNA.[48]

Challenges for Privacy Advocates

As discussed at the beginning of this chapter, there are significant differences between the ways in which genetic privacy is treated in the medical and research context and in the law-enforcement context. There is a growing consensus and near unanimity that, beyond the person's caregivers, the privacy of an individual's medical genetic information should be protected. Many states and the federal government have passed legislation that protects individuals from unauthorized access to and use of medical genetic information. Also, the confidentiality of genetic information, regardless of how it is obtained, is generally accepted among professional societies and bioethicists. The American Society of Human Genetics is on record stating that "genetic information, like all medical information, should be protected by the legal and ethical principle of confidentiality."[49] In a similar vein, bioethicist George Annas noted that "the DNA molecule itself can be viewed as a new form of medical record. It is a source of medical information, and like a personal medical record, it can be stored and accessed without the need to return to the person from whom the DNA was collected for authorization."[50]

However, the standards in forensics for protecting privacy are operating on a different playing field than those in the medical and research communities. While informed consent is the standard for the collection and storage of genetic information in the latter context, the law-enforcement situation can be seen as an evolving free-for-all, where DNA is starting to be collected almost at the whim of a given police officer.

This bifurcated system is not unlike what is seen in the United Kingdom, although, interestingly, a new law in the United Kingdom prohibits non-law-enforcement agents from acquiring and analyzing a person's DNA without consent (see box 14.4). The rationale for the law was articulated by Baroness Helena Kennedy, chairperson of the United Kingdom's Human Genetics Commission:

> Until now there has been nothing to stop an unscrupulous person, perhaps a journalist or a private investigator, from secretly taking an everyday object used by a public figure—like a coffee mug or a toothbrush—with the express purpose of having the person's DNA analyzed. Similarly, an employer could have secretly taken DNA samples to use for their purposes. This sort of activ-

BOX 14.4 U.K. Surreptitious Sampling

Under the United Kingdom's Human Tissue Act of 2004 (which became effective on September 1, 2006) individuals cannot obtain and analyze a person's DNA without his or her consent. Violation of the law is punishable by up to three years in prison or a fine or both. The law does not apply to law enforcement. The British Parliament enacted the law to keep amateurs from breaching the genetic privacy of individuals by analyzing a person's so-called abandoned DNA.

Interestingly, although this law has significantly bolstered privacy protections for individuals in light of a burgeoning genetic testing industry, it further bifurcated the United Kingdom's system of DNA privacy. The law, with its clear exception for law enforcement, underscores that the rules for the police are separate from those for everyone else. In the context of law enforcement, DNA is open for taking.

Source: Authors.

> ity is a gross intrusion into a person's privacy and we are very pleased that the Government has now taken the Human Genetic Commission's advice and made it illegal to take and analyse DNA in this way, without the person's consent.[51]

In contrast, the United States has no restrictions on analyzing DNA from so-called abandoned objects, obtained by stealth or by a ruse, where no informed consent is required. A *New York Times* report titled "Stalking Strangers' DNA to Fill in the Family Tree" describes the current attitude regarding the DNA we continuously shed from our bodies or from personal products we discard. Our laws and policies have not adequately addressed the complex privacy issues that are raised by the growth of medical and forensic interests in people's DNA. The concept of "abandoned DNA" implies that we have no privacy interests in those reservoirs of our genetic code that are discarded or shed continuously and ubiquitously. Currently the default position for police is that DNA unattached to our bodies is unrestricted for anyone's taking:

> They swab cheeks of strangers and pluck hairs from corpses. They travel hundreds of miles to entice their suspects with an old photograph, or sometimes a free drink. Cooperation is preferred, but not necessarily required to achieve these ends. . . . The talismans come mostly from people trying to glean genealogical information on dead relatives. But they can also be purloined from the living as the police do with suspects. The law views such DNA as "abandoned."[52]

Can medical and forensic privacy of DNA be conceptualized into a coherent set of principles, or do we have to live with two independent and incompatible systems?

Part of the challenge for privacy advocates lies in the nature of DNA itself. The fact that DNA is everywhere means that it is fairly easy to collect, whether openly or surreptitiously. The level of intrusion for collection, as a result of the development of buccal swabs, is also minimal, or at least less than for blood collection. DNA's stability means that it can be retained more or less indefinitely so long as it is kept in suitable conditions. The cost of DNA analysis is declining. Generally speaking, we have witnessed a slow and steady erosion of privacy protections in the United States. All these factors make it difficult for privacy activists to effectively oppose the expansion of DNA databases.

At the same time, in the law-enforcement context, very little emphasis and attention have been placed on the informational privacy aspect of DNA collection. Early court cases failed to acknowledge the full scope of privacy concerns that can arise in the amassing and long-term storage of DNA samples, focusing almost exclusively on the initial physical intrusion associated with DNA collection. The "fingerprinting" analogy, meanwhile, has been etched in the minds of policy makers, eager to promote a seemingly high-tech, objective approach to "fighting crime."

Public attitudes and social policies about genetic privacy are evolving. There is ample evidence that people have an expectation of privacy of the genetic code that makes up their cells, at least the fraction of 1 percent of the code that reveals information about their physical or psychological being. A growing list of federal and state privacy protections has made it clear that we have a right to privacy in our genetic makeup in the medical and employment contexts. But in the context of the criminal justice system a relatively new technology of DNA profiling appears to be operating according to a different and shifting set of rules as the reach of DNA data banks is expanded to ever more categories of innocent people.

Public anxiety about open access to genetic information is likely to continue to be exacerbated by the rapidly growing field of behavioral genetics. Two recent studies illustrate this point. One study linked a gene mutation in men to marital discord and infidelity. According to a behavioral geneticist at the Karolinska Institute, "Men with two copies of the allele [gene variant] had twice the risk of experiencing marital dysfunction, with a threat of divorce during the last year, compared to men carrying one or no copies."[53] A second study found a gene that the authors claim can predict voter turnout.[54] Whether or not these research results are replicated and validated, it is reasonably certain that people would not want such information open to anyone who can sequence their DNA from an abandoned paper cup.

Genetics has become the new prism through which science reveals personal identity, behavior, and medical prognosis. Undoubtedly, some of these claims will prove false or simplistic. Others may survive skepticism. But there is little doubt that genetic privacy will become an increasingly important component of personal privacy. Eventually the courts will have to grapple with the discordance that is emerging in our laws and policies and incorporate the public's expectation of genetic privacy within criminal justice under the umbrella of the Fourth Amendment, as Congress has begun to do with medical genetic information. Until then, the question whether we have any privacy at all in our DNA hangs in the balance.

Chapter 15

Racial Disparities in DNA Data Banking

One of the fears is that expanded genome profiling will lead to reification of the belief of the biological basis of race. In particular, the use of expanded genome profiling may lend credence to the opinion that criminal activity is associated with a particular genetic make-up prevalent in certain males and/or individuals.

—Susanne Haga[1]

It is well documented that the American criminal justice system is heavily racialized. By this we mean that racial disparities have been identified in all parts of the system, from arrest, trial, and access to legal services to conviction, sentencing, parole, execution, and exoneration. For example, in regard to prison demographics, New York University sociologist Troy Duster reported that "African Americans are currently incarcerated at a rate approximately seven times greater than that of Americans of European descent."[2] Duster's data show that the disparity in black versus white incarceration has grown significantly in recent years: in the 1930s blacks were incarcerated at a rate that was less than three times that of whites. Explanations that have been provided for this change range from musings on the decline of the moral character of African American males to a society rife with racial prejudice and economic injustice against people of color. Although vigorous controversy remains whether differential criminal involvement or differential criminal justice selection and processing are to blame for racial disparities in the system, most criminologists agree that racial discrimination cannot be dismissed.[3] Indeed, a large body of empirical data supports the notion that racial bias plays a key role in driving these disparities.

Will the technology of DNA analysis and DNA data banks exacerbate or improve the racialized criminal justice system? One response suggests that DNA science trumps racial prejudice. DNA testing provides us with a means to identify suspects that—even if not impervious to error—is more objective than, say, eyewitness identification. The claimed neutrality of forensic DNA technology implies that its introduction would shed light on racial disparities and pave the way toward a more just and equitable criminal justice system. Cases where DNA exclusions have exonerated minorities falsely convicted as a result of racial bias support this view. Another response, however, suggests that race trumps science. In this view science and technology must be understood within their broader social context. In a racially polarized society rife with racial disparities, it is a reasonable expectation that DNA testing will be used to reinforce the bias in criminal justice.

In reality the impacts of DNA technology on racial injustices are most likely a complex combination of each of these responses. This chapter examines the ways in which forensic DNA technology offsets, exacerbates, masks, or highlights racial disparities in criminal justice. Answers to the following set of questions will provide some clarity to these issues.

- What is the current racial composition of the Combined DNA Index System (CODIS), and how will it be affected by broadening the criteria of inclusion to include arrestees?
- Will disproportionately higher minority representation in CODIS result in racial stigmatization or impose relatively greater civil liberties transgressions on minority communities?
- Will uses, other than identification, be made of the forensic DNA databases, and if so, what implications will those uses have for minority populations?

We begin with a brief overview of the way in which racial disparities operate in the criminal justice system.

Racial Disparities in the Criminal Justice System

To understand the nature and persistence of racial disparities in the criminal justice system, one has to begin with the laws that classify and define crimes.

For example, for many years the federal system punished those convicted of crack cocaine offenses much more severely than those convicted of powder cocaine offenses, even though studies showed that the difference in the danger of the drugs is minor.[4] Crack is far more likely to be sold and used by blacks, while powder cocaine is more likely to be sold and used by whites.[5] From 1986 to 1990, the height of the Reagan administration's "War on Drugs," the average prison sentence for blacks compared with that for whites for drug offenses rose from 11 to 49 percent.[6] But the problem goes well beyond sentencing guidelines. The Justice Policy Institute, citing national survey statistics, reported that in 2002, 24 percent of crack cocaine users were African American and 72 percent were white or Hispanic, but more than 80 percent of defendants sentenced for crack cocaine offenses were African American.[7]

Part of the explanation of how these gross disparities arise can be found at the place where individuals tend to first come into contact with the criminal justice system—that of detention and arrest. Study after study has demonstrated that people of color are disproportionately arrested for drug offenses, automobile theft, and driving under the influence.

Data reported by sociologist Harry Levine on marijuana-possession arrests shed light on this phenomenon. From 1997 to 2006, on average, 100 people were arrested every day in New York City for marijuana possession. Each year, on average, New York City police arrested approximately 20,000 blacks for marijuana possession, compared with 5,000 whites. Adjusting for population, arrest rates for blacks have been approximately eight times as high as those for whites. This rate would be acceptable if it were true that blacks were eight times as likely as whites to use marijuana. But, in fact, U.S. national survey data have consistently shown that more whites use more marijuana than blacks.[8] In other words, "Since Whites use marijuana more than Blacks or Hispanics, and since there are more Whites than Blacks or Hispanics in New York City, on any given day significantly more Whites possess and use marijuana than either of the other two groups. But every day the New York Police Department arrests far more Blacks than Whites, and far more Hispanics than Whites, just for possessing marijuana."[9] This pattern is not true only for New York City, nor does it hold only for marijuana. A study of arrestees in Maryland found that while 28 percent of the population is comprised of African Americans, 68 percent of the arrests for drug abuse are African Americans.[10]

Some of the disparity in arrest rates might be explained by disparate policing practices—for example, where police focus their efforts on low-income or ethnic-minority neighborhoods.[11] Data from the Justice Mapping Center show that more than 50 percent of the men sent to prison from New York are from districts that represent only 17 percent of the adult male residents.[12] Pilar Ossorio and Troy Duster observed that the few DNA dragnets carried out in the United States have disproportionately targeted blacks. For example, San Diego police, in search of a serial killer in the early 1990s, identified and genetically profiled 750 African American men on the basis of eyewitness reports that the perpetrator was a black male.[13]

Some of the disparity in arrest rates might also be explained through racial profiling by individual officers. Michael Risher notes:

> Studies have shown that some mixture of unconscious racism, conscious racism, and the middle-ground use of criminal profiles leads law enforcement to focus its attention and authority on people of color. This can include everything from police officers disproportionately selecting people of color to approach, question and ask consent to search, to discriminatory enforcement of traffic laws, and detaining and arresting people of color without sufficient individualized suspicion.[14]

Police officers have significant discretion in making arrests, particularly when it comes to "victimless" crimes, such as drug use or "public order" crimes, and rarely face any consequences in cases where a person is improperly arrested without probable cause. In a study by Aleksandar Tomic and Jahn Hakes, the authors found that when police had broad discretion in making on-scene arrests, blacks were treated more aggressively than whites:

> Our model of racial bias in arrests predicts, and our empirical results suggest, that the policing of blacks has been disproportionately aggressive for crimes associated with high levels of police discretion to make on-scene arrests. . . . The resulting increases in the proportion of Blacks in the criminal justice system can affect the perception of officers in subsequent cases, perpetuating the imbalances even in the absence of racial differences in underlying criminality.[15]

A similar pattern of racial bias is seen in the makeup of the prison population. According to a 2007 report issued by the Justice Policy Institute, despite similar patterns of drug use, African Americans are far more likely to be incarcerated for drug offenses than whites. As of 2003, twice as many African Americans as whites were incarcerated for drug offenses in state prisons in the United States. African Americans made up 13 percent of the total U.S. population but accounted for 53 percent of sentenced drug offenders in state prisons in 2003. In 2002 African Americans were admitted to prison for drug offenses at 10 times the rate of whites in 198 large-population counties in the United States.[16]

According to data from the Innocence Project, an organization dedicated to the use of DNA to exonerate falsely convicted persons, racism is also a significant factor in wrongful convictions. African Americans make up 29 percent of people in prison for rape and 64 percent of those who were found to have been wrongfully convicted of rape. This suggests that there are disproportionately more false convictions for rape of blacks than of whites. Also, most sexual assaults nationwide are committed by men of the same race as the female victim. Only 12 percent of sexual assaults are cross-racial. This fact, in conjunction with data that show that two-thirds of African American men exonerated by DNA evidence were wrongfully convicted of raping white women,[17] reveals the subtext of racial bias in false conviction. The vestigial fears and prejudice held by whites of black-on-white rapes, whose roots go back to slavery, become expressed in our current society by trumped-up accusations, mistaken identification, and false convictions.

Racial Composition of CODIS

Because of the disproportionately high rates of arrest and incarceration of people of color,[18] it can be inferred that the U.S. Combined DNA Index System (CODIS), which, according to the FBI, is not coded by race, is also disproportionately composed of racial minorities. If CODIS, by and large, contains the profiles of past and present incarcerated felons, and minorities are disproportionately represented in prison, then it follows that they will be disproportionately represented in CODIS. Moreover, if Tomic and Hakes's results are corroborated, the mere appearance of innocent blacks in DNA data banks, even if they are never charged with a crime, will impose a penumbra of bias by law-enforcement officials toward those individuals. If DNA databases are superimposed on traditional racial profiling and other forms of bias toward African Americans, the combination could lead to greater degrees of racial disparity in criminal justice. Troy Duster observes:

> Indeed, racial disparities penetrate the whole system and are suffused throughout it, all the way up to and through racial disparities in seeking the death penalty for the same crime. If the DNA database is primarily composed of those who have been touched by the criminal justice system, and that system has provided practices that routinely select more from one group, there will be an obvious skew or bias toward this group.[19]

In chapter 2 we showed that there has been a rapid expansion of DNA data banks in certain states where the criteria have evolved from violent offenders and felons to arrestees who have not been charged with or convicted of crimes. Let us assume that this policy continues beyond the dozen or so states that have lowered the threshold of inclusion in their forensic DNA data banks, and that "being arrested" becomes the standard within the criminal justice system for requiring a biological sample and a DNA profile. How will that affect the racial composition of CODIS?

To contextualize this question, we first examine the racial composition of those who are serving sentences in federal and state prisons. According to a report published by the Pew Charitable Trusts in 2008, 1 in every 15 black males aged 18 or over, compared with 1 in 106 white males of the same age group, is in prison or jail. The highest rate of incarceration is among young black men. One in 9 black men aged 20 to 34 is behind bars.[20] Although African Americans make up 12.8 percent of the U.S. population, they constitute between 41 and 49 percent of the prison population.[21] Sixty-two percent of all prisoners incarcerated in the United States are either African American or Latino, but those groups constitute only one-quarter of the nation's entire population.

There is no published information on the racial and/or ethnic composition of individuals who have DNA profiles contained in CODIS. The FBI does publish annual crime reports broken down into racial categories, but states may upload only names and DNA profiles to CODIS without personal information such as a racial or ethnic identifier. Each state retains the personal information associated with the profiles placed in the national DNA forensic database, but as far as we can tell, that information is not aggregated.

It is reasonable to assume that the racial composition of CODIS mirrors the racial composition of the prison population. The racial composition of CODIS can also be estimated indirectly. Henry Greely and colleagues use a crude measure of felony convictions to estimate the number of African Americans in CODIS, since the vast majority of states currently retain and upload to CODIS DNA from all felons:

> African-Americans constitute about thirteen percent of the U.S. population, or about thirty-eight million people. In an average year, over forty percent of people convicted of felonies in the United States are African-American. . . . We assume, based on the felony conviction statistics, that African-Americans make up at least forty percent of the CODIS Offender Index.[22]

If we assume that the racial composition of CODIS mirrors the current prison population, we arrive at a similar estimate, with African Americans constituting 41 to 49 percent of CODIS.

According to the FBI, the National DNA Index System (NDIS) held 8,201,707 "offender" DNA profiles as of April 2010.[23] On the presumption of 40 percent African American entries, they would constitute 3,280,683 entries in CODIS. This means that approximately 8.6 percent of the entire African American population is currently in the database, compared with only 2 percent of the white population.

Suppose that we asked this question: how would the composition of the database change if all states collected DNA profiles of arrestees? If blacks in American society continue to be stopped, searched, arrested, and charged at a rate in excess of nonminorities, and if collecting DNA samples from arrestees becomes the norm, then it would seem that the racial disparity of blacks in CODIS would continue to grow. D. H. Kaye and Michael Smith make the case for strong racial skewing of U.S. DNA data banks as the system currently exists:

> There can be no doubt that any database of DNA profiles will be dramatically skewed by race if the sampling and typing of DNA becomes a routine consequence of criminal conviction. Without seismic changes in Americans'

> behavior or in the criminal justice system, nearly 30% of black males, but less than 5% of white males will be imprisoned on a felony conviction at some point in their lives. . . . A black American is five times more likely to be in jail than is a white.[24]

The authors estimate that the profiles of about 90 percent of urban black males would be found in DNA data banks if all states required DNA samples from arrestees. At the same time, they contend that racial disparities would be diminished by expanding the databases to include arrestees because many more whites would be brought into the databases: "Racial skewing of the DNA databases will be reduced somewhat if the legal authority to sample and type offenders' DNA continues to expand and come to include the multitudes convicted of lesser, but more numerous, felonies and misdemeanors."[25]

We can use current FBI arrestee statistics to evaluate the claims about racial disparities in expanding CODIS. According to the 2007 FBI data on national arrests, there were 9,014,180 individuals 18 years or older arrested in the United States; 70.2 percent (6,327,954) were white and 27.7 percent (2,496,928) were black.[26] If we use these figures, adding all the arrestees to CODIS would tend to bring down its racial composition from an estimated 40 percent blacks to slightly less because we are diluting what is believed to be a higher relative percentage of blacks with a lower relative percentage. Because DNA profiles might be taken only of arrestees who commit violent crimes, let us consider those figures. The FBI classifies violent crimes as murder or nonnegligent manslaughter, forcible rape, robbery, and aggravated assault. In 2007, 376,745 individuals were arrested for violent crimes as defined by these four categories, of whom 63 percent were white and 36 percent were black. If we added these arrestee profiles to CODIS, the percentage of blacks would be reduced slightly from 40 percent. In other words, adding arrestees who had committed violent crimes to CODIS would increase the racial composition of blacks only if their current composition in CODIS were less than 36 percent.

Of the four categories of violent crimes, only two (murder/manslaughter and robbery) have a higher percentage of black arrestees than whites (50 versus 48 and 53 versus 46, respectively). The racial disparity of blacks in CODIS would increase only slightly if arrestees charged with murder/manslaughter and robbery felonies (and then released) were the only violent perpetrators added to the database because of the relatively small numbers of these crimes.

However, we should not be concerned only with the relative proportion of whites and blacks in the database; we should also consider the proportion of the African American population that is in the database compared with the proportion of the white population. As stated earlier, currently 8.5 percent of African Americans are in the database. If we assume that these are predominantly males (approximately 80 percent of individuals arrested for violent crimes are male, for example), then we are approaching a situation where as many as 14 percent of African American males are in the database, even before we have started to upload arrestees to the system. Of the approximately 2.5 million arrests of blacks that occur each year,[27] a large proportion of those arrests will not result in conviction. The U.S. Department of Justice reports that 47 percent of the more than 140,000 individuals arrested for a federal felony offense in 2004 were not convicted.[28] Similarly, more than 30 percent of the hundreds of thousands of individuals arrested on suspicion of a felony each year in California are never convicted of a crime.[29] Conviction rates for lower-level crimes are even lower; a study in California revealed that 64 percent of drug arrests of whites and 92 percent of those of blacks were not sustainable.[30] Assuming a 50 percent conviction rate for all arrests, this means that of the 2.5 million blacks arrested each year, half will be added to the database who would not have been added under a system that requires conviction for inclusion in CODIS. In just the first year the percentage of the African American population represented in CODIS, from the addition only of unconvicted blacks, would rise to 11.8 percent of the African American population, or close to 19 percent of the male African American population. In comparison, still only 3 percent of whites would be in the database (or as many as 5 percent of white males, under the same assumptions). Predicting how this would play out over time is difficult because one would have to account for repeat arrests or, alternatively, have an estimate for one's lifetime risk of being arrested. Such estimates have been reported as 80 to 90 percent for urban black men.[31] If we layer on top of this the fact that blacks are inappropriately arrested at higher rates than whites, then blacks who are innocent under the law will also be disproportionately represented in the database. This may be especially true where data banks are expanded to arrests for nonviolent ("victimless") crimes, where police have significant discretion in making arrests. As stated earlier, a 1993 California study revealed that while 64 percent of drug arrests of whites were not sustainable, a full 92 percent of the black men arrested on drug charges were subsequently released for lack of evidence or inadmissible evidence. There are nearly 2 million drug-abuse arrests annually.[32] Furthermore, if the use of controlled substances as a percentage of population is equal to or greater in the white community than in the black community while the arrests are disproportionately higher for the latter, then not only are blacks treated unequally by law-enforcement officials but their DNA makes them and their family objects of continued genetic surveillance.

Because blacks are represented in CODIS and in prisons at a much higher percentage than their relative composition in the general population and at a higher percentage relative to nonminorities, if we expand the use of forensic DNA technology to the existing disparity, it will produce downstream effects that will further exacerbate racial prejudice. This is illustrated by the use of familial searching, as in the case when a close, but not exact, match is made between a crime-scene profile and an individual who has a DNA profile in CODIS. When arrestees who have not been convicted of a crime are added, their families are also brought under the surveillance lens of criminal justice. This conclusion was reached by Greely and colleagues:

> Assume that, either using the current CODIS markers or an expansion to roughly twice as many markers, partial matches of crime scene DNA samples to the CODIS Offender Index could generate useful leads from an offender's first degree relatives—parents, siblings and children. . . . Using some additional sampling assumptions, the percentage of African-Americans who might be identified as suspects through this method would be roughly four to five times as high as the corresponding percentage of U.S. Caucasians. . . . More than four times as much of the African-American population as the U.S. Caucasian population would be "under surveillance" as a result of family forensic DNA and the vast majority of those people would be relatives of offenders, not offenders themselves. . . . African-Americans are disproportionately harmed by crimes committed by other African-Americans.[33]

In the United Kingdom, England and Wales have the largest per capita forensic DNA data bank in the world. At the end of March 2009 there were 4.8 million DNA profiles in the database, representing about 7 percent of the population in England and Wales, compared with the national DNA data bank in the United States, which contains 2.6 percent of its population. Recent data reveal that more than one-third of black men in the United Kingdom have DNA profiles in the national DNA data bank, which includes three out of four black males between the ages of 15 and 34.[34] If the United States were to reach that percentage, CODIS would have 13 million DNA profiles of African American males.

Rounding Up the Usual Suspects

Human DNA is shed continuously in all the environments in which people interact. Our saliva is left on paper cups we discard in a coffee shop or empty soda cans we leave behind at a park. We leave hair fragments in bathrooms of restaurants or on shirts we bring to the cleaners. Swabbing for DNA at a crime scene has become almost as common as screening for fingerprints. Newer techniques allow investigators to retrieve trace DNA samples from touched objects. Obtaining profiles of DNA left at a crime scene may sometimes be useful to investigators, but it also may be irrelevant without other forms of evidence. With massive DNA data banks, every DNA sample found at a crime scene that matches someone in the data bank will become grounds for investigation—for no other reason than that the evidence shows that either the person or his or her discarded objects were once at the place of interest.

Let us suppose that there is a robbery and shooting at a Starbucks coffee shop. An employee at the coffee shop believes that the hooded robber had a cup of coffee before the robbery. Police obtain all the discarded cups for DNA samples on the premises. Once the profiling of the DNA is completed, investigators submit the profiles to the national and state DNA data banks, which by 10 years from today may have more profiles of innocent people, including juveniles, than of convicted felons.

If African Americans are disproportionately overrepresented in CODIS, not because of prior convictions, but because blacks are stopped and arrested more frequently than whites, then they are more likely to become suspects in any crime that involves a DNA sweep (see figure 15.1). An old adage claims that crime-scene investigators will go where the evidence takes them. In this technological era we can say with some degree of confidence that crime-scene investigators will go where the DNA takes them. The greater the dis-

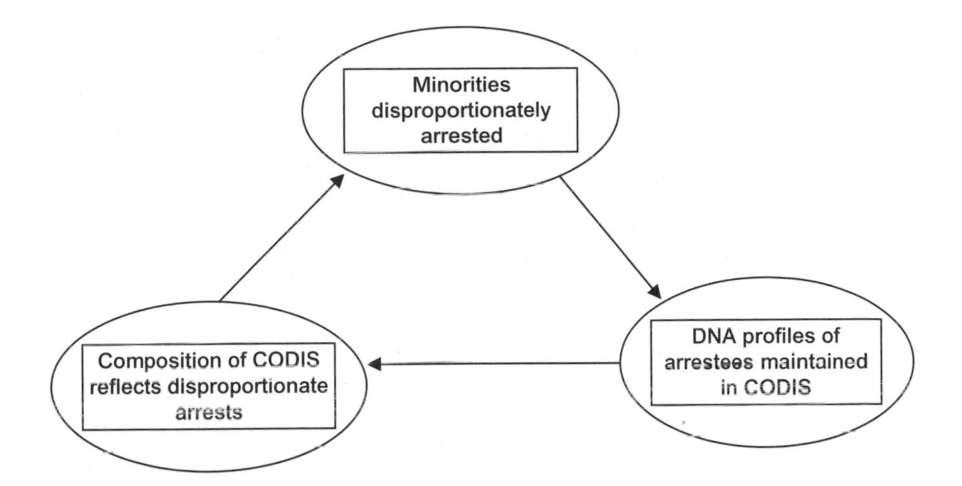

FIGURE 15.1. The cycle of racial disparity in DNA data banks. This flowchart depicts the increasing and compounding racialization of CODIS under a system of expanded criteria for collection and retention of DNA profiles in state and federal databases. Minorities, more likely to be arrested, are disproportionately represented on the database. Racial disparities in the criminal justice system are then further exacerbated, since, once in the database, these individuals are more likely to be arrested again in the future. *Source*: Authors.

parity in the percentage of African American DNA profiles in CODIS, the greater will be the disparity in suspicion of blacks, false arrests, and false convictions.

In the hypothetical Starbucks shooting, perhaps there were a dozen cups in the trash with sufficient cells for DNA analysis. If only two profiles showed up in CODIS that matched the DNA profiles on the cups, those would be the people the police would investigate, interrogate, and follow around for the reason that one pursues the leads one has even though what the police have became potential evidence because the database used to match "abandoned" DNA to profiles in CODIS was weighted disproportionately toward minorities. Given the racial composition of the data bank it would be no surprise to learn that both of the profiles were from black males.

One can argue that if everyone's DNA were in CODIS, at least all the cup users would have an equal chance of becoming a suspect, not just the ones whose DNA can be found in CODIS. Similarly, if all immigrants or illegal aliens were profiled for CODIS, members of those groups would become criminal suspects disproportionately to their ethnic representation in the population.

Another implication of racially imbalanced DNA data banks is stigmatization. It is reasonable to assume that police are likely to be biased toward individuals who have had a prior run-in with the law. To use the language of Bayesian statisticians, such information about the past raises one's prior probability that the individual is guilty. In the future the technology for checking a DNA profile of an individual will assuredly become miniaturized and faster. The digitized profiles can be uploaded to a mainframe computer and accessed with portable devices by police while on patrol.[35] If the profile shows up in the database, it may bias the officer's prior expectation of guilt or innocence. At the very least, an individual may be subject to intensive questioning and treated as a suspect with very little other evidence. As long as the playing field of DNA data banking is not level, where minorities and people of color are overrepresented, for those whose profiles appear in the database, there is a strong probability of stigmatization and the prejudicial use of the information. We should remind ourselves that these consequences will fall on many innocent people, perhaps because they volunteered their DNA in a dragnet, they are part of a minority group that was disproportionately arrested, charged with, or convicted of crimes that they did not commit, or because they entered the country as undocumented immigrants.

Forensic Data Banks in Scientific and Medical Research

Many states have passed laws that give researchers access to DNA information in state data banks. For the time being, scientists cannot learn very much from having access to 13 loci, even if they also have the individual's phenotypic information, that is, medical records. The loci used in DNA profiling were chosen specifically because they are not correlated with physical attributes of individuals. Nonetheless, at least one of these markers has proved to be closely related to the gene that codes for insulin, which itself relates to diabetes,[36] and it is possible that other connections of this sort will occur in time. That said, in order for scientists to acquire extensive knowledge about individuals from the DNA data banks, they will need the entire genome (the biological sample) or at least a portion of it that contains the DNA (functional or coding DNA) that reveals a detectable physical trait. It would also help the research if the scientists had social and medical histories of the individuals.

What type of research is most likely to interest scientists? The prison population consists disproportionately of people of lower socioeconomic status, with poor education, often from broken homes or dysfunctional families, who have committed violent acts, such as child abuse or rape. Historically, researchers have shown an interest in the use of prison populations to study the genetics of aggression, intelligence, and pedophilia. Does someone in prison, who has by law been required to give a biological specimen to a police unit, have any right over potential research uses of his or her DNA profile and/or the original biological sample? Do prisoners retain the right of "informed consent," a fundamental principle of the "common rule" (federal policy regarding human subjects protection), as subjects of research?[37]

Prisoners certainly cannot be forced to serve as human subjects for drug testing. For over 25 years it has been the policy of the American Civil Liberties Union (ACLU) that no therapeutic experiments be conducted on prison populations or any populations of incarcerated individuals.[38] The principle underlying this policy is that people who are in prison or detained cannot exercise their informed consent as free and autonomous human beings.

The issue of informed consent of prisoners for the use of DNA and/or medical information has not been resolved in U.S. policy. In February 2006 the Environmental Protection Agency (EPA) issued its ethical guidelines on human experiments to test pesticides.[39] That policy kept open the possibility that the agency would accept human toxicological data obtained by exposing prisoners to pesticides—in conflict not only with the ACLU's policy but also with well-established international codes of ethics.[40]

Let us suppose that a group of researchers wishes to investigate the genetics of pedophilia. There is no reasonable means of acquiring a study population from pedophiles freely moving about society. Even though released sex offenders are known to society because they are required to register in most states, it is doubtful that this group of individuals would freely join a genetic study of pedophilia. If such individuals were willing to participate, researchers would need a DNA sample and a detailed social and medical history of the subject. Researchers would also be required to fulfill their ethical responsibility of informed consent under the "common rule."

Pedophiles in prison are a captured population whose DNA profiles and biological samples are already on file. It would be far more convenient for researchers to use prisoners for this study than to seek out a study population from individuals living freely, although under public surveillance, in society. But an even easier way to study this population would be to bypass any direct contact with the individuals and instead to go directly to the stored biological samples.

"Criminal" Genes

There is a long tradition in sociology and anthropology and more recently in sociobiology of seeking to discover a biological root cause of criminal behavior or sociopathology in people's blood, brain, or genes. Hereditarian views of behavior and personality saw a resurgence of interest in the post-Darwinian period up to the Second World War. For a time after the war greater emphasis was placed on environmental and social influences on behavior. However, in the late 1960s there was a resurgence of interest in the hereditarian approach to violence. At the same time, new tools were developed for studying human genetics.[41] In a notable case prisoners were chosen to test a theory that an extra Y chromosome in males (XYY males) was a factor in explaining violent behavior.[42] The study was criticized for its methodological flaws and was eventually terminated. More recently scientists have found a region of the chromosome where there are variants of a gene that regulates the production of the enzyme monoamine oxidase A (MAOA), which has been proposed as a possible mechanism for a genetic theory of violence (see the "Behavioral Genetics and Profiling" section in chapter 5). In this theory a variant of a gene either overexpresses or underexpresses a chemical that affects a region of the brain.[43] More recently a study found that individuals with the gene that results in low MAOA activity were twice as likely to join a gang as those with the high-activity form.[44]

A persistent interest in the biological—and, more specifically, genetic—underpinnings of human behavior has made forensic DNA data banks a valuable resource for researchers. The data banks could allow those who seek to find genetic explanations for violent crime (not white-collar crime) a means to pursue this research that avoids the ethical and methodological flaws associated with focusing directly on prisoners. Elisa Pieri and Mairi Levitt discuss the return of behavioral genetics as an explanation for criminal activity and the role of DNA data banking in the United Kingdom to support such research:

> Children as young as ten that are arrested can now be DNA swabbed and entered (for life) on the National DNA database, even if they are never charged and subsequently acquitted. This database, already used for research into ethnic profiling, may potentially provide the DNA data for future behavioural research into criminality, violence or aggressiveness.[45]

A report of the Select Committee on Home Affairs of the United Kingdom's Parliament describes the growing racial disparity of the national DNA data bank (see box 15.1). The availability of a prescreened prisoner database for studying the genetics of human aggression would be highly desirable. However, there are serious methodological as well as ethical concerns with using a forensic data bank for these purposes. DNA samples from convicted felons are not a randomly selected sample. Troy Duster has warned that because African Americans are disproportionately represented in the criminal justice system, any behavioral genetics research that relies on this subset of DNA samples will inevitably be skewed toward that population. Genetic markers may very well be found that are more prevalent among this population than another, but whether these markers explain the causes of criminal behavior is another matter. Nonetheless, we are likely to see what Duster refers to as "the inevitable search for genetic markers and seductive ease into genetic explanations of crime."[46]

BOX 15.1 DNA of Blacks Is Stored Disproportionately in the British Data Bank

Currently, DNA samples can be taken by the police from anyone arrested and detained in police custody in connection with a recordable offence. This includes most offences other than traffic offences. A U.K. parliamentary committee examining the racial disparities in the criminal justice system reported:

> Baroness Scotland confirmed that three-quarters of the young black male population will soon be on the DNA database. Although the Home

(*continued*)

Office has argued in the past that "persons who do not go on to commit an offence have no reason to fear the retention of this information," we are concerned about the implications of the presence of so many black young men on the database. It appears that we are moving unwittingly towards a situation where the majority of the black population will have their data stored on the DNA database. A larger proportion of innocent young black people will be held on the database than for other ethnicities given the small number of arrests which lead to convictions and the high arrest rate of young black people relative to young people of other ethnicities. The implications of this development must be explored openly by the Government. It means that young black people who have committed no crime are far more likely to be on the database than young white people. It also means that young white criminals who have never been arrested are more likely to get away with crimes because they are not on the database. It is hard to see how either outcome can be justified on grounds of equity or of public confidence in the criminal justice system.

Source: Select Committee on Home Affairs, Parliament, United Kingdom, "Nature and Extent of Young Black People's Overrepresentation," in "Second Report," June 15, 2007, http://www.publications.parliament.uk/pa/cm200607/cmselect/cmhaff/181/18105.htm (accessed May 24, 2010).

DNA Profiling for Racial and Ethnic Identification

Crime-scene DNA can now be analyzed by racial-profiling methods. At least, this is the claim. Law-enforcement investigators have an interest in using DNA analysis to develop the physical profile of the perpetrator of a crime from DNA left at the crime scene. To accomplish this, there has to be a correlation that can be made between genetic sequences (genotype) and physical characteristics (phenotype) of an individual. When eyewitnesses provide a profile of a crime suspect, they usually refer to race, height, hair color, weight, or unusual marks on the individual, such as moles or tattoos. What kind of profile of an individual can one's DNA provide?

On the basis of the discovery that several broad racial/ethnic population groups have common and distinguishable clusters of DNA sequences, ancestral DNA analysis has become a new method for reifying racial distinctions. Because it was established that DNA variation within ancestral groups was greater than that between groups, there was reason to believe that there would not be a correspondence between genotype and racial/ethnic self-definition.[47] But recent studies have shown that there are unique genetic clusters that show a high correspondence with racial/ethnic ancestry. Hua Tang and colleagues analyzed a large multiethnic population-based sample of individuals who participated in a study of the genetics of hypertension. The subjects identified themselves as belonging to one of four major racial/ethnic groups (white, African American, East Asian, and Hispanic). The investigators used 326 genetic markers and concluded: "Genetic cluster analyses of the microsatellite markers produced four major clusters, which showed near-perfect correspondence with the four self-reported race/ethnicity categories."[48] The data showed a strong association between a select group of microsatellites and geographical ancestry. But the authors cautioned that "African Americans have a continuous range of European ancestry that would not be detected by cluster analysis but could strongly confound genetic case-control studies."[49]

Classification by racial/ethnic geographical origins appears to be catching on as a popular trend for identifying ancestry. There is also evidence that police investigators have sent crime-scene samples to companies that provide ancestry analysis in order to gain some phenotypic information about crime-scene DNA. One of these companies was the now-defunct company DNAPrint Genomics.[50] It advertised its product, called DNA Witness, as capable of determining race proportions from crime-scene DNA. A company advertisement claimed:

> The new test provides important Forensic Anthropological information relevant for a wide variety of investigative situations. When biological evidence is gathered, an investigative team can use *DNA Witness 2.0* to construct a partial physical profile from the DNA and in many cases learn details about the donor's appearance, essentially permitting a partial reconstruction of their driver's license photo.[51]

But given the admixtures in European and African DNA, the results could be quite misleading and result in mistaken profiling. Probabilistic profiling of crime-scene DNA can lead to harassment of innocent individuals who happen to self-identify with a race or ethnic group. Some experts have argued that because there is no definition of race in genetic terms,[52] genetic analysis of crime-scene data as a surrogate of a racial phenotype has no basis in science and should not be used in criminal investigations.

In his book *Molecular Photofitting: Predicting Ancestry and Phenotype Using DNA*, author Tony Frudakis, a principal in DNAPrint Genomics, cited as the goal of his research (and that of his company) "to establish a method for objectively interpreting an ancestry admixture result in order to safely use the indirect method of anthropocentric trait value inference" such that "knowledge of [genetics of] ancestry can impart information about certain aspects of physical appearance which is what we are after if we are attempting to characterize DNA found at a crime scene."[53] Frudakis's working definition of "molecular photofitting" is "methods to produce forensically (or biomedically) useful predictions of physical features or phenotypes from an analysis of DNA variations."[54]

At first glance, this is an idea backed by extraordinary hubris, namely, that we can translate the genetic code into physical appearance. This has been referred to as "DNA reverse engineering" or "DNA photofit."[55] Those who have attended their thirtieth-anniversary high-school or college reunion understand that there are some individuals whom we cannot currently identify from the picture that was taken at their graduation. Nevertheless, although the facial characteristics, hair color, or body shape of an individual might have changed radically over several decades, there may be some physical characteristics that remain invariant, such as eye color, skin tone, or hair type. If there are invariant phenotypic characteristics, can they be strongly correlated with DNA alleles? That was the project undertaken by Frudakis in DNAPrint Genomics. He argues that some phenotypes are highly heritable, such as skin color. He uses ancestry data where alleles are selected that have been invariant with respect to certain geographically based populations.

Admixture defines the percentage of the selected alleles in an individual's genome. When there is a strong correlation between a phenotype and the alleles that define ancestry, Frudakis maintains that a prediction of the phenotype from the percentage of the alleles can be made. One trait that lends itself to this analysis is pigmentation. He claims that "pigmentation traits are under the control of a relatively small collection of highly penetrant gene variants."[56] Thus, when an individual can be characterized by greater than 75 percent West African admixture, it can be inferred that the person has a darker skin shade than someone with less than 35 percent West African ancestry. He developed the melanin (M) index, which is a measure of the skin's melanin content. Frudakis writes: "With individual admixture estimates and M values it was then possible to search for correlation between admixture and pigmentation among individuals."[57] He claims a "significant correlation between constitutive pigmentation and individual ancestry" in a group of admixed samples.[58] If these results are validated in larger databases, then one should be able to predict melanin content (and therefore skin pigmentation, an observable trait) from the percentage of a person's African admixture (a genetic composition of alleles).

What, if anything, would be problematic about developing a set of statistical correlates between a selection of alleles and a physical trait? When the DNA profile of biological material found at the scene of a crime has no match in the database, criminal investigators could benefit from having some clues in the genetic code about the physical traits (red hair, green eyes, light skin) of the person who left his or her DNA at the scene. We have already noted that the rarer a genetic mutation is, the more helpful it might be in criminal investigation.

DNA Witness was not able to garner general scientific support for its methods. Anthropologist Duana Fullwiley wrote: "DNA Witness falls short of legal and scientific standards for trial and admissibility, while it eludes certain legal logics with regard to the use of racial categories in interpreting DNA."[59]

CODIS was established on the premise that its value was in the comparison of two sources of DNA and therefore in the concept of "identity." The alleles used in the STRs for DNA profiles have no significance for a person's physical characteristics. At best, even with the full DNA of the biological sample, police will have a probability estimate that the DNA found at the crime scene came from a person with red hair and green eyes, or that a DNA sample consisting of 80 percent African and 20 percent European admixture probably came from an individual with a melanin index (a proxy measure of skin color) of 35–50.[60]

Those who transition from haplotype ancestry maps of geographically isolated populations to techniques for drawing human physical properties from a DNA sample are making many scientifically contested assumptions. They use a process called Ancestry Informative Marker (AIM) technology. According to Fullwiley,

> AIMs-based technologies, like DNA Witness, are attempts to model human history from a specifically American perspective to *infer* present-day humans' continental origins. Such inferences are based on the extent to which any subject or sample shares a panel of alleles (or variants of alleles) that code for genomic function, such as malaria resistance, UV [ultraviolet] protection, lactose digestion, skin pigmentation, etc. There is a range of such traits that are conserved in, and shared between, different peoples and populations around the globe for evolutionary, adaptive, migratory, and cultural reasons. To assume that people who share, or rather co-possess, these traits can necessarily be "diagnosed" with a specific source ancestry is misleading.[61]

Without a cold hit in CODIS the police must resort to traditional policing techniques, which certainly would not include rounding up all red-headed persons with green eyes. But if familial searches are used in CODIS to troll for suspects (see chapter 4), then these gene-to-phenotype correlations may, in the minds of criminal investigators, play a role in narrowing the search. Suppose that a familial search yielded 20 current or past prisoners or, in some states, arrestees. If all the partially matched profiles had additional genetic information accessible to police, then the police could troll through the 20 saved biological samples for a person with red hair and green eyes who was a partial match in the familial search. If the technique described by Frudakis shows increasing promise, there will be pressure to maintain biological samples and to transform CODIS from a DNA profile identity database to a data bank for building physical profiles of individuals from their DNA. Fullwiley believes that these techniques are substantiated neither in ancestry testing nor in forensic phenotyping of DNA: "As a forensics market version of the AIMs technology, DNA Witness may offer precise mathematical ancestry percentages, but the accuracy of that precision remains debatable."[62]

Because the national forensic DNA databases have a disproportionately high percentage of African Americans, the traits most common to this group will be sought out in developing profiles of suspects. Moreover, there will be pressure to use the sample population of African Americans in CODIS to refine the genotype-to-phenotype correlations. Although the developers of "molecular photofitting" agree that it is not foolproof, they believe that "there is information about M [pigmentation] we gain by knowing an individual's genomic ancestry and as long as it is communicated responsibly, this information would be more useful for a forensic investigator as a human-assessed measure of skin color would be."[63] In other words, Frudakis claims that this method is more reliable than eyewitness testimony of skin color while also providing data with a very poor correlation between pigmentation and ancestry.[64]

The forensic DNA data banks in the United States and the United Kingdom contain a disproportionately high number of people of color because of the demographics of crime and criminal arrests. Even if the purpose of DNA data banks remains true to its original mission of matching the identity of samples, the significant racial skewing of the database, combined with a well-established systemic racial bias in the overall system, means that people of color who commit crimes are more likely to be identified in a DNA match, while white individuals will be more likely to escape detection. The expansion of databases to arrestees will worsen this situation, since decisions concerning who is arrested are highly discretionary and therefore especially prone to bias. Since minorities are arrested disproportionately to their contribution to criminal activity, these communities will be placed under greater DNA surveillance and subjected to greater stigmatization.

In an effort to develop "profiles" of perpetrators from DNA left at a crime scene, the criminal justice community will inadvertently reify race as a scientific term, even though it has been widely discredited and undefined scientifically. From this, new consequences of abuse are likely to result. One of these consequences involves picking up the usual minority suspect from weak probabilistic assignments of skin color from DNA through ancestry analysis. Skin pigmentation is a continuous spectrum from albino to dark brown and every shade in between, with no clear breaks. Race, however, is socially constructed as a two-value concept—black or nonblack (white). When DNA is used to determine pigmentation, the outcome, as interpreted by police investigators, will most likely be that the perpetrator was black or nonblack. If the science remains weak, this type of inference can exacerbate the racialization of criminal justice because society still operates under the one-drop rule. If the forensic investigator estimates that a DNA sample from an individual shows 90 percent European and 10 percent African ancestry, does the one-drop rule imply that he or she should be looking for a person of color? Or does the melanin index (with its poor predictive power) trump ancestry? Although the introduction of these techniques in the courtroom is unlikely, there are no restrictions on their use by police in generating suspects. That is where we can see the confluence of expanded data banks of innocent people, disproportionate numbers of minorities in the data banks, and the use of not-ready-for-prime-time science to invade the privacy of people's lives.

Chapter 16

Fallibility in DNA Identification

> It will not be easy to dispute the prevailing wisdom, fed by CSI-style media fantasies, that forensic science is virtually infallible. Yet, the intellectual weaknesses of many of the "forensic sciences" are now becoming increasingly apparent.
>
> —William Tobin and William Thompson[1]

It can justifiably be argued that, to date, the greatest contribution to civilization arising directly from the genetics revolution has been the exoneration of hundreds of falsely convicted individuals who have logged thousands of years in prison, some on death row. The failure to match the DNA of the alleged perpetrator with the DNA at the crime scene has, in most cases, trumped other physical and circumstantial evidence, including eyewitness testimony, that juries deemed "beyond a reasonable doubt" in deciding the guilt of an innocent individual. For many, DNA exculpatory evidence has come to mean "beyond a conceivable doubt."

Although DNA exonerations are hailed as one of the most important achievements of forensic science, the primary goal of the government in applying DNA to criminal justice has been one of finding guilt, not innocence. District attorneys have deployed DNA as indisputable evidence to argue that a suspect was the perpetrator of a rape or was at the scene of another type of violent crime. DNA evidence has sent thousands of individuals to prison, but it has only helped exonerate around 250 falsely convicted individuals.[2] However, in the process of vetting suspects and evaluating alibis, police also use DNA to exclude individuals in routine casework before they are drawn too deeply into the web of investigation and surveillance. Early exclusions are a win-win situation for criminal justice and those individuals who might otherwise have to experience protracted periods under the cloak of suspicion and with fear of prosecution.

When DNA evidence is used to support a hypothesis either of guilt or of innocence, and all other hypotheses that contradict the proffered one are either patently false, are highly improbable, or lack any substantial evidentiary support, DNA evidence takes on authoritative preeminence. It is under these conditions that DNA evidence is presented in the popular *Crime Scene Investigation (CSI)* vernacular as infallible. But in reality, ideal circumstances are rarely the way in which criminal cases get played out. DNA infallibility is a myth, even though DNA evidence can be highly authoritative and effective in identifying and prosecuting criminals. The following describes eight central myths of forensic DNA evidence. These myths have been identified from media accounts and in exaggerated claims made by individuals who view DNA as the ultimate and incontestable authority within forensic science.

Myth of DNA Consistency

DNA Is DNA Is DNA. All DNA Evidence Is Strong Evidence; There Is Not Much Difference in Quality from One Sample to the Next.

Although DNA is revered as the "gold standard" of forensic science, not all DNA is the same. William C. Thompson has pointed out that there is considerable case-to-case variation in the nature and quality of DNA evidence.[3] For example, it is not uncommon that biological evidence collected at the scene of a crime does not provide a full DNA profile when analyzed. Sometimes only a few cells are picked off a piece of clothing or an object. Also, in many cases DNA is analyzed or reanalyzed many years after a crime has been committed, and by that time it has further degraded. DNA, like any other chemical, will break down over time. The rate of degradation depends on the type of cells (saliva, blood, semen, skin), as well as the conditions under which the samples are stored. Improper storage or exposure to sunlight, moisture, bacteria, or other unfavorable conditions can accelerate degradation rates. Where significant degradation has occurred, the DNA analysis may not result in a full DNA profile.

Some DNA evidence is also less probative than other DNA evidence. For example, a DNA sample taken from a cigarette butt on the street where a crime was committed is less likely to have come from the perpetrator than DNA extracted from a vaginal swab of a rape victim.

DNA evidence presented in a case is only as good as the DNA found at the crime scene. When DNA evidence is compromised because the biological sample produces less than a full profile or because it may be unrelated to the crime in question, that should be acknowledged from the start. Partial DNA matches that occur as a result of degradation must be understood to carry far less weight than full matches where DNA is being used to establish that an individual was at the scene of a crime. Likewise, matches with DNA that might have been left by an innocent passerby should not carry the same relevancy as those where the DNA evidence is more likely to have come from the perpetrator.

Myth of Infallible Matches

There Are No False Positives in DNA Testing. If Two Samples of DNA Are Found to Match, Then the Samples Must Have Come from the Same Individual.

Although current scientific consensus supports the conclusion that, except possibly for identical twins, no two individuals have an identical genome, the conclusion of individuated genomes does not imply that a match of two DNA *profiles* indicates that the samples came from the same individual. False positives can and do occur in forensic DNA analysis. They can happen because of error, contamination, interpretation of the output of DNA analyzers, and chance profile matches that can be expected in a sufficiently large population.

We have explained in some detail in chapter 1 the stages involved in performing DNA analysis for identification. The reliability of the process depends on the quality of the DNA obtained at the crime scene; the care with which it is collected, labeled, and transported; the standards and quality-control procedures of laboratories performing the DNA profile analysis; and the interpretation of the DNA analyzer data, including whether a partial profile or a mixed profile is obtained. There are a number of opportunities where errors can occur in the collection, handling, and storage of DNA samples that can result in false positives and, therefore, constitute a risk of incriminating an innocent person.

Sample mix-ups and mislabeling in rape cases, where biological evidence that is being compared contains mixtures of the perpetrator's DNA with that of the victim, can lead to the incrimination of the wrong person. If the reference DNA samples are switched, then the DNA that is believed to have come from the suspect (but in actuality is that of the victim) will invariably be found to be included in a vaginal swab. In 2000 the Philadelphia City Crime Laboratory admitted to having accidentally switched the reference samples of the defendant and the victim in a rape case. As a result of the sample switch, the lab issued a report that stated that the defendant's profile was included in a mixed sample taken from vaginal swabs. The report also stated that the defendant was a potential contributor of what the analysts took to be "seminal stains" on the victim's clothing, which they later realized were in fact bloodstains from the victim.[4]

BOX 16.1 The Case of Lazaro Soto Lusson

In 2002 it was discovered that 26-year-old Lazaro Soto Lusson was mistakenly charged with multiple felonies because the Las Vegas Police crime lab switched the labels on two DNA samples. While in jail on an immigration hold, Lusson's cellmate, Joseph Coppola, accused him of rape. Police took DNA samples from both men to investigate the allegation. While they were conducting the analysis, they ran the samples against the state database and matched Lusson's mislabeled DNA to two unsolved sexual assaults. Lusson faced life in prison and was incarcerated for over a year before this mistake was discovered.

Source: Glenn Puit, "Wheels of Justice Turn Slowly," *Las Vegas Review-Journal*, July 6, 2002.

DNA samples can also be contaminated, either before or after collection, especially if they are not stored under proper conditions. Samples can be contaminated by the inadvertent transfer of trace amounts of DNA. Ironically, this error type is of increasing concern as DNA tests become more sensitive.[5] Lab analysts have cross-contaminated samples by not properly sterilizing lab equipment between cases or with their own DNA by not wearing the proper protective clothing while conducting DNA analyses.

Even trace amounts of outside DNA can complicate a DNA analysis. Where more than one source of DNA is present in a mixture, the results of the DNA analysis can be difficult to interpret. It can be difficult to tell how many individuals contributed to the source of the DNA sample, let alone which alleles are associated with each of those contributors.[6] The presence of one source of DNA can also mask another, and degradation might cause one source to go undetected. Any of these situations can lead to a false positive match.

BOX 16.2 The Case of Timothy Durham

In 1993 Timothy Durham was convicted of raping an 11-year-old girl and sentenced to 3,000 years in prison despite having produced 11 alibi witnesses who placed him in another state at the time of the crime. The prosecution's case rested almost entirely on a DNA test, which showed that Durham's genotype matched that of the semen donor. Postconviction DNA testing showed that Durham should have been excluded as a possible suspect, and reanalysis of the initial test showed that the misinterpretation arose from the difficulty of separating mixed samples. The lab had failed to separate completely the male and female DNA from the semen stain, and the combination of alleles from the two sources produced a genotype that could have included Durham's. Durham was released in 1997 after serving four years in prison.

Source: W. C. Thompson, F. Taroni, and C. G. G. Aitken, "How the Probability of a False Positive Affects the Value of DNA Evidence," *Journal of Forensic Sciences* 48, no. 1 (January 2003): 47–54, information at 50.

Sample cross-contamination appears to be a surprisingly common lab error. Thompson has found that these errors are chronic and occur even at the best-run DNA labs.[7] Under a guideline issued by the FBI, DNA laboratories are required to maintain corrective-action files to keep track of discrepancies that arise in casework. Many laboratories do not adhere to this guideline, but Thompson has reviewed corrective-action files for some labs where a file is maintained. A small laboratory in Bakersfield, California, for example, documented

> multiple instances in which blank control samples were positive for DNA, an instance in which a mother's reference sample was contaminated with DNA from her child, several instances in which samples were accidentally switched or mislabeled, an instance in which an analyst's DNA contaminated samples, an instance in which DNA extracted from two different samples was accidentally combined into the same tube, falsely creating a mixed sample, and an instance in which a suspect tested twice did not match himself.[8]

Thompson worries that this and other similar examples are only "the tip of the iceberg," especially since these represent only the errors that the lab itself has caught, corrected, and documented.

In 2004 the *Seattle Post-Intelligencer* reported that forensic scientists at the Washington State Patrol Laboratory had made mistakes while handling evidence in at least 23 major criminal cases over three years. Most of these mistakes involved contamination by DNA from unrelated cases, from the lab analysts themselves, or between evidence in the same case.[9]

Cross-contamination of DNA samples in laboratories has led to false cold hits in several cases. For example, in Washington State a cold hit turned up when DNA from a rape case was compared with the state database. However, the juvenile offender who appeared to be the source of the DNA would have been only 4 years old at the time the rape was committed. In realizing that this individual could not have been connected with the crime, the Washington State Patrol Laboratory came to understand that the juvenile's sample had been used as a training sample by another analyst when the rape case was being analyzed.[10]

Processing DNA samples requires that humans collect and handle biological samples, which are then subjected to laboratory techniques run by human technicians. DNA testing is only as reliable as are the people overseeing each of these processes, and infallibility simply cannot be achieved. Therefore, forensic scientists must depend on quality control, retesting, troubleshooting, and transparency of every decision made in the process to achieve reliable, trustworthy forensic evidence every time.

BOX 16.3 Gary Leiterman: Murderer or Victim of Cross-Contamination?

In March 1969 Jane Mixer, a 23-year-old University of Michigan law student, was murdered. No leads were generated for the case until 2002, when the Michigan State Police Crime Laboratory in Lansing processed the DNA evidence from the crime and found DNA from two men. A drop of blood from Mixer's hand was found to match an individual named John Ruelas, who was in the database because he had been convicted of killing his mother. DNA taken from the pantyhose of the victim was found to match Gary Leiterman, who was in the database for having previously been convicted of fraud involving prescription drugs. Ruelas, it turned out, was only 4 years old at the time of the murder. Police could not find any link between the child Ruelas and Mixer, and no explanation was provided why his DNA was found on the victim's hand 33 years after the crime was committed. At the same time, a review of the lab records revealed that DNA samples from both Ruelas and Leiterman were being processed for submission to CODIS in connection with other cases on the same day on which the old samples from the Mixer case were being analyzed. In addition, DNA from the victim was barely detectable on her pantyhose, indicating that significant degradation had occurred over the 33 years since the crime had been committed. However, Ruelas's and Leiterman's profiles did not show a similar level of degradation, indicating that they were unlikely to have been deposited at the same time as the victim's DNA.

When these facts are taken together, it seems plausible that the detection of DNA from both Ruelas and Leiterman could have been due to contamination of the Mixer crime-scene evidence that occurred in the laboratory while the DNA samples were being processed. Certainly this seems the only likely explanation for the presence of Ruelas's DNA. Nonetheless, Leiterman was convicted of Mixer's murder in 2005.

Sources: Amalie Nash and Art Aisner, "DNA Evidence Key in 1969 Slaying Trial," *Muskegon Chronicle*, July 22, 2005; William C. Thompson, "The Potential for Error in Forensic DNA Testing (and How That Complicates the Use of DNA Databases for Criminal Identification)" (paper produced for the Council for Responsible Genetics [CRG] and its national conference, "Forensic DNA Databanks and Race: Issues, Abuses and Action," New York University, June 19–20, 2008); Theodore Kessis, "Report of Findings, *People v. Gary Leiterman*, No. 04-2017-FC," http://www.garyisinnocent.org/web/CaseHistory/NewDNAFindings/tabid/58/Default.aspx (accessed April 28, 2010); *People of the State of Michigan, Plaintiff-Appellee v. Gary Earl Leiterman, Defendant-Appellant*, Court of Appeals of Michigan, no. 265821, decided July 24, 2007.

Myth of Objectivity

Two DNA Samples Either Match or They Do Not Match; DNA Analysis Is Not Subject to Interpretation.

Even if DNA samples are collected, handled, and processed with the greatest care and errors are minimized to the highest extent possible, the results of DNA analysis are still subject to interpretation. The subjectivity of DNA analysis is largely unacknowledged in popular accounts.

In the early development of DNA forensic analysis, the output came in the form of an autoradiogram, which had some resemblance to a supermarket bar code. The bars, which represented the appearance of an allele at a specific locus, showed up in different intensities. When bar segments of the autoradiogram were very light, some interpreters of the data might neglect the bar, believing that it was an imperfection or an artifact. By neglecting the faded bar, the forensic DNA specialist could conclude that there was an exact or partial match; in the latter case he might report that the individual's DNA profile was consistent with the profile of the crime-scene sample.

Advances in DNA testing methodology replaced autoradiograms with electropherograms that generally give a cleaner visualization of the alleles in a DNA profile. These graphs have peaks and numbers associated with each peak that identify the locus where the peak is found and give the number of STRs within an allele at the peak, and the height of the peak indicates the amount of DNA associated with the peak (see chapter 1). Although the new output from the DNA analyzers is a marked improvement over the output in autoradiograms, discretionary factors remain in the interpretation of the data.

First, there is no standard rule of thumb for how an analyst should report ambiguous results of DNA analyses. For example, where degradation has occurred, peak heights might be very low, and the profile might be considered incomplete. One analyst might decide that these measurements are spurious and unreliable and might report this result as "inconclusive," while another might report a partial profile. Partial matches may not provide sufficient information to support the conclusion that it is extremely unlikely that someone other than the suspect would have the identical partial match, but they may provide sufficient information to exclude a person from a crime. If even a subset of alleles is found not to match that of a suspect, he or she cannot be the source of that sample. Unfortunately, there have been cases where lab analysts have failed to report analysis information that might have led to a different outcome in a case.

BOX 16.4 The Case of Robin Lovitt

In September 1999 Robin Lovitt was convicted and sentenced to death for the murder of a pool-hall manager in Arlington, Virginia. His conviction rested heavily on DNA evidence. A bloodstain was found on a jacket that Lovitt was wearing when he was arrested, several days after the crime was committed. The lab analyst reported that the results of the analysis of the stain were "inconclusive," but the prosecutor argued that the stain came from the victim. In addition, bloodstains on the murder weapon matched the blood of the victim but also contained a single, additional allele that was shared by Lovitt. A DNA expert testified that this allele was shared by 19 percent of African Americans.

A closer look at the DNA evidence revealed that the lab analyst failed to report that the DNA analysis of the sample from Lovitt's jacket produced a partial profile of five loci. All five of those loci matched Lovitt's DNA. Therefore, there was no evidence that the blood on his jacket matched that of the victim, and if anything, the evidence that it matched Lovitt was exculpatory. In addition, the single extra allele that was found on the murder weapon was far more common in the population than was reported to the jury; according to the FBI's population data, that allele is found among 33 percent of African Americans, 46 percent of Caucasians, and 40 percent of Hispanics.

Lovitt's death sentence was reduced by Governor Mark Warner in 2005 to life imprisonment without the possibility of parole. The problems with the interpretation and presentation of the DNA evidence were never fully considered in any of the appeals.

Source: William C. Thompson and R. Dioso-Villa, "Turning a Blind Eye to Misleading Scientific Testimony: Failure of Procedural Safeguards in a Capital Case," *Albany Law Journal of Science and Technology* 18 (2008): 151–204.

Cases that involve mixtures of DNA from two or more sources provide the most opportunity for ambiguity. In the face of a mixture, a forensic analyst usually attempts to separate the alleles so that the profiles of each of the contributors can be determined. Even for experienced forensic analysts, however, there may be several ways to sort the alleles among two or more contributors. If all the contributors are available for DNA testing, the sorting process can be carried out with a reasonable degree of accuracy. Some forensic scientists believe that the sorting of the alleles must be done before one has information about the profiles of possible suspects. To do otherwise could bias the interpretation.[11] When the contributors to the DNA profile of interest are not all available or are in dispute, then certain assumptions can be made about the likelihood that the suspect's profile is included among the evidence. The forensic investigator may choose one hypothesis that fits the suspect's profile in the assemblage of alleles without disclosing to the jurors the other possible allele assortments that do not match the suspect. As Thompson and colleagues have noted:

> By their very nature mixtures are difficult to interpret. The number of contributors is often unclear. Although the presence of three or more alleles at any locus signals the presence of more than one contributor, it often is difficult to tell whether the sample originated from two, three, or even more individuals because the various contributors may share many alleles.[12]

Misinterpretation of mixtures has resulted in false cold hits and even wrongful convictions. In analyzing mixed samples it is critical that the person engaged in the DNA analysis not have an interest or a stake in the outcome of the case and not be seeking to find a match among the alleles with a preexisting suspect's DNA profile.

BOX 16.5 The Case of Josiah Sutton

In 2004 Josiah Sutton was exonerated after spending four and one-half years in prison for a rape he could not have committed. Sutton's conviction rested almost entirely on the basis of DNA tests performed by the Houston

(*continued*)

Police Crime Laboratory. The lab claimed that a semen stain found in the back of the car where the rape occurred contained two profiles—Sutton's and that of an unidentified man. In addition, the lab analyst testified that the DNA found in the sperm fraction of vaginal swabs and on the victim's jeans "matched" Sutton's. Reanalysis of the lab report showed that the semen sample came from a single source, and not from Sutton. In addition, the lab analyst exaggerated the significance associated with the inclusion of Sutton's DNA profile in the mixed evidentiary sample by not reporting any statistics and repeatedly testifying about the uniqueness of each DNA pattern. It turned out that the chance of a coincidental match in this case was quite high: the frequency in the African American population of men who would be "included" in the vaginal sperm fraction was 1 in 15. Exposure of the errors in Sutton's case led to a full-scale investigation of the Houston Crime Lab and the review of hundreds of cases involving DNA evidence.

Source: William C. Thompson, review of DNA evidence in *State of Texas v. Josiah Sutton* (District Court of Harris County, Case No. 800450), February 6, 2003.

Myth of Individuality

No Two People Can Have the Same DNA Profile. The Probability of a Coincidental Match Is Zero or Infinitesimally Small.

When the DNA profiles of two pieces of biological evidence "match," it is often presumed that they must have come from the same source. But could the match have been coincidental? Do the police have an innocent person whose DNA profile happens to match perfectly the profile of the DNA left at the crime scene? What are the chances that two or more people share the same DNA profile and that the matching profiles do not represent DNA from the same individual?

There is a generally recognized assertion that, except possibly for monozygotic (identical) twins, no two people can have identical sets of 3 billion base pairs of DNA.[13] In forensics, however, no person's complete DNA is sequenced. In the United States 13 loci, as well as markers on the X and Y chromosomes for gender determination, are selected for DNA analysis. The question about coincidental matches reduces to this: what are the chances that more than one person (such as sibling pairs) will have the exact number of short tandem repeats (STRs) in the 26 alleles of the 13 loci used to profile their DNA?

There are three important principles used in developing the probability statistic that prosecutors use in court. The first principle states that the 13 loci are independent and thus are not linked in the population. This is based on a testable assumption that the loci are assorted randomly. The second principle, derived from the first, is that the probability that any individual has a particular array of STRs is given by the product of the frequency with which each allele appears in the population. The third principle states that individuals who are identified with similar racial, ethnic, or ancestral groups have a greater likelihood of allelic similarity in their DNA profiles, including the number of STRs at a locus in the chromosome, than individuals associated with other population groups. When a DNA profile match is found, the population frequencies of the alleles are most commonly determined from one of three population reference groups, Caucasian, African American, or Hispanic, on the basis of the perpetrator's closest "racial" identity. The third principle allows forensic scientists to estimate the likelihood that two unrelated individuals have the same DNA profile for 13 loci or fewer.

Dan Krane describes a three-stage process that forensic laboratories use to determine the probability that the DNA taken from a random, unrelated individual in the population has the same profile as the evidence sample (the random-match probability, RMP).[14] In step 1 the frequency of each allele in the DNA profile of interest is estimated in the reference database. As an example, at a particular locus, allele 1 (7 STRs) appears at a frequency of 3 percent, and allele 2 (12 STRs) occurs at a frequency of 4 percent. In the second step the frequency of each genotype is calculated by the formula 2 times p times q, where p and q are the frequencies of the two alleles in the genotype. The multiplier 2 comes from the fact that each allele can come from either the mother or the father. In the example here, the frequency of the genotype with alleles of frequencies .03 and .04, respectively, is $2 \times .03 \times .04 = .0024$. The frequency of the overall genotype of 13 loci is obtained by multiplying the frequencies of each locus.

To highlight the principles of probability underlying forensic DNA, consider the following example. Suppose that we have an urn filled with 1,000 balls, some red and some blue. Now imagine that we have selected 10 balls and found that 7 were red and 3 were blue. Can we assume that our next pick of 10 balls would give us the same number of red and blue ones? Obviously not. For one thing, we do not know whether we took a random sample of the balls in the urn. We also do not know whether the balls are distributed homogeneously, or whether all of the red balls are stacked at the bottom of the urn. But if we repeatedly selected 10 balls, we could calculate the average number of blue and red picks. If there were in fact 700 red balls and 300 blue ones, then our ratios in the picks of 10 would cluster around a mean of 7 red and 3 blue, although we would not get that ratio in every selection.

How do scientists know what the allele frequencies are in the population? How do they know how many people have a particular allele at locus 2? There is no direct way because neither the government nor scientists have the DNA profile of everyone in the world, and therefore they cannot calculate the exact frequency of the STR alleles of interest that exist in the entire population. Instead, scientists use convenience databases rather than a random sample of the population. The databases from which forensic scientists draw allelic frequencies could be a few hundred people whose DNA happened to be on hand when allele frequencies first needed to be determined. Because the databases are not a random sample of the population, in theory the probability estimate could either overestimate or underestimate the frequency of the alleles and thus give a false value for the chances of a coincidental match. As in the urn example, even without a random sample of the population, if we kept taking samples of allele frequencies from the population, we would eventually get an average that approaches the real frequency of the allele in the entire population.

But, unlike the example of balls in the urn, where we may know the distribution of red and blue balls, we do not know what the exact allele frequencies are in the population. Forensic scientists infer the actual allele frequencies from the small sample of people who do not represent a random sample of the population. If we drew the allele frequencies from a completely different population, it is likely that we would get a different set of frequencies. As in the case of the urn, if we chose enough population samples, we would expect to obtain a distribution of allele frequencies whose average would approach the average of the entire population.

Some forensic scientists argue that a random selection of the population for the purpose of obtaining allele frequencies is not necessary because even small reference groups will have allele frequency distributions at specific loci that are similar to those of the larger population. The analogy for the urn is that with sufficient mixing we can select 10 balls from the top (not random, but a convenience sample) and get 7 red and 3 blue balls.

The racial or ethnic background of someone who is the source of an evidence sample is often unknown or in dispute. As a result, allele frequencies from the three common racial groups named earlier are commonly used to attach a weight to any matches to such a profile. It is plausible to assume that the greatest chance of a coincidental match of a DNA profile would come from individuals who have similar phenotypes and therefore come from the same "racial" lineage. When an exact match is found between a DNA profile obtained from a biological specimen left at the crime scene and a DNA profile obtained from a suspect, the forensic investigator then determines the likelihood that some randomly chosen unrelated person whose DNA was not left at the crime scene would exhibit an identical DNA profile (the RMP).

Even with frequencies for each STR sequence in the range of 1 in 10 (1 person in 10 has the same number of repeats), the product rule rapidly yields a very low probability [$(1/10)^n$, where n is the number of alleles]. The conventional wisdom within the forensic field is that the likelihood of a random match from 13 loci is inconceivable (perhaps one in a trillion) so long as the DNA is properly handled, no laboratory errors occur, and the individuals involved are not identical twins.

For example, if the frequency of each of the 26 alleles in a DNA sample as determined by the relevant population database is 1 out of 10, then the RMP that two nonidentical twins would have the identical profile is $2 \times (1/10)^{26}$, or 1 in 50 septillion (1 septillion $= 10^{24}$ or a trillion trillion). This point was made in *Discover Magazine*:

> Reports on DNA matches . . . include scientifically rigorous probabilities of the likelihood of finding the same DNA profile in a random, unrelated indi-

> vidual. The chances are typically far less than 1 in 10 billion for a full DNA profile from a single individual. It is that degree of improbability that forms the basis for the common perception that DNA testing is foolproof.[15]

Suppose that the crime-scene sample is profiled and compared with that of the suspect, and on the basis of the population database used by police, the RMP is determined to be 1 in 10 million. The prosecution tells the jury that the suspect is the likely offender because if we choose an unrelated person at random in the population, the chance that this individual would have the exact profile of that found at the crime scene is 1 in 10 million. But with 6 billion people on the earth, there could, on average, be 600 people with the identical profile. The defense can justifiably say that the suspect is 1 out of 600 people who could have the same DNA profile as that found at the crime scene.[16] An RMP of 1 in 10 million does not necessarily mean that you will find one and only one such profile in a population of 10 million. On average an event with a probability of 1 in 10 million occurs once in every 10 million trials, but in some instances it might occur more than once in 10 million trials.[17]

But what about people who are related or who are from a relatively isolated or highly inbred population? Could there be an exact match for a DNA profile of 13 loci of two individuals who are not identical twins? To date, no such case has been recorded. According to Dan Krane,

> The crux of the problem is simply that the RMP delivers pretty much what it says that it will (the chance that a randomly chosen, unrelated individual from a particular population has a perfectly matching DNA profile) and that it is completely silent on the chance that a close relative (or that one of a very large number of relatively close relatives) would have [identical] DNA profiles.[18]

Krane notes that the assumptions behind the product rule (random assortment of all alleles) do not apply for relatives of individuals. The chances for a coincidental match, then, even if small, are not zero.

Myth of the Infallibility of a Cold Hit

A Cold-Hit Match Made Against a Large Database Has the Same Weight as a Match Between a Person Suspected of a Crime and Evidence from a Crime Scene.

Generally there are two ways in which police seek to find DNA profile matches with crime-scene evidence. First, when they have a suspect, they obtain a biological sample from that individual and compare it with the profile derived from the crime-scene sample. If police get an exact match (all 26 alleles are identical), it usually comes with other evidence linking the suspect to the crime. Otherwise they would not have had reason to obtain the DNA profile of the suspect in the first place.

Second, when police have no suspect, they may compare the DNA profile from the crime scene with all the profiles that have been entered into a DNA offender database or a DNA database consisting of offenders, arrestees, and/or volunteers. This is a fishing expedition using computer technology to make comparisons between one DNA profile (from the crime scene) and the more than 8 million profiles that have been banked in the national Combined DNA Index System (CODIS).[19] If they get a match in this case, it is called a "cold hit" because they are operating blindly, without any evidence linking a suspect to the crime or any a priori suspicion.

Do both of these kinds of DNA profile matches—a match that occurs by comparing a known suspect's DNA with that of the crime scene and a match that occurs as a cold hit—merit the same statistical weight? Keith Devlin, a statistician at Stanford University, argues that a 13-locus match would be a definitive identification provided that "the match is arrived at by comparing a profile from a sample from the crime scene with a profile from a sample from a suspect who has been identified by means other than his or her DNA profile."[20] Devlin argues that the chance that the match is coincidental is higher, however, when a given sample is compared with many samples in a database. In cold-hit cases the investigation involves searching a database of hundreds of thousands or even millions of genetic profiles for a match. Each individual comparison increases the chance that a match will occur with an innocent person.

David Kaye uses the "birthday problem" in statistics to illustrate this point.[21] If you are in a room with a group of people and you choose one, then the chances that the two of you have the same birthday is 1 out of 365. But if we ask what the chances are that you have the same birthday as anyone in the room, that will depend on how many people are in the room. Moreover, if you asked what the probability is of one birthday match (not necessarily yours) in the room, the probability would even be greater because you are making pairwise comparisons with everyone in the room.

This example has been used to illustrate the point that RMPs can underestimate the chances of a coincidental match in a cold-hit case, where no other evidence but a DNA profile match is found. Even with their aggressive collection of DNA from citizenry, good practice guidelines adopted by the British police state clearly that because of chances of a coincidental match and other limitations of DNA evidence, individuals should not be convicted exclusively on DNA evidence (i.e., a cold match in a database).[22]

The National Academy of Sciences recognized that the method of determining the RMP from a suspect sample (where there is prior evidence of suspicion) should not be identical with that from a cold hit (where there is no prior evidence of suspicion). In the latter case the RMP should depend on the size of the database. The chance of finding a random match is greater with a very large database. The academy wrote: "If the only way that the person becomes a suspect is that his DNA profile turned up in a database, the calculations [of RMP] must be modified. . . . Multiply the match probability by the size of the database searched. This is the procedure we recommend."[23]

Although it is true that the larger the database, the greater are the chances of finding a match, including a random match, for crime-scene DNA, it is also true that finding a single match of a suspect in a large database improves the chances that the suspect was at the crime scene because it rules out all the other people in the database.

The reliability of the calculation of the RMP is dependent on the reliability of the independence of the genetic loci used in the calculation. But the independence principle remains an assumption or idealization. Devlin has argued for an empirical method of calculating RMPs that requires large data sets and not simply 200 to 400 data points. If we use the product model for calculating RMPs, we could validate it by comparing its results with the frequency of matches found in a large database.

One such test was run in 2005 on the Arizona convicted-offender database containing approximately 65,000 entries, which was analyzed for profile similarities. Approximately 1 in every 228 profiles in the database matched another profile in the database at 9 or more loci; approximately 1 in every 1,489 profiles matched at 10 loci; 1 in 16,374 profiles matched at 11 loci; and 1 in 32,747 matched at 12 loci (both were siblings). Devlin opined: "How big a population does it take to produce so many matches that appear to contradict so dramatically the astronomical, theoretical figures given by the naive application of the product rule?"[24] About 1 in 1,489 profiles matched at 10 loci. If we calculated the RMP based on STR frequencies of a very conservative 1 of 5, the theoretical answer would be 1 in 11 million, a much lower probability than was actually found in Arizona. On the basis of this empirical result Devlin concludes:

> It is not much of a leap to estimate that the FBI's national CODIS database of 3,000,000 entries will contain not just one but several pairs that match on all 13 loci, contrary (and how!) to the prediction made by proponents of the currently much touted RMP that you can expect a single match only when you have on the order of 15 quadrillion profiles.[25]

The debates among statisticians and forensic scientists on RMPs play out in the courtroom as well. The same information can be packaged and presented differently to a panel of jurors, one framing that supports the prosecution and another that supports the defense.

Let us suppose that the calculated frequency of an individual's 26 alleles is 1 in 6 billion. This means that when you multiply the frequencies of the individual alleles in the relevant population, the product of the frequencies yields a frequency of 1 in 6 billion. This could be presented to the jury as follows: "There is only one person in 6 billion with this DNA profile and that is our suspect, because there are only 6 billion people on the earth. If there were 12 billion we would have to conclude that there might be another person with the same DNA profile."

But "1 in 6 billion" is a theoretical calculation based on databases that have not been chosen randomly to determine allele frequencies. So there is still a chance that more than one person on the planet will have the same DNA profile. If we had a DNA profile for every living person on the planet, we could ascertain definitively whether more than one exact profile match occurs.

In cold-hit matches the profile is uploaded to a database where, let us assume, one match is found. There are two ways of thinking about the probability of this being a coincidental match. On the one hand, if the database is very small, we might think that this could be a coincidental match because we have not seen a large-enough population from which to judge the profile. On the other hand, since the database is small, the likelihood of getting a coincidental match should be small because it increases with the size of the comparison population (a world database would increase the chances of a coincidental match). Even though the theoretical calculation gives us an RMP of 1 in 6 billion, we know that the assumptions behind the calculation do not take account of close family relations; those have to be analyzed using kinship statistics. Thompson notes, "These estimates understate the probability of a coincidental match in actual cases because they take no account of the possibility that the pool of possible suspects contains the relatives of the perpetrator, who would be more likely to have the same profile due to common ancestry."[26]

Now suppose that the database from which police obtained the cold hit was very large. An actual DNA database of 6 million profiles that yielded one cold hit tells us that 5,999,999 people have been excluded from the crime-scene match. The larger the database, the more confidence we can have that our cold hit—with no other evidence—is not a false match because we are approaching the true population size. By imagining a database with 60 million people and one cold hit, we gain even more confidence, given that 59,999,999 people are excluded. But if we found 1 match in 60 million people, then there could be 100 matches in 6 billion people (1 match per 60 million, using a kind of inductive logic).

So another narrative that could be presented to the jury is that the chance of a coincidental match for a cold hit in a database of 60 million people is 1 out of 100. Telling a jury that there could be another 99 people on the planet with the same DNA profile presents a very different statistic that could change its psychology when it is trying to determine the grounds for "beyond a reasonable doubt."

Would higher probability statistics in cold-hit cases make a difference in their probative value or how juries relate to the evidence? Thus far, in cold-hit cases the courts have opted for the RMP estimates from forensic statisticians over the mathematical statisticians. Juries typically do not get to hear the controversy because it is often resolved before experts appear before the jury.

BOX 16.6 The Case of John Puckett

In 1972 a 22-year-old nurse was sexually assaulted and stabbed to death in San Francisco. More than 30 years later a swab that had been taken from the victim's mouth in 1972 containing a degraded sperm sample and at least one other person's DNA produced a partial DNA profile of 7 markers. When the profile was compared with California's DNA databases of 338,000 profiles, it matched with that of John Puckett. Puckett, then 70, denied ever knowing the victim, and there was virtually no other evidence linking him to the crime, aside from the fact that he lived in San Francisco in 1972 and had a previous rape conviction. During his trial the jury was provided a random-match probability of 1 in 1.1 million, based on population statistics. During pretrial hearings Bicka Barlow of the San Francisco Public Defender's Office argued that this figure did not take into account the size of the database. Following the NAS recommendation to multiply the RMP by the number of profiles in the database, she argued that the chances were in fact 1 in 3 that the database search had resulted in linking an innocent person to the crime. The judge did not allow this statistic to be presented to the jury. Puckett was convicted and sentenced to life in prison.

Source: Jason Felch and Maura Dolan, "DNA Matches Aren't Always a Lock," *Los Angeles Times*, May 4, 2008.

Increasingly, police are trolling their databases for partial matches of DNA profiles. This means that they might be interested in a cold hit with 20 out of 26 matched alleles. It is possible to generate a fairly high RMP with fewer than 13 loci that could sound convincing to a jury. In 1999 police in the United Kingdom found an exact match of 6 loci between the profile of crime-scene DNA from a burglary and a profile logged into the United Kingdom's national databank. The frequency of a random match was calculated by law enforcement to be 1 in 37 million, which is persuasive evidence in a country of 60 million people. When the suspect was arrested, it soon became obvious that the match was a coincidence because the man was disabled and was physically incapable of carrying out the crime. The coincidental match could have been corroborated by testing more alleles in the biological samples.[27]

According to Thompson, "The British Home Office has reported that between 2001 and 2006, 27.6 percent of the matches reported from searches of the U.K. National DNA Database (NDNAD) were to more than one person in the database,"[28] largely because police were uploading partial DNA samples where degradation of the crime-scene sample had taken place or because a number of individuals were entered into the database more than once. The current interest in familial DNA searching has resulted in greater interest among criminal investigators in partial matches. Although forensic scientists have made efforts to develop statistical models that predict the probability that a partial match of an individual implicates that person's family as the source of the DNA, the results have been highly problematic and contested (see chapter 4).[29]

Myth of Infallible Rape Evidence

If the DNA of a Suspected Rapist Is Found in the Vaginal Smear of the Victim, Then the Suspect Must Be the Rapist.

DNA testing has been responsible for a high conviction rate in crimes involving rape. It is widely assumed that if a suspect's DNA is found in a vaginal smear of the rape victim, then the suspect's guilt has been established beyond a reasonable doubt. There are two separate issues. First, does the DNA of the suspect in the vaginal smear prove beyond a reasonable doubt that the suspect had sexual intercourse with the victim? Second, does the DNA match prove that the suspect raped the victim?

The answer to the first question must most probably be in the affirmative. It seems extremely unlikely that a suspect's DNA could enter the vaginal canal without intercourse. The victim could surely set the record straight in such an event. That said, there was one case where a woman implanted a sperm sample in order to thwart law enforcement. In 1999 a convicted rapist named Anthony Turner smuggled a sample of his semen out of prison, concealed in a ketchup packet. Turner's family members paid the woman $50 to use the sperm to stage a phony rape as a way of casting doubt on the DNA evidence that placed him in prison.[30]

The second question asks whether a DNA match implies a rape. There are cases where a victim has had multiple sexual partners, one or more of whom may have been consensual, where a vaginal smear by itself may not reveal the actual rapist. This is where forensic investigators can use elimination samples in mixed DNA samples where there have been consensual partners. It is also possible that in violent crimes against women involving more than one man, one of the perpetrators did not penetrate the victim or did not ejaculate. Thus the nonappearance of sperm is not by itself conclusive evidence that the suspect was not involved in violence or a rape against the woman. DNA evidence, separated from its context, is never solely definitive for either conviction or exoneration, although the burden for the former is much higher.

Myth of DNA Detection Equaling Physical Presence

When the DNA of a Suspect Is Found at a Crime Scene, Then the Suspect Must Have Been Present at the Crime.

The fact that an individual's DNA is found at the scene of a crime does not indicate that he or she committed the crime in question or even was present at the crime scene. There are many ways in which a person's DNA can wind up at the scene of a crime. As discussed in "Myth of DNA Consistency," DNA in the form of a vaginal swab found on a rape victim might be far more useful to investigators than DNA lifted from a cup or a cigarette butt.

Moreover, even if a person's DNA is reportedly found at the scene of a crime, it is not necessarily the case that the person deposited it there. There is always the possibility that the DNA could have appeared as result of secondary transfer, that the DNA could have been planted, or that the results of the DNA analysis were fabricated.

Secondary transfer refers to the phenomenon where DNA deposited on one item winds up on another. The individual does not have direct contact with that item (primary transfer); instead, his or her DNA is transferred by way of an intermediary, which could be either another person or another object. For example, if person A shakes person B's hand, they are each likely to have trace amounts of the other's DNA on their hand. If A then takes out a kitchen knife and cuts vegetables, it is quite possible that the DNA of both A and B could be found on the knife handle, even though B never touched the knife. Ironically, the potential for inadvertent transfer of DNA to muddy an investigation has increased over time as DNA testing techniques have become more sensitive and able to type the DNA of samples of only a few cells.

BOX 16.7 *Massachusetts v. Greineder*

In 1999 Mabel Greineder was found beaten to death in a wooded area in Wellesley, Massachusetts. Her husband, Dr. Dirk Greineder, a prominent physician and adjunct Harvard professor, was arrested after a DNA profile similar to his was found, mixed with his wife's profile, on gloves and a knife found near the scene of the crime. Some of Greineder's alleles were not found, and additional alleles that did not belong to him or his wife were also found on the items. Greineder challenged the DNA evidence in the case. He argued that his DNA could have appeared on those objects as a result of *tertiary transfer*. He claimed that because he and his wife had shared the same towel that morning, his DNA could have been transferred from his face to the towel, and then from the towel to his wife's face. Then, in the process of her murder, his DNA could have been transferred again to the knife and the gloves. This theory, he claimed, was also consistent with the fact that additional alleles had been found on the gloves and the knife that matched neither him nor his wife.[a] Greineder hired a private DNA lab to test his hypothesis. The lab ran an experiment and presented testimony in the case that tertiary transfer could indeed have occurred as he described it.[b] The jury ultimately convicted Greineder of murder in 2001. In 2005 Greineder's lawyers filed a motion with the Supreme Judicial Court requesting a new trial. They argued that DNA testing crucial to the prosecution's case had been conducted improperly and that Greineder had been deprived of effective legal counsel. Arthur J. Eisenberg, the director of the DNA Identity Laboratory at the University

(*continued*)

of North Texas Health Science Center and then chairman of the U.S. DNA Advisory Board, submitted an affidavit stating that the genetic testing conducted for the trial by the forensic laboratory Cellmark "was contrary to what is generally accepted in the science community. . . . There was no scientifically reliable evidence that Dirk Greineder was a potential contributor to the DNA obtained from any of the three key pieces of evidence." Eisenberg said that too little DNA was found on the items to obtain reliable results and that, furthermore, the profiles of both Dirk and Mabel Greineder were ascertained by Cellmark before interpreting the key evidentiary samples, potentially biasing the analyst's interpretation of the results. The motion was denied. In October 2009 his lawyers filed another motion for a new trial.

[a] William C. Thompson, Simon Ford, Travis E. Doom, Michael L. Raymer, and Dan E. Krane, "Evaluating Forensic DNA Evidence, Part 2," *The Champion* (April 2003): 16–25, at 24.
[b] Suzanna Ryan, "Transfer Theory in Forensic DNA Analysis," *LawOfficer.com*, January 20, 2009, http://www.lawofficer.com/news-and-articles/columns/ryan/transfer_theory_in_forensic_dna_analysis.html (accessed April 28, 2010).
Sources: Rachel Lebeaux, "Greineder Appeals Murder Conviction," *Wellesley Townsman*, August 3, 2005; Denise Lavoie, "Wellesley Doctor Seeks New Murder Trial," *Associated Press*, October 8, 2009, http://www.boston.com/yourtown/news/wellesley/2009/10/wellesley_doctor_seeks_new_mur.html (accessed April 28, 2010).

In science, misconduct, including outright fraud, rises to the level of high crimes and misdemeanors. A special federal office called the Office of Research Integrity was established in March 1989 to investigate scientific misconduct. Among the most blatant and reviled forms of misconduct is the "cooking of data," a term that means that the investigator discards or fabricates data to conform to a hypothesis. There have been cases of data fabrication so egregious that even seasoned observers found them difficult to comprehend. In one case biologist William T. Summerlin used a felt pen to mark a mouse and claimed that it expressed a skin transplantation without immunosuppression.[31] Others have been known to doctor photographs or reuse old photographs.[32] Arthur Koestler's classic book *The Case of the Midwife Toad* tells the story of the highly acclaimed early twentieth-century biologist Paul Kammerer, who used india ink to fabricate darkened nuptial pads in the toad in order to support a Lamarckian theory that inherited characteristics can be acquired from environmental conditions.[33]

The idea of fabricating evidence is not unique to science. It is a well-documented practice in law enforcement, where criminal investigators are either so confident that the suspect is guilty or are so pressured to solve a crime that they feel justified in "cooking the evidence" by planting drugs, a gun, or other incriminating items in the home or car of a suspect. In 1995, six Philadelphia police officers pleaded guilty to charges of planting illegal drugs on suspects, the theft of more than $100,000, and the falsification of reports. The investigations into the officers' actions led to the release of hundreds of defendants whose convictions were overturned by the appeal courts. Also in 1995, two other officers from Philadelphia received prison sentences of 5 to 10 years for framing young men. Since 1993 the city of Philadelphia has paid out approximately $27 million in more than 230 lawsuits alleging police misconduct.[34]

In one analysis of the O. J. Simpson case the author noted, "Evidence presented later at the trial showed that the officer had used racist language in an interview with a writer, that he described police beating a Black suspect and that he asserted that the police planted evidence against Black suspects."[35] Merrick Bobb reports that the Los Angeles Police Department (LAPD) suffered "embarrassment and opprobrium" when it was disclosed that "LAPD officers were shown to have planted evidence and guns and wrongfully shot young Latinos suspected of gang activity."[36] If police can plant drugs, they can certainly plant DNA. Another account of the Simpson case noted that sloppy handling of DNA evidence—including an inability to account for missing blood from a reference sample collected from Simpson and the discovery of several bloodstains at the crime scene several weeks after the crime had been committed—supported a theory that Simpson's blood had been planted after the murders had taken place.[37]

DNA can also be planted at a crime scene by a criminal in an attempt to thwart the police or to frame someone else for a crime. Several instances have already been reported where criminals have planted or tampered with evidence or have paid inmates to take DNA tests as a way of confusing investigators or evading prosecution. Prisoners have also been overheard coaching each other on how to plant biological evidence at a crime scene and how to avoid leaving their own DNA behind.[38] We have seen how DNA was smuggled out of prison to cast doubt about a conviction.[39] An elaborate scheme is hardly needed, of course; more simply, DNA evidence can be deposited at a crime scene by way of discarding DNA-carrying items, such as used cups, cigarette butts, a hair sample, or other items likely to contain testable amounts of DNA.

In 2009 a study published by scientists in Israel demonstrated that a somewhat more motivated criminal with access to a single hair strand, cigarette butt, or dry saliva stain and some basic laboratory equipment (a polymerase chain reaction [PCR] analyzer and a testing kit that is commercially available) could amplify DNA and spread it around a crime scene. Similarly, an artificial DNA profile could be assembled and amplified on the basis of a reference profile, without the need for any source DNA. In either case the amplified DNA can then be applied to objects and planted at the crime scene.[40] Although neither of these approaches is likely to be pursued by an average criminal, neither of them would require significant resources or more than a basic knowledge of molecular biology. Dan Frumkin, lead author of the article "Authentication of Forensic DNA Samples," has stated, "You can just engineer a crime scene. Any biology undergraduate could perform this."[41] Laboratories often update their equipment and sell off their PCR analyzer machines on the Internet; when the authors last checked, there were two such analyzers for sale for approximately $500 each on eBay.

Finding someone's DNA at a crime scene may be a prima facie reason to consider that the person was at the location at some time, but it is certainly not definitive or infallible evidence of this. Given the history of misconduct in criminal justice, planting of DNA evidence by police seeking to close the case or perpetrators seeking to divert police cannot be left out of the equation. Finding someone's DNA at a crime scene is not infallible evidence either that they were there or that they committed the crime. DNA typing helps determine the source of the biological material at a crime scene; other evidence is needed to determine whether the true donor of the sample committed a crime.

Myth of the Infallible Mismatch

If Two Samples of DNA Are Found Not to Match, Then the Samples Cannot Have Come from the Same Individual.

This claim appears, at first glance, to be well grounded in science. Textbooks report that all our cells contain the identical string of DNA molecules. If two DNA samples do not match, then it would seem that they surely cannot come from the same individual.

It is true that a nonmatch is more definitive than a match. In other words, a nonmatch offers more conclusive evidence that two samples did not come from the same individual than does a match in showing that the source of DNA of two samples is the same. As an analogy, a single black swan falsifies the statement "all swans are white," whereas a white swan (or many white swans) supports the statement but does not prove it. Nonetheless, even a nonmatch has its limitations.

Lydia Kay Fairchild, a resident of Washington State, was pregnant at age 26 with her third child in 2002. She had an on-and-off relationship with the putative father of her children, a man named Jamie Townsend. They were separated during her pregnancy. Without a job or means of support, Fairchild applied for welfare benefits. The state welfare agency required proof that Townsend and Fairchild were indeed the biological parents of the children. DNA tests were performed. The results confirmed that Townsend was the children's father. But there was a wrinkle. Fairchild's DNA was found not to match that of her children. Ordinarily there would be a 50 percent similarity between the DNA of a child and each biological parent. The court ruled that Fairchild was not the biological mother of her two children, a son aged 4 and a daughter aged 3, as she had claimed, and considered this a case of welfare fraud. The judge discounted the hospital birth records as forgeries and accepted the DNA evidence as indisputable.[42]

The state prosecutor for the case wanted Fairchild's two children to be placed with guardians while the investigation continued. She was charged with attempting to defraud the state and was denied public assistance. Her insistence that she was the biological mother of her children convinced the judge to offer her a last opportunity to prove her case. The judge ordered someone to be present during the birth of Fairchild's third child; the court-appointed witness would take blood samples of the newborn immediately after delivery and have them analyzed. Like his siblings, the newborn's DNA was found to be different from that of his mother. The state could no longer claim that, despite the DNA conundrum, Fairchild did not gestate her children. Other explanations were sought. Fairchild's lawyer characterized the response of the prosecutor to this result:

> The questions that have gone through the prosecutor's mind include whether or not she [Fairchild] was involved in being a surrogate mother. If the egg and sperm had been planted then she wouldn't have a [genetic] relationship

> to the child. Maybe she'd abducted the children from somewhere or was involved in some other criminal activity.[43]

The explanation for the dissimilarity of DNA between mother and child was eventually solved: Fairchild is a chimera. This means that some of her cells have one DNA type, while other cells have an entirely different DNA type. Fairchild's skin and hair-root DNA did not match that of her children, while the DNA from her cervical cells did match their DNA.

Chimerism occurs during the development of a blastocyst in the womb. Two fertilized eggs, either implanted by in vitro fertilization or dropped from the ovaries, fuse in the early stages of development, creating an embryo with cells that have different DNA profiles. Another route to chimerism is the vanishing-twin thesis. Somewhere between 20 and 30 percent of pregnancies start out as fraternal twins but end up as single babies. One of the early-stage fraternal embryos is absorbed by the mother, while some of its cells enter the body of the remaining embryo and remain there throughout development. These embryonic anomalies occurring after in vitro fertilization are sometimes referred to as embryo amalgamation.[44] Alternatively, chimerism can also arise from cells that routinely pass from mother to fetus and get integrated into the fetus.

There are no clear estimates of the rate of chimerism in the population. Howard Wolinsky, who estimates as many as 1 in 8, believes that it is "not rare, but rarely discovered, because it seldom generates any observable anomalies."[45] Catherine Arcabascio reports chimerism figures ranging from as high as 1 in 10 to 1 in 2,400 persons.[46] *New York Times* science writer Gina Kolata reported the following on chimerism:

> Dr. Ann Reed, chairwoman of rheumatology research at the Mayo Clinic, who uses sensitive DNA tests to look for chimerism, finds that about 50 to 70 percent of healthy people are chimeras. The more scientists look for chimerism, the more they find it. It seemed not to exist in the past, she said, because no one was explicitly looking for small amounts of foreign cells in people's bodies. "Some believe that if you look hard enough you can find chimerism in anybody," said Dr. Reed. . . . It is so common that she thinks there must be a biological reason for it. It also may cause problems, she and others say.[47]

There are insufficient empirical data to narrow the uncertainty about chimerism incidence. Chimerism can have serious implications for individuals undergoing blood transfusions or organ transplantation. It has also emerged as a defense on the part of professional athletes who have been accused of transfusing themselves to boost their endurance.[48] The implications for paternity testing and forensic analysis are significant:

> Take, for example, the hypothetical case of a chimeric criminal who leaves DNA at the scene of the crime. The suspect may leave a sample of hair, semen, saliva, perspiration, urine, ear wax, mucus, bone, fingernail scrapings, blood, or skin. He may even leave a combination of those forensic clues at the scene. If he is a chimera, however, the DNA from his saliva could, in theory, differ from the DNA in his semen, skin, blood, or some other sample left at the scene.[49]

Criminal chimeras could be mistakenly exonerated if DNA served as the definitive evidence. In addition, those who are falsely convicted of a crime and whose only chance at exoneration is the submission of the crime-scene DNA for a cold hit in CODIS could also be stymied by the actual perpetrator if he or she were a chimera. If chimerism occurs at a higher rate than the lower estimates predict, the entire project of forensic DNA would have to be reconsidered for fallibility of identification.

Where does this leave us? There is nothing infallible about DNA. DNA evidence can be strong or weak or anything in between. Human error can and does occur in the collection, analysis, and interpretation of DNA results. Samples can be switched, cross-contamination can occur, analyses can be improperly interpreted, and the results can be poorly communicated. Any errors of this sort can lead to the false incrimination and wrongful conviction of an innocent person. The possibility that chimeras are a rule rather than a rare exception could undermine the very basis of the forensic DNA system.

In the meantime, are the myths or exaggerations of infallibility obstructing the cause of justice? Is too much power attributed to DNA as truth telling? Would higher probabilities in the estimate of RMPs in cold-hit cases make a difference in their probative value or how juries relate to the evidence? Is contamination of evidence going unnoticed? How often are people being wrongfully arrested, tried, and convicted of crimes on the basis of flawed DNA evidence? These questions illustrate the human dimension in the use of forensic DNA. Human judgment is notoriously fallible, but it remains our only guide as long as we understand its limitations.

Chapter 17

The Efficacy of DNA Data Banks: A Case of Diminishing Returns

> The more complete the database the better the chance of detecting criminals, both those guilty of crimes past and those whose crimes are yet to be committed.
>
> —Lord Brown, House of Lords, British Parliament[1]

Forensic DNA data banks have been the subject of many narratives, but none more forceful in its advocacy and more universally held than the one that claims that, ceteris paribus, the larger the data banks, the more crimes will be solved and the more crimes will be prevented. When the New York State Senate was debating whether to expand the state's forensic DNA databases to arrestees, then Senator Joseph L. Bruno said, "Expanding the scope of the DNA databank means expanding the ability of our law enforcement officials to solve both new crimes and old ones."[2]

There are certainly intuitive aspects to this narrative. Forensic DNA data banks of felons overall have been a good thing for law enforcement, both in prosecutions of murder-rapists who might have escaped being caught were it not for DNA evidence and in early DNA exclusions of suspects who otherwise might have run up high investigative costs. But is more of a good thing necessarily better? Examples abound in daily life where this is not the case. Sometimes "more" can go hand in hand with "more complicated," resulting in diminishing returns and increasing inefficiencies. We all like choices when we go to the grocery store, but the more choices we have, the more time we spend making our way down the aisle. Sometimes expanding the good beyond the boundaries of its initial intent can have unforeseen consequences.

Soon after estrogen therapy proved successful in helping women who lacked normal estrogen levels to conceive, physicians began using estrogen as a replacement therapy for women who lost estrogen in the aging process under the theory that, since estrogen was important for women's health, more of it would make women healthier. Instead, this continuous, long-term use[3] of synthetic estrogen proved to carry with it substantial risks, requiring special medical oversight. Similar fallacies have been made in law enforcement. Few would deny that police should be adequately armed against violent criminals. But the fact that an armed police force is a good thing does not imply that more powerful arms, such as a 12-gauge shotgun or an AK-47 rifle, makes for better policing and would not result in the abuse of deadly force.

This leads us to the central question of this chapter: what is the relationship of the size of a country's forensic DNA database to its rate of solving crimes? As noted, it is frequently argued that the larger the database, the greater the opportunity police have to solve new crimes and to clear their cold-case files of unsolved crimes. Is there any evidence that supports these claims? And if there is evidence, does it apply to databases that have been expanded to include arrestees who have not been convicted of a crime and volunteers who gave a DNA sample to police (so-called elimination samples) to rule themselves out as suspects in a highly publicized case?

Are there any reasons that expanded forensic DNA databases might impede law enforcement? Are there diminishing marginal returns for criminal justice in the expansion of DNA data banks? If there are errors in DNA analysis that may lead to false convictions, is there reason to believe that the rate of false convictions could increase by expanding DNA databases to arrestees who have not been convicted of a crime or to those convicted of minor offenses?

Current Standards for Measuring Database Success

The U.S. Combined DNA Index System (CODIS) database currently holds more than 8 million "offender" profiles (including over 100,000 profiles from arrestees)[4] and over 300,000 forensic profiles from crime scenes.[5] The FBI measures the success of the database by the number of matches or "hits" made against the database, as well as the number of "investigations aided," defined as criminal investigations where CODIS has added value to the investigative process. As of March 2010 CODIS had produced 114,300 hits assisting in 112,300 investigations.[6]

Neither of these pieces of information—number of "hits" or "investigations aided"—is particularly helpful in determining the efficacy of the database. A "hit" is only a match; it tells us nothing about whether those hits were followed up, how many resulted in arrests, and how many arrests resulted in convictions. The fact that an individual's DNA is found at a crime scene does not tell us whether his or her DNA was left at the time the crime was committed, let alone whether he or she was responsible for the crime in question. There have also been reported cases of a DNA cold hit leading to a suspect who was found to be too young (e.g., an infant) at the time of the crime.[7] The match was most likely the result of cross-contamination in the forensic laboratory. In addition, multiple "hits" can occur for single cases or investigations. Likewise, "investigations aided" include those cases where an offender was already a suspect and would have been convicted regardless of DNA evidence; they might even include cases where the individual identified turned out not to have been related to the crime in question at all and just happened to have his or her DNA at the scene of the crime. These cumulative figures also tell us nothing about whether the rate of "hits" or "investigations aided" has increased in proportion to the growth of the database. Last, it is not possible to know from these figures what the actual contribution of the database has been to crime investigations overall; for example, what proportion of investigations conducted consists of those involving "hits" that occurred through the DNA database?

In the United Kingdom more information is routinely published by the government that sheds some light on database efficacy, although these data are also limited. Official reports from the United Kingdom offer evidence that the collection of DNA samples at crime scenes has contributed to crime detection. As noted by Helen Wallace, "Detections are crimes that have been recorded as 'cleared up' by the police. This includes crimes where [a] person has been charged, cautioned or warned, and some crimes that are not proceeded against (for example, because the victim is unwilling to give evidence)."[8] The "crime-detection rate" is defined as the percentage of crimes recorded by police that are cleared up. Police in the United Kingdom have reported that the clear-up rates for crimes where DNA evidence is available are significantly higher than for those crime scenes where no DNA is found.

In 2004 clear-up rates for crimes where DNA was successfully recovered were between 38 and 43 percent, compared with the overall detection rate of 23.5 percent.[9]

One might argue that efficacy ends at detection. However, conviction is the standard endpoint for success of criminal justice techniques. If DNA can only lead to a suspect but plays little or no role in conviction, then the efficacy of DNA, despite the *Crime Scene Investigation* (*CSI*) effect, is diminished. It is difficult to isolate the role of DNA exclusively as a contributor to conviction rates since DNA is rarely the only or even the primary factor in a conviction (indeed, it should not be).

What is of additional interest is whether the growth in the database has improved crime-solving capacity. According to data presented by Carole McCartney, the proportion of matches resulting in a successful detection in the United Kingdom rose from 37 percent (1999–2000) to 48 percent (2000–2001) but then fell in 2002–2003 to 42 percent.[10] The size of the United Kingdom's DNA database rose significantly from 1999 to 2003, but the proportion of matches has not followed that trend.

Another factor to consider in evaluating the use of DNA databases to solve crimes is the number of crime scenes that produce DNA evidence. A report by the United Kingdom's National DNA Database (NDNAD) stated that a mere 0.85 percent (85 in 10,000) of all recorded crimes produce a DNA sample that can be tested.[11] In other words, in more than 99 out of 100 cases, DNA cannot be collected from the crime scene and uploaded to the database in the first place. McCartney concludes: "Objective assessment of the 'success' or otherwise of the DNA Expansion Programme on the basis of crime detection rates is therefore near impossible. Even some police officers concede that perhaps the ability of DNA to solve the crime problem has been overstated."[12]

Another factor British criminologists look at to measure the impact of the database is the "match rate," which is simply the proportion of crime-scene profiles that are found to match profiles of individuals in the database. The "match rate" is considered a useful indicator for determining whether enlarging the databases will yield more cold hits when crime-scene data are loaded into the national database. The "match rate" is defined as 100 times the number of crime-scene profile matches made when profiles are entered into the database divided by the number of crime-scene profiles collected in that year. In the United Kingdom, between 2004 and 2006 the match rate rose slightly but then flattened out, even though the total number of profiles retained in the database rose each month. As of March 2006 the match rate was reported as 52 percent, up 4 percent from the previous year.[13]

Although the match rate is often used to promote data-bank expansion, it does not give us insight into the efficacy of DNA data banks. Suppose that for a particular year there were 1,000 crime-scene profiles, and 600 of them were matched to a profile in the national forensic DNA database. That would yield a "match rate" of 60 percent. But this indicator does not take account of the size of the database. Suppose that the database doubled and the "match rate" remained the same even though one expects more matches in a larger database. "Match rate" is not a surrogate for efficacy.

The U.K. NDNAD report for 2005–2006 stated: "Of the 200,300 or so profiles on the NDNAD that have been retained under the CJPA 2001 [profiles mainly of arrestees] and would previously had to have been removed, approximately 8,500 profiles from some 6,290 individuals have been linked with crime scene sample profiles from some 4,000 offenses."[14] The claim seems to be that were it not for the expansion of the database to arrestees, these individuals would not have been connected with these crimes. However, these figures do not tell us how many of these "matches" were followed up, resulted in convictions, or were associated with crimes that otherwise would not have been solved. They also do not tell us how many individuals linked to crimes were convicted for other offenses and thus would have been linked regardless at the point where their DNA was entered into the database upon conviction. Moreover, again, it does not tell us whether these additional matches are in proportion to the increase in the database over this period.

The U.K. NDNAD also publishes data about the match rate as they relate to the number of "subject" profiles (offenders, arrestees, and others qualified for database inclusion). Data from 2004 to 2006 actually show a decline in the crime-to-subject match rate; as additional subject profiles were added to the database, the match rate actually went down from about 2 percent to 1.2 percent.[15] Although it is difficult to tease out the factors that might have contributed to this pattern,[16] it certainly does not support the notion that expanding the database necessarily leads to more effective or efficient crime solving.

Evaluating DNA Database Efficacy

In evaluating the efficacy of DNA databases for solving crimes, one has to go beyond simply looking at the number of "hits," "investigations aided," "detection rate," or "match rate" and explore, to begin with, the following factors.

The percentage of crime scenes that are screened for DNA. DNA profiles can contribute to solving crimes only if crime scenes are screened for DNA evidence. Therefore, the effectiveness of the database will, in part, depend on the extent to which DNA is collected at crime scenes.

The number of crime-scene DNA samples with complete profiles that are loaded into the database. The fact that crime scenes are screened for DNA samples does not necessarily mean that there will be one or more viable DNA profiles obtained that meet the criteria for uploading them to the database. The DNA might be too degraded, or there might be mixtures that cannot be individuated into discrete profiles.

The number of uploaded crime-scene DNA profiles that match identified profiles already in the DNA data bank. The matches of crime-scene DNA profiles with profiles in the database must produce one or more suspects who can be investigated for the crime in question.

The number of detections (people charged with a crime) based on those matches. When a match is found between a crime-scene DNA profile and a person whose profile has been entered into the national data bank, the police may have a suspect, but they have not solved the crime. There may be several explanations of why that individual's DNA was left at the scene of the crime. This is why, at least in the United Kingdom, additional corroborating evidence is required before an individual can be charged with a crime.[17] It is imperative that we know how many of these matches result in individuals who are charged with a crime if we are to evaluate the efficacy of enlarging these databases.

The number of convictions based on the DNA crime-scene matches. The true test of the efficacy of DNA data banks is measured by how effective they are in contributing to convictions. Moreover, in how many cases is the DNA match a necessary condition for a conviction, as opposed to ancillary or secondary evidence?

For an overall index of DNA database efficacy with respect to solving all reportable crimes, we construct the following indicator that we call the

DNA Crime-Solving Efficacy (CSE) Index (see figure 17.1 for components of the CSE Index).

$$\text{CSE (DNA crime-solving efficacy)} = P \times Q \times R \times S \times T,$$

where

P = Rate of crime scenes analyzed for DNA
Q = Rate of crime-scene DNA profiles loaded into the DNA database per crime scenes analyzed
R = Rate of matches per loaded DNA profiles
S = Rate of detection per matches
T = Rate of conviction per unit of detection

The rate of crime scenes analyzed is simply the number of crime scenes (from all recorded crimes) analyzed for DNA divided by the total number of recorded crimes. In the United Kingdom 17 percent of crime scenes are examined for DNA evidence.[18] Not all the crime scenes investigated for DNA will yield viable profiles that can be uploaded to the national database, as discussed earlier; in the United Kingdom, of the 17 percent of crimes scenes that are examined for DNA evidence, only 5 percent result in a successful DNA sample being collected, profiled, and loaded into the database. Taken together, this means that a testable DNA sample is obtained in only 0.85 percent of all recorded crimes.[19]

The third factor in the equation is the rate of matches for all the crime-scene profiles loaded into the database. We are assuming in this case that the matches are determined through the database, as opposed to a preexisting suspect. Obviously, many crimes are solved by matching the DNA at the crime scene with the DNA of a suspect, where the national DNA database is not involved. If we are seeking an indicator of DNA efficacy, then we will include detections both with and without the database. On the other hand, if we are seeking an indicator of DNA database efficacy, we will include only the matches obtained by uploading crime-scene DNA to the database (so-called cold hits). The match rate in the United Kingdom was reported at 52 percent as of March 2006; in general it averaged around 50 percent over the course of that year.

The fourth factor is the rate of detection (charges filed) per matches. For this indicator, we make the distinction between "detection" and "conviction."

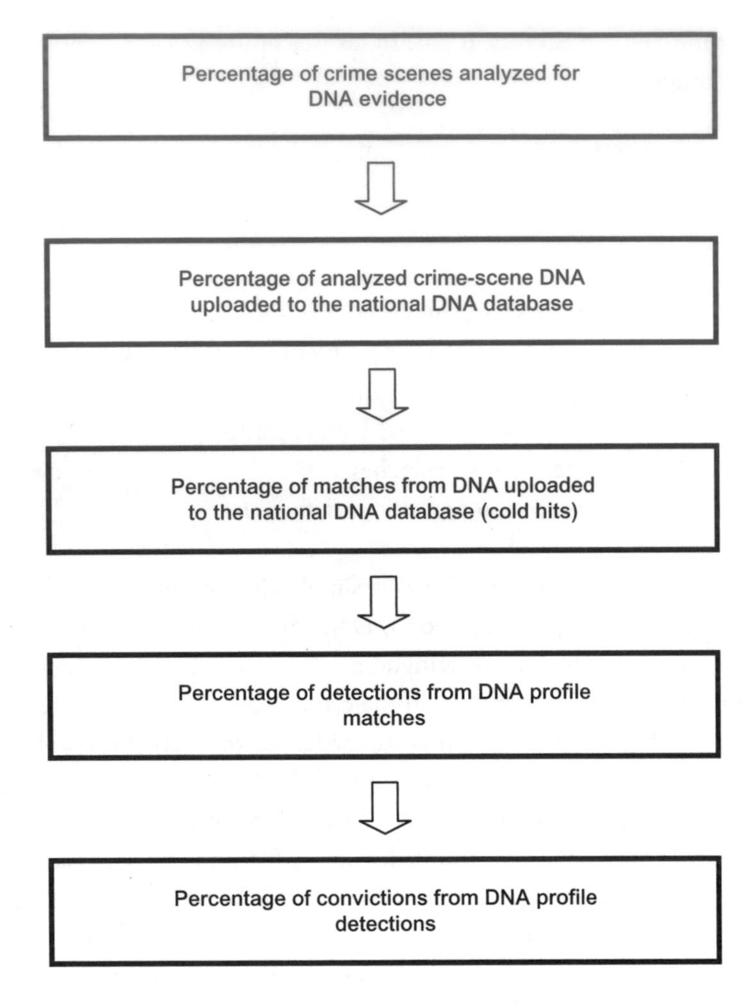

FIGURE 17.1. Flowchart of the components of the Crime-Solving Efficacy Index (CSE) from crime scenes analyzed to convictions. *Source*: Authors.

As discussed earlier, the detection rates for crimes involving DNA matches in the United Kingdom ranged from 37 percent to 48 percent for the period 1999–2003. For illustration purposes, we estimate this at 50 percent.

The fifth and final factor is the rate of conviction from the number of detections. The fact that someone's DNA profile in the database is matched with crime-scene DNA by no means necessarily results in a conviction.

Someone's DNA may have been left at the scene before the criminal event, or, as in some notable cases, the DNA match arose from contamination of the evidence. With no reliable data, for the purpose of this example, let us assume that the conviction rate is 50 percent.

From the numerical values of these percentages, we can calculate the values for the CSE Index and assess the role of DNA databases in solving recordable offenses as follows:

$$CSE = .17 \times .05 \times .50 \times .50 \times .50 = .0010625, \text{ or } 0.1 \text{ percent.}$$

This tells us that the role DNA plays in convictions of all crimes is minuscule—DNA plays a role in only about 1 out of every 1,000 convictions.

But is it fair to compare minor robberies, where DNA is not even sought, with the efficacy of DNA in solving violent crimes? Shouldn't we define efficacy of DNA data banks in terms only of those crime scenes that have been examined for DNA? We can begin with the assumption that there will be DNA screenings in all serious crimes. Thus factor P will be 100 percent ($P = 1$). Overall detection rates in the United Kingdom are about 23 percent, whereas detection rates for violent crimes are about 50 percent, about the same as the rate for cases involving DNA. These rates are similar to those in the United States. Figure 17.2 shows that the clearing rates for forcible rape and murder in the United States in 2006 are just over 40 and 60 percent, respectively.

For purposes of illustration, assuming $P = 1$ and using the same values as in the previous example ($Q = .05$; $R = .50$; $S = .50$; $T = .50$), the four factors will yield a Crime-Solving Efficacy Index of 0.63 percent, still less than 1 percent. The number of cold-hit matches does provide some indication of the role of DNA data banks. Disaggregated national data on cold-hit matches versus suspect-to-crime-scene profile matches are not always available. If we exclude from the "frequency of matches" those that arise from preexisting suspects and leave only cold hits, then we have a better indicator of the efficacy of the data bank by itself.

Suppose that for one year 100 percent of all violent crimes had DNA samples taken ($P = 1$), 20 percent of all crime-scene DNA was loaded into the database ($Q = .2$), 40 percent of the DNA loaded yielded cold matches ($R = .4$), 30 percent of the matches resulted in charges filed (detection) ($S = .3$), and 20

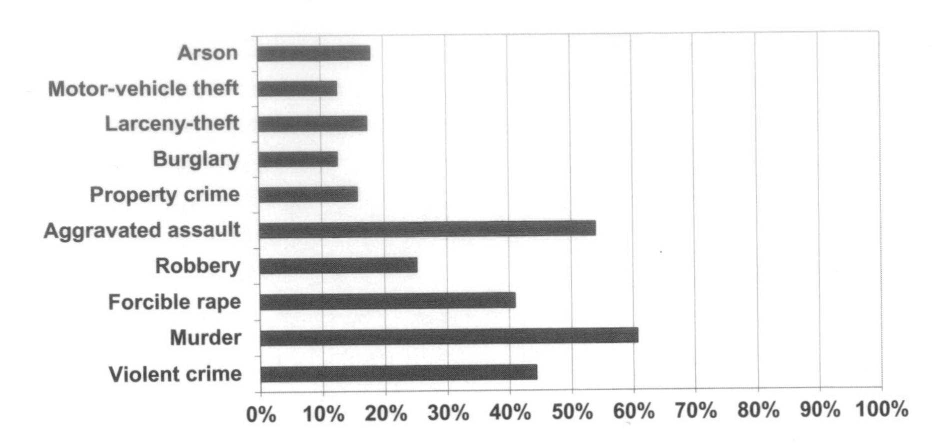

FIGURE 17.2. The clearing rate of felony crimes in the United States, 2006. *Source*: FBI, *Crime in the United States*, table 25, http://www.fbi.gov/ucr/cius2006/data/table_25.html, compiled by Tania Simoncelli, Harry G. Levine, and Jon B. Gettman, March 2008.

percent of those charged were convicted ($T = .2$). The Crime-Solving Efficacy Index for the database would be given by

$$\text{CSE} = 1 \times .20 \times .40 \times .30 \times .20 = .0048, \text{ or } 0.48 \text{ percent.}$$

We still need a factor that accounts for the size of the data bank. If the data bank expands by 100 percent but the CSE rises by 5 percent, then the increased size of the data bank is not having a major impact on the match, detection, and conviction rates, or at the very least there are diminishing marginal returns from the expansion of the DNA database. We can choose a base year for setting the size of the database N_B. Let the database for any other year be designated as N. Our modified indicator of DNA database efficacy would be given as follows:

$$\text{CSE} = P \times Q \times R \times S \times T \times N_B / N.$$

Thus, if in one year the number of DNA cold-hit convictions (T) goes up 20 percent, while the database increases twofold (100 percent or $N = 2N_B$), then the efficacy for crime solving as determined by our index would go by about 40 percent from what it was when the database was half its present size. Our index for efficacy of DNA data banks grows when DNA plays a stronger role in cold-hit convictions and shrinks when a larger database does little to improve the conviction rate. This indicator tells us that when all the factors of the index remain constant (*P*, *Q*, *R*, *S*, and *T*) while the size of the database grows, the efficacy of the DNA data bank decreases.

Size of a DNA Data Bank Versus Solving Crimes

Why would a larger database not necessarily lead to more crimes being solved more efficiently? As previously noted, it is taken for granted by many police authorities that the larger the database, the more crimes will be solved. As an example, consider the following statement published in the annual report of the U.K. NDNAD for 2005–2006: "The additional CJ [criminal justice] arrestee sample profiles to the NDNAD have brought significant benefits, including direct police savings through speedier investigations, quicker apprehension of offenders, earlier elimination of suspects and greater victim reassurance."[20]

There are many reasons that a larger database does not necessarily deliver more crime control. First, as we have seen from the previous exercise, the factor currently limiting database efficacy is the number of crime-scene profiles, not the number of offender or subject profiles. Although numbers of subject profiles have increased exponentially in both the United States and the United Kingdom from 2000 to 2010, the number of crime-scene samples added to the database has risen gradually (see figures 17.3 and 17.4). To improve the efficacy of DNA databases in solving crimes, one has to increase the number of crime-scene DNA profiles loaded into the database in the first place.

This is not an easy undertaking. The proportion of crime scenes where DNA has been recovered has increased since 1988 from 4.5 percent to 17 percent in the United Kingdom.[21] However, there are limitations on crime-scene profiling that are likely to prevent this number from getting much higher. First, DNA is most readily obtainable from scenes of violent crimes—crimes where biological evidence is likely to be left behind in the form of blood, semen, or skin cells. The majority of crimes that are committed, however, are property crimes. This is true even for the most serious crimes, which are tallied annually by the FBI in its "Crime Index." Out of the 11.9 million index offenses reported to the FBI in 2002, 1.4 million constituted violent

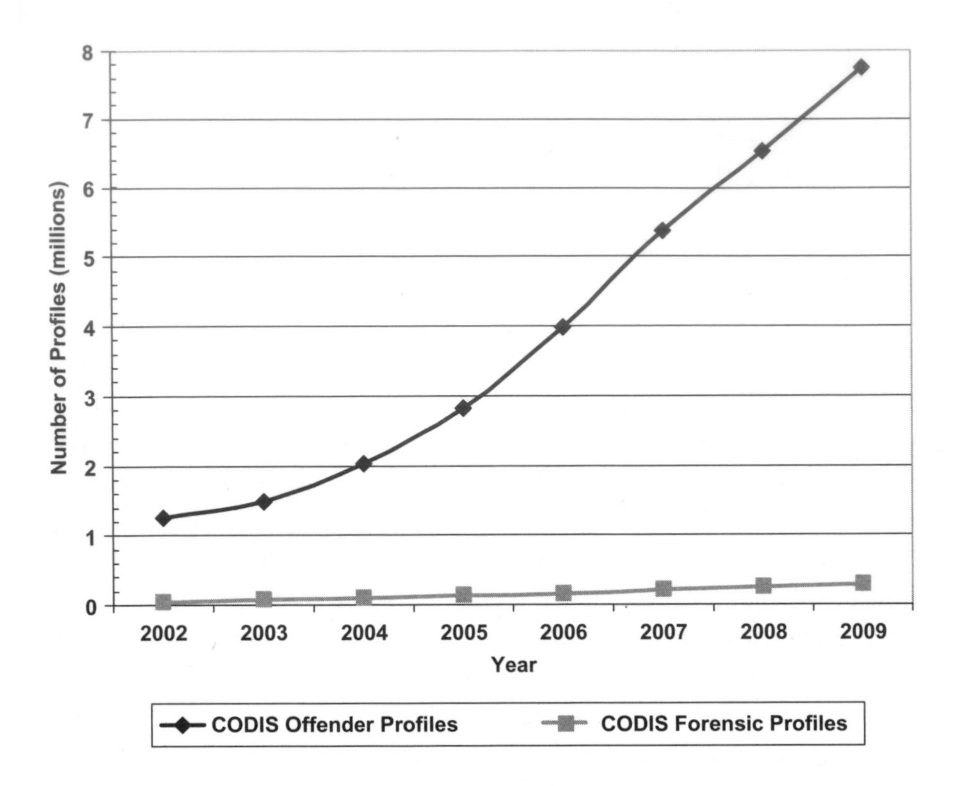

FIGURE 17.3. Diminishing returns: CODIS offender profiles (in millions) relative to forensic profiles. *Source*: CODIS Combined DNA Index System (FBI brochure); CODIS Program Office, FBI, personal communication. Compiled by Tania Simoncelli, Harry G. Levine, and John Gettman, March 2008. Updated by the authors May 2010.

crimes, while 10.5 million were property crimes.[22] To look for DNA where it is not obviously present requires painstaking and costly crime-scene investigation, during which forensic technicians scour the scene looking for trace evidence that may or may not carry the DNA of the perpetrator. Moreover, DNA found in these situations may not be of sufficient quality and quantity to permit testing.

It is possible that DNA technology will continue to change in ways that will allow us to detect even smaller amounts of DNA. But improved DNA detection will not contribute to solving property crimes, where multiple sources of DNA will be found at the same crime scene, and regardless of the technology it will not be possible to sort out which samples belong to the perpetrator. In fact, a concern with overreliance on DNA in criminal inves-

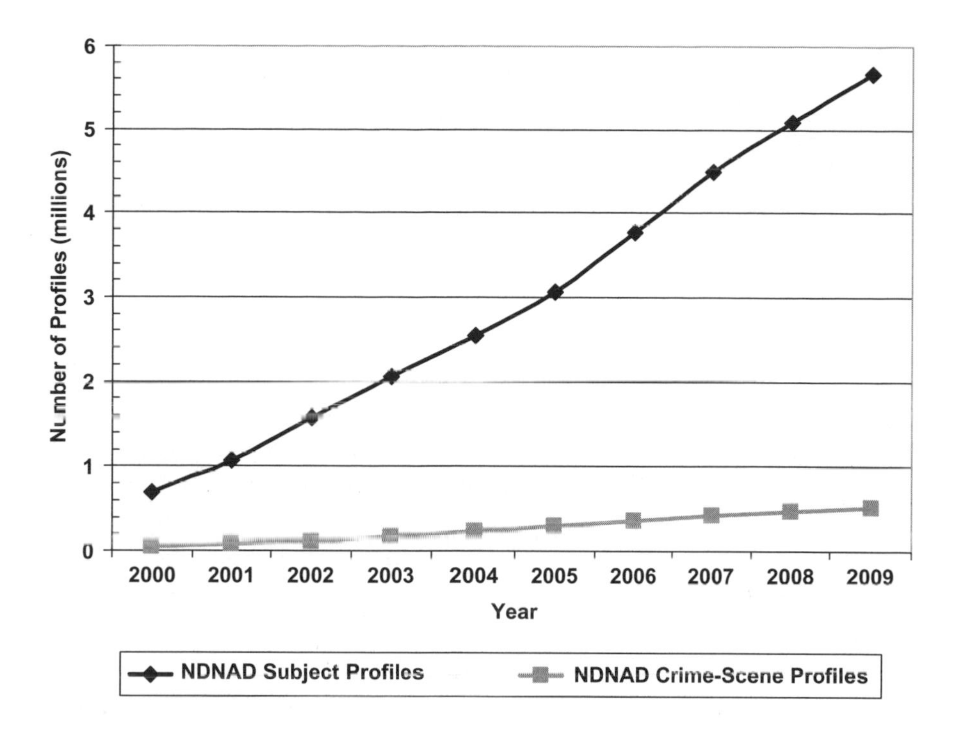

FIGURE 17.4. NDNAD offender profiles (in millions) relative to forensic profiles. The number of subject profiles held on NDNAD includes profiles from arrested individuals as well as volunteers and is higher than the number of individuals represented in the database because some of the profiles are replicates. *The National DNA Database Annual Report 2007–09* estimates that 13.5 percent of the subject profiles are replicates, such that the number of individuals in the database was approximately 4.9 million as of March 2009. Similarly, the number of crime-scene profiles reported here do not reflect those that have been removed from the database. As of March 2009 there were a total of approximately 350,000 crime-scene profiles retained on the NDNAD (*Annual Report*, p. 25). *Source*: Authors. *Date Source*: *National DNA Database Annual Report 2007–09*, http://www.npia.police.uk/en/docs/NDNAD07-09-LR.pdf (accessed May 14, 2010).

tigations is that by attempting to increase the input of crime-scene samples to generate more "hits," we may end up generating higher numbers of spurious matches. This may in turn decrease the crime-detection and conviction rates and also runs the risk of leading to more wrongful convictions.

Even within the category of violent crime the use of DNA is limited. In most cases of rape the rapist is known to the victim. DNA is seldom useful in those cases, since what is most often in dispute is whether the act was forced or consensual (see chapter 8).

In the overall scheme of solving crimes, Wallace notes, "The number of cases that can be solved using DNA will always be limited by the number of crime scenes from which DNA profiles can be collected and the need for corroborating evidence."[23] She adds that even if the DNA match rate (number of DNA matches per crime-scene sample) were 100 percent (for example, if the database were truly universal), the DNA detection rate (or conviction rate) would never be that high.[24]

There are other reasons to be skeptical about the notion that a larger database is a better one. When DNA databases are expanded to include samples and profiles taken from petty criminals or arrestees who are not charged with a crime, they are including individuals who are less likely to go on to commit future crimes, especially the types of crimes where DNA can be collected. Some argue that adding arrestees to databases will either hinder the conviction rate or not help at all. In 2005 the police liaison officer of the Scottish DNA Database was quoted as saying, "It is arguable that the general retention of profiles from the un-convicted has not been shown to significantly enhance criminal intelligence or detection."[25]

Beyond this, there are ways in which DNA expansions can even interfere with or undermine crime-solving efficiency. DNA database expansions create backlogs of unprocessed crime-scene DNA samples, which may remain in a queue for months or even years. As a result, when serious crimes are committed, the crime-scene DNA samples await their turn to be profiled. In 2003 the National Institute of Justice estimated that more than 350,000 rape and homicide cases awaited DNA testing and that crime labs were deluged with analysis requests for convicted-offender samples.[26] If we add to the convicted felons individuals arrested but not charged or those detained by federal agencies, the backlog could inhibit the role of DNA in convicting serious felons. As an example, in 2001 police in Wyoming took a DNA sample from an individual convicted of kidnapping. The sample languished in the laboratory without being processed or uploaded to the national database. When it was finally processed, it matched the profile found at a 1997 murder of a University of Colorado student. This case reveals that when biological samples are taken from too many people and scheduled to be processed and uploaded to the forensic DNA database, fewer crimes are actually solved. This may sound counterintuitive. As P. Solomon Banda explained, "The nation's DNA tracking system is beset with a huge backlog that could take years to clear. And in the meantime, law enforcement officials say, crimes are going unsolved."[27]

DNA expansions also are likely to result in more errors in profiling, labeling, and transcription of samples that will add to administrative costs, needless lawsuits, and human hardships for people falsely accused and perhaps even convicted by DNA. Indeed, errors of these types have already occurred, and some have been attributed to the problem of mounting backlogs.[28] The error rate will undoubtedly increase as the backlog of DNA profiles rises. Cross-contamination of samples, which has explained past errors and remains a serious quality-control problem in sequencing DNA samples, is more likely to occur when lab technicians are under increased pressure to speed up processing.

The larger the database, the more opportunity there is for a person to be falsely accused and incriminated for a crime. As the database grows, so too does its potential use as a resource for criminals to frame others for crimes by obtaining and planting DNA evidence from others. Already there have been cases where criminals have planted or tampered with evidence or paid inmates to take DNA tests as a way of confusing investigators or evading prosecution (see the case of Anthony Turner, discussed in chapter 16). Unfortunately, it is not only criminals who might plant DNA in an attempt to frame someone for a crime; there have been a surprising number of police frame-ups reported over the last 10 to 15 years (see chapter 16 for a discussion of Philadelphia police who pleaded guilty to planting illegal drugs on suspects). William C. Thompson notes, "If your profile is in a DNA database you face a higher risk than other citizens of being falsely linked to a crime. You are at higher risk of false incrimination by coincidental DNA matches, by laboratory error and by intentional planting of DNA."[29]

The expansion of DNA databases to include DNA profiles of petty criminals and innocent people requires resources, which may come at the expense of traditional criminal investigation techniques. In other words, a transfer of resources takes place from shoe-leather investigations to DNA profiling and database matches. Jenny Rushlow argues that "all of the resources devoted to DNA testing and database management, which are extremely costly and time-intensive, diverts necessary resources away from other aspects of investigations, like tracking down witnesses, victims and suspects."[30] Rockne Harmon, a senior deputy district attorney from Alameda County, California, has complained that the costs associated with building California's database as a result of the passage of Proposition 69 have taken away resources that are needed to place officers on the streets.[31]

Finally, thus far we have been discussing the efficacy of DNA data banking in solving crimes. It has also been argued that DNA data banks will prevent or deter crimes from taking place. The presumed conventional wisdom is that people who are in the database will be deterred from committing a crime out of a fear of getting caught. This argument does not apply to the vast majority of innocent people who have not yet and may never commit any serious crimes. Also, there is no evidence that crime rates have declined as a result of the growth of the DNA databases. Furthermore, there is no reason to think that crimes of passion will be deterred by having one's profile on a DNA data bank. Finally, sophisticated criminals will not be deterred from committing crimes; rather, they will seek to confound the criminal justice system, for example, by introducing foreign DNA into the crime scene.

We return to an earlier discussion of efficacy. By overextending the function of a reliable technology, we can undermine its efficacy. When DNA profiling and data banking are used in connection with violent crimes, there is ample evidence that they can be efficacious and cost effective for generating suspects and evidence that an individual was at the crime scene. However, when the technology is extended to petty crimes or innocent and suspicionless individuals, the evidence leads to the conclusion that there will be a rapid decline in efficacy; the marginal benefits of loading names into a national database decline rapidly. Moreover, there is no evidence that posting people's DNA profiles on a national database will deter or prevent crimes.

Chapter 18

Toward a Vision of Justice: Principles for Responsible Uses of DNA in Law Enforcement

I don't know how we, as members of the community, who are concerned with public safety, can look victims and families in the face knowing we can do things in the lab to help bring perpetrators of serious crimes to justice. To me, it's unconscionable not to use these methods to solve serious crimes and to prevent future crimes.

—Fred Bieber[1]

Scientific truth-making . . . is always a social enterprise. . . . As such, even scientific claims are subject to distortion, through imperfections in the very human systems that produced them. In attempting to render justice, the law's objective should be, in part, to restore to view these potential shortcomings, instead of uncritically taking on board a decontextualized image of science that ignores its social and institutional dimensions. Doing justice, after all, demands a complex balancing of multiple considerations.

—Sheila Jasanoff[2]

Over the last 20 years we have witnessed an extraordinary explosion in the development and use of DNA technology in the criminal justice system. Initially DNA was relied on as an occasionally useful tool in investigating very serious crimes. In those cases DNA, sought from suspects by way of a court-issued warrant supported by probable cause, was compared with DNA left behind at the scene of a heinous crime. But as we have described throughout this book, this fairly circumscribed use of DNA analysis has given way to a massive and ever-expanding system of collecting and permanently retaining DNA for ongoing investigation and use. Most recently a number of law-enforcement policies, techniques, and practices have expanded to allow police to take DNA from innocent people—people who have never been convicted and in some cases never even suspected of a crime.

As described in chapters 2, 3, 4, and 6, the expansion of DNA databases to arrestees, familial searching, the increasing use of DNA dragnets, and surreptitious DNA collection are all ways in which innocent people are increasingly being brought into the criminal justice system by way of their DNA. The mass expansion of DNA collection to innocent people marks a radical shift in the way in which DNA is used in the criminal justice system, one where DNA is starting to look much more like a surveillance tool than a tool for criminal investigation. Throughout this book we have explored multiple dimensions of these developments, including their impact on crime solving, their role in exonerating the wrongly convicted, their implications for privacy, and their potential for error. But perhaps the overriding issue at stake is whether DNA data banks, now a mainstay in all modern industrial societies, are bringing greater justice to the policing functions of civil society, while minimizing injustice. Will ever more DNA expansion necessarily advance the cause of justice?

The answer may very well depend on how one perceives "justice" and its role in the criminal justice system. When we speak of the "criminal justice system," it usually refers to a set of practices and institutions of government directed at maintaining social order, defining, deterring, and mitigating crime, and establishing sanctions, penalties, and rehabilitation for those convicted of crimes. The system includes the perpetrators and victims of crime, the falsely accused and wrongfully incarcerated, the courts, the police, prosecutors, the defense bar, the written body of legal precedents established in case law, and the complex tapestry of regulations and rules that set standards of behavior and accountability within the criminal justice subsystems. But how does "justice" enter into "criminal justice"?

Concepts of Justice in Criminal Law

In law enforcement and its ancillary institutions the concept of justice is expressed in several distinct forms. *Retributive justice* (also *punitive justice*) is based on the idea that the guilty must be apprehended and punished for their crimes. Without some form of retribution, it is often argued, there would be no deterrence against violating the laws of civil society and committing injustices against persons or the state. The victims of crimes are often the most vocal about seeking retribution. Retributive justice draws from the ethics of "an eye for an eye" or the philosophy of the "just desert," namely, that "you reap what you sow."

The concept of *restitutive* (restorative) *justice* signifies that the victim's losses shall be restored or that some effort is made to pay restitution for one's crimes. In the criminal justice system restitutive justice is introduced in particular cases where a convicted felon is obliged to repay the victim with some form of restitution for causing diminished capacity or to repay the state for crimes against property. Tort law is premised on elements of restitutive and retributive justice, for example, when courts award punitive and compensatory damages for losses. Some states provide financial relief for those wrongfully convicted and incarcerated for a crime, where the amount of compensation depends on the length of the term served.

Procedural justice seeks to ensure that those charged with a crime obtain adequate counsel and are afforded equal opportunity (compared with anyone else charged with a similar crime) to prove their innocence. The term "due process" or "due process of law" implies that the government must respect all the legal rights that are owed to an individual, and that a law shall not be unreasonable, arbitrary, or capricious. The constitutional guarantee of due process of law is found in the Fifth and Fourteenth Amendments to the U.S. Constitution. The Fifth Amendment prohibits the federal government from arbitrarily or unfairly depriving individuals of their basic rights to life, liberty, and property; the Fourteenth Amendment applies these same limitations to the states. The Fourteenth Amendment has been interpreted by the U.S. Supreme Court to incorporate the basic civil liberties protections of the Bill of Rights, so that those protections also apply to the states, as well as to the federal government. Injustices include unwarranted intrusion into one's privacy, incarceration without being charged of a crime, and the failure to afford an individual charged with a felony the right to a trial by an impartial jury of one's peers.

Procedural justice also implies that justice for an individual is associated with fair and equal treatment under the law. This includes the idea of "fair sentencing." It would be unjust if some people received disproportionately harsh sentences for the same crime. Procedural justice also includes the ideas of "presumed innocence" and "burden of proof." Innocence is the default state unless one can demonstrate otherwise. In addition, it is the state's burden to demonstrate guilt rather than the suspect's burden to prove innocence.

Justice and Forensic DNA

Law enforcement's primary role is to investigate crime and prosecute criminals. The Combined DNA Index System (CODIS) in the United States and similar systems in other countries have been explicitly created as tools toward these ends. This vision of justice that DNA is intended to serve—at least from the perspective of those who employ it—is perhaps narrower than the one discussed in the previous section because it focuses more heavily on retribution and restitution and much less on procedural fairness and equality. A narrow vision of justice such as this might very easily embrace the maximum expansion of DNA technology in the criminal justice system, which elevates the resolution of a crime and punishment of the guilty above all other matters.

On the other hand, a vision of justice that values not only crime solving but also notions of fairness, equity, and privacy might view expansions of DNA technology in a different light. The often-debated question "How many offenders might we tolerate escaping in order to avoid an innocent person being wrongfully condemned?" becomes central when procedural fairness is valued. Questions of wrongful identification as a result of error or abuse associated with DNA testing also become relevant in this formulation of justice. Similarly, questions about the disproportionate impacts of long-term genetic surveillance on minority communities and the broad societal implications of widespread data banking for privacy and autonomy take on special significance.

There is no question that debates about the appropriate role of DNA in the criminal justice system are directly tied to competing notions of justice. One who believes that criminals must be caught at any expense might subscribe to the approach "the bigger the database, the better" to DNA data banking. But even for those of us who agree that procedural fairness and equity are of extraordinary importance in achieving justice, the question whether more DNA will further that cause is still murky at best. As Simon Cole writes, "Seemingly egalitarian proposals, such as a universal DNA database, will not necessarily have equalizing effects. Whether DNA profiling technology as a whole mitigates or exacerbates inequality does not yield a single, simple answer. Rather, the answer depends on the 'equality of what?' . . . and the equality of whom."[3]

Differences of opinion about the use and expansion of forensic DNA technology are also entrenched in different understandings of the role and limitations of science. When DNA was introduced into the criminal justice system, a number of experts testified that false positives were impossible in DNA testing.[4] A defense attorney in Virginia characterized the perceived infallibility of DNA as follows: "If you put God on the witness stand . . . and God's testimony conflicted with the DNA evidence, everyone would automatically say, 'Why is God lying like this?' "[5] No credible scientist would claim that DNA is infallible today, and the laboratory scandals that have arisen around the country have made clear that human error can and does occur in the processing and typing of DNA samples (see chapter 16). Nevertheless, there is still a great divide between those who see these errors as exceptions to the rule of science as an objective and truth-telling endeavor and those who instead view it as an inherently social enterprise open to human fallibilities.

There is a very clear difference in how these perspectives become operationalized in policy. For those who believe that all the scientific and technical issues have been resolved, the focus is on the truth revealed by the science. Moreover, DNA science becomes the ultimate and unquestioned authority for discovering guilt or innocence. Other sources of knowledge are secondary and defer to DNA. It was perhaps this type of near-blind faith in DNA to deliver justice that led to the embarrassing investigational blunder in Germany discussed in the introduction to this book. On the other hand, for those who embrace the idea that, even with strong consensus, science is a social enterprise, then its vulnerability to error, bias, and misuse must always be a continuous presence in law enforcement. Moreover, the complexity of DNA testing relative to nontechnological methods of criminal investigation, such as eyewitness identification, means that DNA procedures and processes are left to a narrow class of "experts" and are not subject to public scrutiny. The question whether justice is advanced by DNA must take into account the potential that error, abuse, or misuse can occur unchecked or unreported as a result of this limitation.

Previous chapters have discussed the impact of forensic DNA databases on three groups of individuals: those convicted of and serving sentences for a crime, those arrested for or charged with a crime, and those who come under surveillance by way of their DNA (for example, through a dragnet or through a familial relation with an individual on the database). Certainly questions of fairness weigh most heavily when DNA is taken, stored, and analyzed from the innocent, or when the innocent, simply by nature of their DNA, fall under the lens of suspicion.

In cases where police, in their effort to find suspects and solve crimes, suspend or exceed the protections that suspicionless people traditionally have been afforded, the question whether DNA is advancing justice becomes especially relevant. We have illustrated violations of personal liberty in cases where police, in pursuit of an unknown perpetrator, undertake a DNA dragnet in a manner whereby individuals are coerced or pressured to give a DNA sample when there is no probable cause that links them to the crime. Efforts by law enforcement to coerce, intimidate, scare, or penalize individuals who choose not to participate in a DNA dragnet constitute a form of injustice, despite the best motives of police and prosecutors to bring about retributive justice on behalf of the crime victims and the state. Convicting alleged felons by "any means necessary" is a form of "frontier justice" that violates safeguards guaranteed by the Constitution.

The surveillance, intimidation, or coercion of innocent family members of a convicted felon whose DNA fulfilled the criterion of a low- or moderate-stringency match in a forensic DNA database against a crime-scene sample violates the presumption of innocence afforded to the members of that family. The mere probability of "DNA resemblance" without any other evidence of probable cause or suspicion is not sufficient to violate a person's presumed innocence or civil liberties. Although the science of DNA identification has an important role to play in affirming or disaffirming a match between crime-scene biological evidence and the DNA of a suspect, there is nothing in the science or in our accepted principles of social justice that justifies using DNA science as a "net" to collect suspects based on a blunt notion of "DNA resemblance." Once the idea of familial searches is accepted, it can easily entangle larger numbers of innocent people in the web of criminal police investigations under the rationalization that genetic resemblance raises the probability that one is implicated in criminal activity. As was pointed out in chapter 4, familial searches disproportionately affect certain ethnic, racial, and socioeconomic groups that tend to have larger families, as well as people who happen to have a relative who was convicted of a felony (and in some states those who have been arrested for, but not convicted of, a crime). The practice of familial searches in DNA databases reinscribes into our cultural mind-set the idea of "crime families" or groups whose members have a genetic propensity toward criminal behavior.

Of what relevance to our sense of justice are the collection and deposition in a database by law enforcement of the forensic profiles of individuals who have never been convicted of a felony offense or even charged with one? Criminal investigations will inevitably lead to some blind alleys, turn up refuted initial hypotheses about suspects, or incur investigational errors and forensic misconduct. As a matter of fairness, why should errors or false hypotheses about a crime suspect, who is presumed innocent, place that individual under permanent DNA surveillance? That is precisely the injustice committed by maintaining the forensic DNA profiles and biological samples of some selected class of innocent individuals. Under the Equal Protection Clause of the Fourteenth Amendment, it would seem reasonable to presume that all innocent people (people who have never been convicted of a felony) should be treated equally with respect to the collection and banking of their DNA by the government. If police are more likely to bring in suspects from certain neighborhoods for questioning in states that include arrestee forensic DNA profiles in state databases, citizens in those communities will be disproportionately represented in CODIS, and as a result their DNA profiles remain under the constant eye of criminal justice. A person picked up for exceeding the speed limit may someday be routinely checked to see if his or her DNA profile is in CODIS. The mere appearance of that person's profile in CODIS can have a stigmatizing impact, especially when racial minorities and individuals in lower socioeconomic strata are more likely to show up.

Another consideration of justice involving DNA forensic science arises in the proper balancing of societal interests in solving crimes against the privacy interests of individuals (chapter 14). The courts have a long tradition of balancing these interests. Thus when the use of surveillance technology has a proven record of preventing or solving crimes, the courts have sometimes ruled in favor of the state's interest over the protection of individual privacy. Even in balancing, the courts place controls on an overzealous state. Certainly the technology must be shown to work effectively, such as radar used by police on highways. But even in cases where the technology is efficacious, such as infrared detectors, courts have required a warrant when the technology is used to detect criminal activity in one's home.[6] In chapter 17 we have cited evidence that expanding DNA databases to include innocent arrestees reaches a point of diminishing returns in the state's interest in solving crimes. If we consider that "balancing" of interests is a form of social justice, we should recalibrate the balance when arrestee databases are not achieving their goals and at the same time are placing increasing numbers of innocent people under "genetic surveillance." Bruce Budowle, one of the original architects of CODIS and a strong supporter of the use of DNA databases to develop investigational leads in criminal cases, has pointed out that the actual utility of the database system is currently not known:

> There is no indication if the tax payer has gotten his/her money's worth regarding solving crime or whether a victim's case will be resolved because sufficient resources and processes are not in place to assess the overall performance of CODIS. Simply put, the actual numbers of success are not known. Therefore, we are left only with balancing decisions of expansion and privacy on the value of individual victims, the number of hits, and the assumption that most hits translated into successful investigative leads.[7]

As for the category of individuals who have been convicted and sentenced to prison terms for felony offenses, as we have discussed in chapter 14, the courts have made clear that the balancing of interests points in the direction of the state in questions whether convicted felons should have to turn over their DNA. That balancing becomes less clear when the question turns to individuals convicted of minor crimes or crimes where DNA evidence is generally not of interest or use (e.g., low-level drug possession), or when the individuals involved are juveniles.

We have also discussed the role of DNA in those cases where incarcerated individuals have sought to prove "actual innocence" (see chapter 7). In these cases individuals with a claim of innocence may obtain access to appropriate DNA testing that might exclude them as the perpetrator of a crime for which they have been tried and convicted. Over 250 cases of "actual innocence" have been accepted by the courts and have resulted in exonerations. Several principles of justice are intertwined with these types of cases.

There are a number of reasons that the American judicial system has, to a large degree, adopted a principle of finality in regard to litigation. Once an individual has been afforded all the legal remedies available to him or her, including the possibility of entering new exculpatory evidence after conviction, an incompetent defense attorney, or procedural errors in the trial, the system begins the process of finalizing the decision, making it increasingly difficult to reopen the conviction. Without finality, it is argued, the courts would be overburdened with spurious postconviction claims. The system seeks to frontload all the opportunities for defending one's innocence and for making appeals, leaving very few opportunities beyond the last appeals process. But forensic DNA has resulted in a reconsideration of judicial finality because of its exculpatory power.[8] Where an individual's DNA profile is not included in the profile of well-preserved DNA evidence known to have come from the perpetrator, it means that this individual did not leave behind his or her DNA and someone else did. Unlike the case of a so-called match where *inclusion* of the profile in the crime-scene DNA could be explained by contamination, mixtures, or other potential complications, a nonmatch of quality DNA samples is a definitive exclusion. Particularly where there is no other concrete evidence linking that individual to the crime, it is hard to imagine how a state can possibly allow a person to continue to serve time for a crime under this type of scenario. Nonetheless, as we saw in chapter 7, while state authorities have fully embraced the use of DNA to place individuals behind bars, some have been far more reluctant to open the door to post-conviction DNA testing.

The principle of "DNA exculpatory justice" may be stated as follows: if a claim of "actual innocence" can be definitively resolved at a reasonable cost by DNA testing, then it would be an injustice if the state refused to provide the defendant the preserved crime-scene DNA evidence for comparison with his or her DNA. Not all states provide the same opportunities for post-conviction testing. DNA exculpatory justice is also compromised when local police departments exercise different standards for preserving postconviction DNA evidence. Those who have been falsely convicted, where exculpatory DNA evidence is available, even after having served their sentence, cannot feel whole again until they have been afforded the opportunity to use DNA to prove their innocence. This raises the question whether the scales of justice can be balanced when the protection of DNA evidence varies across jurisdictions.

The concept of justice in the criminal justice system must include the opportunity to correct mistaken convictions when it is generally recognized that wrongful convictions do sometimes occur. No society can aspire to an ideal of justice if it does not take every reasonable precaution to prevent innocent people from having their liberties taken away by the state, or, when it has done so, it is recalcitrant to correcting its mistakes.

Toward a Vision of Justice

The vision of justice to which the "criminal justice system" aspires should be based on a proper balance between the protection of the civil liberties, presumed innocence, and procedural rights of persons and the needs of the state to apprehend, punish, and rehabilitate perpetrators of crime. Ideally this is achieved by a type of maximin principle where retributive justice (convicting perpetrators of crime) is maximized under the constraint of minimizing intrusions into the protections of individual privacy, autonomy, and the presumption of innocence.

On the basis of this general vision of justice, we offer the following set of axioms and commentaries on the responsible uses of forensic DNA in law enforcement. These axioms seek to balance the concept of justice for victims of crime with the ideals of justice for the presumed innocent and convicted felons who seek to prove "actual innocence" through postconviction DNA testing. They are also designed to bring the medical and forensic systems into greater concordance by adopting the principle that all people have an expectation of privacy of their genetic information.

Axiom 1. Genetic information collected in the law-enforcement context should be protected in a manner that is consistent with the protections afforded to medical information.

Forensic biological samples are analogous to medical records in that they contain highly intimate and personal information, including information pertaining to disease predisposition, genetic abnormalities, recessive genes, paternity, and immunological sensitivities. Therefore, the privacy of genetic information protected in medical records should carry over to other sources and uses of one's genetic code. One should be guided by the sensitivity of the source information and not by how the information was acquired or the purpose for which it is intended to be used. As we discussed in chapter 14, while the DNA profiles (noncoding regions of DNA) contain only limited information about an individual, the source biological samples that are collected in order to profile the DNA are permanently retained by law enforcement.

There is near unanimity that, beyond the person's caregivers, the privacy of an individual's medical genetic information should be protected, and as a result, many states and the federal government have passed legislation to protect their residents from unauthorized access to and use of medical genetic information. However, quite the opposite is taking place in the criminal justice system, where the collection of people's DNA without their consent has expanded. As long as agencies of criminal justice maintain databases that include the complete genome of individuals who are charged with, arrested for, and/or convicted of crimes, there will be a bifurcated system in which the same information available through DNA analysis will be protected under one set of public and private institutions while being unprotected under another.

Axiom 2. People have an expectation of privacy in the informational content of their DNA regardless of where it has been obtained or acquired (on their person, shed, within medical records).

A corollary to this axiom is that it is an invasion of one's privacy when his or her DNA is analyzed outside the scene of a crime. Police should not have the right to analyze DNA for medical, paternity, ancestry, or behavioral information without a warrant. Covert involuntary DNA sampling by amateurs, private investigators, and police should be prohibited unless there is a court order.

Axiom 3. People have a prima facie but not fundamental right to withhold their identity.

In *Hiibel v. Nevada* the U.S. Supreme Court ruled that a person does not have a constitutional right to withhold his or her identity. On the other hand, police cannot stop a person without reasonable suspicion simply to acquire the individual's identity.[9] Thus, if we extrapolate from *Hiibel v. Nevada*, even if DNA were used exclusively for "identification purposes," there are still limits on what police can do to obtain DNA identity. Law enforcement may have to meet a just-cause or reasonable-suspicion requirement to create a DNA profile.

It is important to note that DNA is not simply being collected "for identification purposes." The identity of individuals arrested or convicted is already known or can be made known through means far simpler and more efficient than DNA testing. The purpose of collecting DNA from known individuals is not principally one of identification, but rather of investigation and inculpation.

Axiom 4. The taking of DNA constitutes a search. Therefore, in order for the police to forcibly collect DNA from an individual suspected of a crime, they must have a warrant supported by probable cause.

In *United States v. Mitchell* the U.S. District Court (Western District of Pennsylvania) addressed whether the government could collect a sample of the defendant's DNA before trial and without a warrant. This case directly challenged the federal law enacted in 2006 that granted the U.S. attorney general the authority to collect DNA from individuals arrested or non-U.S. persons detained under federal authorities. In its ruling the court noted:

> A DNA profile generates investigatory evidence that is primarily used by law enforcement officials for general law enforcement purposes. To allow such suspicionless searches, which are conducted in almost all instances with law enforcement involvement, to occur absent traditional warrant and probable cause requirements will intolerably diminish our protection from unreasonable intrusion afforded by the Search and Seizure Clause of the Fourth Amendment.[10]

Axiom 5. DNA data banks should be limited to DNA profiles from persons who are convicted of felonies.

The United States and other countries should follow the wisdom of the European Court of Human Rights in rejecting the blanket and indiscriminate collection of DNA from persons suspected but not convicted of crimes. Individuals detained or arrested are presumed innocent and should not have their DNA and accompanying profiles shared in a database.

The state has an obligation under its public safety mandate to maintain accurate records of convicted felons, for it is not unusual for such felons to commit additional crimes or to take on new identities to mask their criminal record. Nevertheless, convicted felons, whether or not they are incarcerated, do not lose all their privacy rights under the Fourth Amendment. For example, informed consent may still apply to a prisoner's DNA when it is sought for research purposes. Beyond the storage of the DNA profile in the database for use in linking the offender to additional or future crimes, convicted felons should still retain privacy rights over the coding sequences of their genome without the state's overriding interest being demonstrated.

In cases where the DNA from a suspect is collected by way of a warrant, and the charges against that suspect are dropped or the individual is not convicted, the individual's DNA profile should be expunged automatically from the police record, and its biological source should be destroyed. Responsibility for expungement should rest with law enforcement; no petition or written request from the individual should be required.

Axiom 6. Written informed-consent procedures and proper protections against coercion should be in place for warrantless searches of nonsuspect DNA samples when police engage in voluntary DNA dragnets.

As noted in chapter 3, DNA dragnets have been conducted with varying degrees of police neglect of informed voluntary consent by suspicionless individuals who are asked to submit a DNA sample. In the very least, procedural guidelines are needed that establish a proper balance between individual privacy, informed consent, and law-enforcement goals.

DNA dragnets should be used by police only as a last resort and should be limited in scope to those who had access to the victim or who match a detailed description of the perpetrator. Those approached to provide DNA samples should be informed of their rights of refusal. Samples and profiles should be destroyed upon close of the investigation.

Axiom 7. Police seeking to acquire and analyze the DNA of family members of an individual identified through a partial match must obtain a warrant.

There are currently no consent procedures, warrants required, or national guidelines for familial searches. States, including New York and California, are beginning to introduce rules that allow law-enforcement authorities to conduct familial searches under their own standards.[11] DNA sweeps carried out through familial searches inculpate suspicionless individuals. In some families this can include parents, children, and siblings. Therefore, by Axiom 4, police must obtain a warrant if they seek to acquire and analyze DNA of a family member of someone identified through a partial match. Being a family member of someone whose DNA has been partially matched with DNA left at a crime scene should not, by itself, constitute probable cause.

Axiom 8. Surreptitious taking, testing, or storing of DNA from suspects or their relatives is a violation of a person's privacy and should be prohibited.

The current dominant framework that assumes that DNA collected from coffee cups, cigarette butts, and saliva samples is "abandoned" allows police to pick up DNA anywhere, from anyone, at any time, and is in direct conflict with the most fundamental notions of genetic privacy. As discussed in chapter 6, thus far no court has ruled against obtaining DNA from discarded objects through surreptitious means. As legal analyst Elizabeth Joh states, "The collection of abandoned DNA by police threatens the privacy rights of everyone. The law permits it, and the police seek it. Advances in molecular genetics will permit ever greater exploitation of that personal information once it is acquired."[12] Joh refers to sampling DNA on discarded objects as "covert involuntary DNA sampling."[13]

Axiom 9. The analysis of crime-scene DNA should be limited to identity and to those externally perceptible traits whose DNA markers have been scientifically validated.

At a crime scene all materials, including DNA, that could help police determine the identity of the victim and perpetrators and/or the methods used in the criminal activity are open to forensic investigation without warrants. However, such investigation must be strictly limited to standard forensic DNA analysis of the 13 STRs (or equivalent in other countries) and nonsensitive, nonstigmatizing, externally perceptible traits, such as hair color and stature. Any attempts to mine crime-scene DNA to make predictions about the medical or behavioral characteristics of the alleged perpetrator should be prohibited.

Axiom 10. Offender or suspect biological samples should be destroyed after DNA profiling so that the encoded information cannot be accessed for information beyond the DNA profile.

The most significant privacy concerns with DNA collection relate to the stored biological samples. The only way to ensure that misuses of the samples do not occur is to destroy the biological source of DNA after a DNA profile is generated. The biological sample is not necessary to link the source individual to DNA left behind at a crime scene—all the information that is needed is contained in the DNA profile. Since the individuals are known, another sample can always be collected if needed. If for some reason additional coding sequences of a felon's DNA are needed for an investigation, a warrant can be sought to obtain that information.

The development and implementation of policies for collecting DNA profiles in a national data bank has followed a path that some political scientists and policy experts call "disjointed incrementalism." Each state sets its own guidelines and rules in the context of de minimis federal standards. Step-by-step expansions of the reaches of the database have served to mask the long-term downsides to a system that places inadequate emphasis on overarching principles of privacy and justice in the quest to solve crime. The pitfalls of piecemeal policymaking have been worsened by the fact that many of the decisions with regard to DNA collection and use by law enforcement have been made by default and have not been reviewed in the courts.

In this book we have tried to provide a holistic and coherent picture of forensic DNA data banks. Building on data and studies from the United States, Asia, Australia, and Europe, we have sought to illustrate both the assets and liabilities of the technology for its use in criminal justice. The principles we have outlined in this chapter arise out of our exploration of the impacts of DNA expansion on privacy and justice and our examination of questions of efficacy and fallibility. Much remains to be done to establish this powerful technological tool in a manner that conforms to our sense of fairness and justice and balances the values to society of solving felony crimes, freeing innocent and wrongly convicted individuals, and preserving our rights of privacy.

APPENDIX: A COMPARISON OF DNA DATABASES IN SIX NATIONS

	United States	England and Wales	Japan	Australia	Germany	Italy
National system	Combined DNA Index System (CODIS).	National DNA Database (NDNAD).	DNA Profile National Database System.	National Criminal Investigation DNA Database (NCIDD).	DNA-Analyse-Datei (DAD).	The national DNA database is not yet operational.
Relevant national law	DNA Identification Act (1994).	Police and Criminal Evidence Act 1984 (PACE).	No legislative authorization; National Public Safety Commission Regulation no. 15.	Crimes Act 1914 (Part 1D) (as amended by the Crimes Forensic Procedures Act 1998 and subsequent amendments).	Statute on Identification through DNA Testing; Code of Criminal Procedure (both enacted in 1997).	Law No. 85 (2010); Pisanu Law (2005).
Agency operating database	Federal Bureau of Investigation (FBI).	Forensic Science Service (FSS).	National Police Agency.	CrimeTrac Agency.	Bundeskriminalamt (BKA).	RIS (Reparto Investigatorivo Speciale) and the Scientific Police currently operate data banks. The national data bank will be housed within the Department of Public Security in the Home Office with a Central Laboratory in the Ministry of Justice.
Number of loci	13	10	15	9	8	9

(continued)

APPENDIX (CONTINUED)

	United States	England and Wales	Japan	Australia	Germany	Italy
Categories of individuals included	Anyone convicted of a crime; anyone arrested or non-U.S. persons detained by federal authorities.	Anyone arrested for any recordable offense.	Anyone suspected of an offense where DNA evidence has been obtained from the crime scene.	Those convicted of offenses carrying a penalty of five years imprisonment or more.	Individuals convicted of serious crimes, where there is probable cause to assume that the persons will be involved in similar crimes in the future.[1]	Individuals convicted of any crime punishable by three or more years of imprisonment, as well as arrestees and juveniles charged with similar offenses.
DNA collection from juveniles	Yes.	Juveniles aged 10 and older.	15 and older (14 and younger are not punishable for a crime).	DNA collection requires a magistrate's order or consent by a parent or guardian. Consent by a parent or guardian is insufficient if the child objects to the procedure.	Yes, 14–17 years old, but only with guardians' consent.	Yes.
DNA collection from suspects	Requires informed consent.	With consent.	Requires a warrant to obtain a sample.	Requires probable cause and reason to believe that the forensic procedure is likely to produce evidence tending to confirm or disprove the suspect's guilt.	Requires a court order; suspects must be deemed at risk for repeat offenses before their profile can be entered into the database.	Anyone arrested and charged for a crime punishable by three years imprisonment or more.

Sample retention and destruction	Indefinite retention.	Retention for 100 years.	Biological samples destroyed after successful profiling.	Forensic material must be destroyed if proceedings are not initiated against a suspect within 12 months after collection of forensic material, if charges are dropped, or if an individual is acquitted or a conviction is overturned.	Samples are routinely destroyed after successful profiling.	Samples may be retained for a maximum of 20 years and be immediately destroyed if the person is proved innocent; individuals must petition to request removal of DNA samples.
Profile retention	Indefinite.	100 years.	Retained regardless of conviction. Profiles expunged after suspect has died or is determined to be no longer useful.	Government is under no obligation to remove data entries; proposals have been put forth to require destruction of both the samples and profiles in cases where a suspect has been eliminated or where charges are dropped.	Profiles are destroyed upon acquittal or discontinuance of criminal proceedings. All database records are subjected to a case review after 10 years to determine whether they should be retained in the database or can be removed.	Profiles may be kept for a maximum of 40 years and destroyed upon acquittal or discontinuance of criminal proceedings; individuals must petition to request removal of profiles from database.

(*continued*)

APPENDIX (CONTINUED)

	United States	England and Wales	Japan	Australia	Germany	Italy
Surreptitious sampling	Yes.	Yes.	Yes; under debate whether this requires a warrant.	Unknown.	No.	Yes.
Familial searching	Yes.	Yes.	No.	Unknown; DNA tests have been performed on relatives of absent suspects as a way of building a case against that suspect.	No; would likely violate existing law.[2]	Unknown.

[1] Alexander Hanebeck, "DNA Analysis and the Right to Privacy: Federal Constitutional Court Clarifies Rules on the Use of Genetic Fingerprints," *German Law Journal*, no. 3 (February 15, 2001), 2, http://www.germanlawjournal.com/index.php?pageID=11&artID=50 (accessed June 10, 2010).

[2] Peter M. Schneider, Institut für Rechtsmedizin, Universität Mainz, Mainz, Germany, personal communication with Tania Simoncelli, August 28, 2008.

NOTES

Introduction

1. Tristana Moore, "Germany Hunts Phantom Killer," *BBC News*, April 11, 2008, http://news.bbc.co.uk/2/hi/europe/7341360.stm (accessed May 17, 2009).

2. "DNA Bungle Haunts German Police," *BBC News*, March 28, 2009, http://news.bbc.co.uk/2/hi/europe/7966641.stm (accessed May 17, 2009).

3. Roger Boyes, "Phantom of Heilbronn: Hunt for the Killer Who Leaves Clues and Bodies," *The Times*, April 10, 2008, http://www.timesonline.co.uk/tol/news/world/europe/article3715800.ece (accessed May 17, 2009).

4. "DNA Clues in Hunt for 'Faceless' Serial Killer," *Telegraph*, April 14, 2008, http://www.telegraph.co.uk/news/worldnews/1584625/DNA-clues-in-hunt-for-faceless-serial-killer.html (accessed May 17, 2009).

1. Forensic DNA Analysis

1. Committee on DNA Forensic Science, National Academy of Sciences, *The Evaluation of Forensic DNA Evidence* (Washington, DC: National Academy Press, 1996), 36.

2. Barry Commoner, "Unraveling the DNA Myth," *Harper's* 304, no. 1821 (February 2002): 44.

3. Saurabh Asthana, William S. Noble, Gregory Kryukov, Charles E. Grant, Shamil Sunyaev, and John A. Stamatoyannopoulos, "Widely Distributed Noncoding Purifying Selection in the Human Genome," *Proceedings of the National Academy of Sciences of the United States of America (PNAS)* 104 (July 17, 2007): 12410–12415.

4. John M. Greally, "Encyclopedia of Humble DNA," *Nature* 447 (June 14, 2007): 782–783.

5. Colin Nickerson, "DNA Study Challenges Basic Ideas in Genetics: Genome 'Junk' Appears Essential," *Boston Globe*, June 14, 2007, A1.

6. Oak Ridge National Laboratory, U.S. Department of Energy, "Human Genome Project Information," http://www.ornl.gov/sci/techresources/Human_Genome/elsi/forensics.shtml (accessed March 27, 2010).

7. William C. Thompson and Dan E. Krane, "DNA in the Courtroom," in *Psychological and Scientific Evidence in Criminal Trials*, ed. Jane Campbell Moriarty (Minneapolis: West Publishing, 2003), 11-4, 11-5.

8. Reprinted from National Institute of Standards and Technology, Chemical Science & Technology Laboratory, http://www.cstl.nist.gov/strbase/fbicore.htm (accessed March 27, 2010).

9. L. A. Zhivotovsky, "Population Aspects of Forensic Genetics," *Russian Journal of Genetics* 42, no. 10 (2006): 1426–1436.

10. William C. Thompson, Simon Ford, Travis Doom, Michael Raymer, and Dan Krane, "Evaluating Forensic DNA Evidence," *Champion* (April 2003): 23.

11. Erin Murphy, "The Art in the Science of DNA: A Layperson's Guide to the Subjectivity Inherent in Forensic DNA Typing," *Emory Law Journal* 58 (2008): 496–512, quotation at 503.

12. Michael R. Bromwich (Independent Investigator), "Final Report of the Independent Investigator for the Houston Police Department Crime Laboratory and Property Room," June 13, 2007, http://www.hplabinvestigation.org (accessed May 22, 2010), quotation at 5.

13. Simon A. Cole, "How Much Justice Can Technology Afford? The Impact of DNA Technology on Equal Criminal Justice," *Science and Public Policy* 34, no. 2 (March 2007): 95–107, quotation at 95.

14. Michael Lynch, Simon A. Cole, Ruth McNally, and Kathleen Jordan, *Truth Machine: The Contentious History of DNA Fingerprinting* (Chicago: University of Chicago Press, 2008), 23.

2. The Network of U.S. DNA Data Banks

1. John M. Butler, *Forensic DNA Typing* (Burlington, MA: Elsevier, 2005), 445.

2. Stephen R. Reinhardt, Dissenting Opinion, *United States v. Kincade*, U.S. Court of Appeals, Ninth Circuit, No. 02-50380, D.C. No. CR-93-00714-RAG-01, filed August 18, 2004.

3. Federal Bureau of Investigation, CODIS Combined DNA Index System, http://www.fbi.gov/hq/lab/html/codisbrochure_text.htm (accessed April 19, 2010).

4. Dwight E. Adams, deputy assistant director, Laboratory Division, Federal Bureau of Investigation, "The FBI's DNA Program," testimony before the House Committee on Government Reform, Subcommittee on Government Efficiency, Financial Management and Intergovernmental Relations, June 12, 2001, Attachment A, http://www.fbi.gov/congress/congress01/dwight061201.htm (accessed February 9, 2007).

5. M. Dawn Herkenham, "Retention of Offender DNA Samples Necessary to Ensure and Monitor Quality of Forensic DNA Efforts: Appropriate Safeguards Exist to Protect DNA Samples from Misuse," *Journal of Law, Medicine and Ethics* 34, no. 2 (Summer 2006): 380–384, at 380.

6. Dwight E. Adams, deputy assistant director, Laboratory Division, Federal Bureau of Investigation, "The FBI's DNA Program," testimony before the House Committee on Government Reform, Subcommittee on Government Efficiency, Financial Management and Intergovernmental Relations, June 12, 2001, http://www.fbi.gov/congress/congress01/dwight061201.htm (accessed April 19, 2010).

7. Herkenham, "Retention of Offender DNA Samples," 382.

8. Butler, *Forensic DNA Typing*, 439.

9. Adams, "The FBI's DNA Program."

10. 42 U.S.C. § 14135(a); see also Public Law No. 107-56 § 503, 115 Stat. 272, 364 (2001).

11. See 42 U.S.C. § 14135(a)(1).

12. Justice for All Act, Title II, Section 203 (H.R. 5107, P.L. 108-405), http://www.ojp.usdoj.gov/ovc/publications/factshts/justforall/content.html (accessed May 23, 2010).

13. See "Combined DNA Index System," http://en.wikipedia.org/wiki/Special:Search?search=Combined+DNA+Index+System%2C&go=Go (accessed April 17, 2010).

14. Public Law No. 109-162, Section 1004.

15. Department of Justice, "DNA-Sample Collection and Biological Evidence Preservation in the Federal Jurisdiction," *Federal Register* 73, no. 238 (December 10, 2008): 74932–74943.

16. Ibid., 74935.

17. Department of Justice, "DNA-Sample Collection Under the DNA Fingerprint Act of 2005 and the Adam Walsh Child Protection and Safety Act of 2006 (Proposed Rule)," *Federal Register* 73, no. 76 (April 18, 2008): 21083–21087.

18. Ibid., 21084.

19. Ibid.

20. J. Luttman, FBI Laboratory, DNA Unit 1, Federal Convicted Offender Program, "Implementation of Database Expansion" (presentation for Annual CODIS Conference, Federal Bureau of Investigation, October 24, 2006).

21. Department of Justice, "DNA-Sample Collection Under the DNA Fingerprint Act of 2005," 21084.

22. Ibid., 21087.

23. American Civil Liberties Union, "Comments on RIN 1105-AB24 Proposed Rule, DNA-Sample Collection Under the DNA Fingerprint Act of 2005 and the Adam Walsh Child Protection and Safety Act of 2006," prepared by Caroline Fredrickson, Michael Macleod-Ball, Tania Simoncelli, and Michael Risher, May 19, 2008, http://www.aclu.org/racial-justice_prisoners-rights_drug-law-reform_immigrants-rights/aclu-comments-justice-department-r (accessed April 18, 2010).

24. Federal Bureau of Investigation, CODIS Bulletin, "Interim Plan for the Release of Information in the Event of a Partial Match at NDIS," July 20, 2006,

http://www.bioforensics.com/conference08/Familial_Searches/CODIS_Bulletin.pdf (accessed April 12, 2010).

25. Ibid., 1.

26. Susan Price Livingston, "DNA Database of Convicted Felons," *OLR Research Report* (December 20, 2002), http://www.cga.ct.gov/2002/olrdata/jud/rpt/2002-R-0984.htm (accessed August 4, 2008).

27. Steven Messner, "Comment: Law Enforcement DNA Database: Jeopardizing the Juvenile Justice System Under California's Criminal DNA Collection Law," *LaVerne Law Review: Journal of Juvenile Law* 28 (2007): 159–173, quotations at 172–173.

28. *Haskell v. Brown,* Class Action Complaint for Declaratory & Injunctive Relief (submitted on behalf of Plaintiffs), United States District Court for the Northern District of California, filed October 7, 2001, ¶ 6 (citing California Department of Justice).

29. American Civil Liberties Union of Northern California, "ACLU Lawsuit Challenges California's Mandatory DNA Collection at Arrest," October 7, 2009, http://www.aclunc.org/news/press_releases/aclu_lawsuit_challenges_california's_mandatory_dna_collection_at_arrest.shtml?ht=dna%20dna (accessed April 18, 2010).

30. South Carolina General Assembly, 117th Session, 2007–2008, passed H.3304; ratified June 12, 2007. Under the bill, section 2, A. Section 23-3-620 of the 1976 Code of South Carolina would have been amended to read: "Section 23-3-620. (A) Following a lawful custodial arrest or a direct indictment for a felony offense or an offense that is punishable by a sentence of five years or more, either of which is committed in this State, the person arrested must provide a saliva or tissue sample from which DNA may be obtained for inclusion in the State DNA Database." http://www.scstatehouse.gov/sess117_2007-2008/bills/3304.doc (accessed May 23, 2010).

31. Governor Mark Sanford, South Carolina, letter to Robert W. Harrell Jr., Speaker of the House of Representatives, South Carolina, June 18, 2007.

32. Kevin Johnson, "DNA Not Kept in Half of States," *USA Today*, August 5, 2008.

33. Betty Layne DesPortes, Richmond, VA, lawyer with Benjamin and DesPortes, P.C., personal communication with Sheldon Krimsky, July 27, 2009.

34. Maria Gold, "Letters to Inform 400 Felons of DNA Evidence Retesting," *Washington Post*, August 7, 2008, B04.

35. Maria Gold, "Va. DNA Project Is in Uncharted Territory," *Washington Post*, August 17, 2008, C01.

3. Community DNA Dragnets

1. Joseph Wambaugh, *The Blooding* (New York: Morrow, 1989), 168.

2. Aaron B. Chapin, "Arresting DNA: Privacy Expectations of Free Citizens Versus Post-convicted Persons and the Unconstitutionality of DNA Dragnets," *Minnesota Law Review* 89 (June 2005): 1842–1875, quotation at 1867.

3. Amanda McElfresh, "Man Sues to Retrieve DNA Sample," *Daily Reveille* (Baton Rouge: Louisiana State University), August 24, 2003, http://media.www.lsureveille.com/media/storage/paper868/news/2003/08/24/news/man_sues.To.Retrieve.DNA.Sample-2046570.shtml (accessed October 30, 2007).

4. Kevin Bersett, "Victims Challenge Police Use of Controversial DNA Dragnet," *New Standard* (Syracuse, NY), September 27, 2004.

5. According to Merriam-Webster's dictionary, a dragnet is "a network of measures for apprehension (as of criminals)"; http://www.merriam-webster.com/dictionary/dragnet (accessed December 10, 2008).

6. Samuel Walker, "Police DNA 'Sweeps' Extremely Unproductive" (report by the Police Professional Initiative, Department of Criminal Justice, University of Nebraska, Omaha, September 2004).

7. Wambaugh, *Blooding*.

8. Mark A. Rothstein and Meghan K. Talbott, "The Expanding Use of DNA in Law Enforcement: What Role for Privacy?" *Journal of Law, Medicine & Ethics* 34 (Summer 2006): 153–162, reported that the German DNA dragnet collected samples from 16,400 men. See page 156. *Science* reported that 18,000 men were tested. See "Gene Hunt," *Science* 313, no. 5789 (August 18, 2006): 897.

9. "Random Samples," *Science* 313, no. 5789 (August 18, 2006): 897.

10. A. R. T. Bates, M. Hequet, and R. Laney, "The DNA Dragnet," *Time* (January 24, 2005), 39–40.

11. Carole McCartney, *Forensic Identification and Criminal Justice: Forensic Science, Justice and Risk* (Cullompton, UK: Willan Publishing, 2006), 5.

12. Chapin, "Arresting DNA," 1858.

13. Adam M. Gershowitz, "The iPhone Meets the Fourth Amendment," *UCLA Law Review* 56 (October 2008): 27–58, quotation at 27.

14. Glynn Wilson, "In Louisiana, Debate Over a DNA Dragnet," *Christian Science Monitor*, February 21, 2003.

15. *Kohler v. Englade*, 470 F. 3d 1104, No. 05-30541, U.S. Court of Appeals, Fifth Circuit, November 21, 2006.

16. Rosemary Roberts, "Open Your Mouth for a DNA Swab," *News and Record* (Greensboro, NC), January 14, 2005.

17. Jeffrey S. Grand, "The Blooding of America: Privacy and the DNA Dragnet," *Yeshiva University Cardozo Law Review* 23 (2002): 2277–2323, quotation at 2307.

18. Ibid., quotation at 2317.

19. Peter J. Neufeld, member, New York State's Forensic Science Review Board, testimony at the Subcommittee on Crime, Terrorism, and Homeland Security, July 17, 2003, http://commdocs.house.gov/committees/judiciary/hju88394.000/hju88394_0f.htm (accessed March 25, 2010).

20. Associated Press, "DNA Dragnet Raises Concern," *St. Petersburg Times*, June 17, 2003.

21. Kathy Marks, "Christmas Day Killer Trapped by DNA Test," *Independent* (London), February 10, 1998.

22. Stephen Wright, "Trapped by a DNA Dragnet," *Daily Mail*, February 10, 1998.

23. Sepideh Esmaili, "Searching for a Needle in a Haystack: The Constitutionality of Police DNA Dragnets," *Kent Law Review* 82 (2007): 495–523, quotation at 520.

24. Ibid., 522.

25. Rothstein and Talbott, "The Expanding Use of DNA in Law Enforcement, " 156.

26. Wilson, "In Louisiana, Debate Over a DNA Dragnet."

27. Marc Rotenberg, Marcia Hoffman, and Melissa Ngo, Electronic Privacy Information Center, Brief of Amicus Curiae, *Shannon Kohler v. Pat Englade, et al.*, U.S. Court of Appeals, Fifth Circuit, Case No. 3:03-cv-00857-JJB-CN (2005), 6–7.

28. Maria Gold, "Police in Charlottesville Suspend 'DNA Dragnet,'" *Washington Post*, April 15, 2004, B01.

29. Lorraine M. Blackwell, "Virginia U. Suspends DNA Dragnet Locating Serial Rapist," *The Daily Texan*, May 4, 2004.

30. Esmaili, "Searching for a Needle in a Haystack," 516.

31. Grand, "Blooding of America," quoted at 2301.

32. Walker, "Police DNA 'Sweeps' Extremely Unproductive."

33. Ibid., 3. As of 2008 there were at least 20 DNA dragnets conducted throughout the United States since 1990. "Kohler v. Englade: The Unsuccessful Use of DNA Dragnets to Fight Crime," http://epic.org/privacy/kohler (accessed March 26, 2010).

34. Walker, "Police DNA 'Sweeps' Extremely Unproductive," 14–15.

35. Ibid., 4.

36. Chapin, "Arresting DNA," 1863.

37. Frederick R. Bieber and David Lazer, "DNA Sweep Must Be Accompanied by Informed Consent," *Provincetown Banner*, January 20, 2005.

38. Pam Belluck, "Slow DNA Trail Leads to Suspect in Cape Cod Case," *New York Times*, April 16, 2005, http://www.nytimes.com/2005/04/16/national/16arrest.htm?ref=christa_warthington (accessed August 13, 2010).

39. Rotenberg, Hoffman, and Ngo, Brief of Amicus Curiae.

40. Samuel Walker and Michael Harrington, "Police DNA 'Sweeps': A Proposed Model Policy on Police Requests for DNA Samples," Police Professionalism Initiative, University of Nebraska at Omaha, July 2005, http://www.unomaha.edu/criminaljustice/PDF/dnamodelpolicyfinal.pdf, 6 (accessed March 26, 2010).

41. Nuffield Council on Bioethics, *The Forensic Use of Bioinformation: Ethical Issues* (London: Nuffield Council on Bioethics, September 2007), 56, http://www.nuffieldbioethics.org/fileLibrary/pdf/The_forensic_use_of_bioinformation_-_ethical_issues.pdf (accessed March 26, 2010).

42. Rotenberg, Hoffman, and Ngo, Brief of Amicus Curiae, 7.

43. Chapin, "Arresting DNA," 1859, quoting Judge Stephen Reinhardt.

4. Familial DNA Searches

1. Daniel J. Grimm, "The Demographics of Genetic Surveillance: Familial DNA Testing and the Hispanic Community," *Columbia Law Review* 107 (June 2007): 1164–1194, quotation at 1164.

2. Tony Lake, chief constable, Lincolnshire Police, presentation before the FBI Symposium on Familial Searching and Genetic Privacy, Arlington, VA, March 17–18, 2008.

3. "Brick Thrower Jailed Over Death," *BBC News*, April 19, 2004, http://news.bbc.co.uk/2/hi/uk_news/england/3639057.stm (accessed April 12, 2010).

4. Forensic Science Service, "Case Files: Craig Harman—Family DNA Link Offers Crime Breakthrough," 2008, http://www.forensic.gov.uk/html/media/case-studies/f-39.html (accessed April 12, 2010).

5. Of course, it is also possible that "tipping off" a relative may warn the suspect that the police are onto him or her, which happened in the "stiletto shoe case" (see box 4.1).

6. European Network of Forensic Science Institutes (ENFSI), ENFSI DNA Working Group, "DNA-Database Management Review and Recommendations," April 2009, http://www.enfsi.eu/page.php?uid=98 (accessed April 12, 2010).

7. See Frederick R. Bieber, "Science and Technology of Forensic DNA Profiling: Current Use and Future Directions," in *DNA and the Criminal Justice System: The Technology of Justice,* ed. David Lazer (Cambridge, MA: MIT Press, 2004), 23–62. See also Ben Mitchell, "Police Warning to Criminals Over DNA Breakthrough," *Scotsman*, November 19, 2004.

8. Federal Bureau of Investigation, CODIS Bulletin, "Interim Plan for the Release of Information in the Event of a Partial Match at NDIS," July 20, 2006,

http://www.bioforensics.com/conference08/Familial_Searches/CODIS_Bulletin.pdf (accessed April 12, 2010).

9. Henry T. Greely, Daniel P. Riordan, Nanibaa' A. Garrison, and Joanna L. Mountain, "Family Ties: The Use of DNA Offender Databases to Catch Offenders' Kin," *Journal of Law, Medicine and Ethics* 34, no. 2 (Summer 2006): 248–262.

10. Ibid., 252.

11. Michael Chamberlain, deputy attorney general, DNA Legal Unit, California Department of Justice, Memorandum to Attorney General Jerry Brown, California, June 6, 2007, 3.

12. Interpol DNA Unit, *Global DNA Databank Inquiry: Results and Analysis 2002*, International Criminal Police Organization (I.C.P.O.), Interpol, General Secretariat, 2003, http://www.interpol.int/Public/Forensic/dna/inquiry/default.asp (accessed August 8, 2008).

13. Canada, National DNA Data Bank, "Technology" (last updated April 23, 2007), http://www.nddb-bndg.org/techno_e.htm (accessed September 8, 2008).

14. Kristen Lewis, University of Washington, Department of Biostatistics, statement at American Association of Forensic Science Annual Meeting, Washington, DC, February 18–23, 2008.

15. Ted Staples, Scientific Working Group on DNA Analysis Methods (SWGDAM) Ad Hoc Group on Partial Matches, presentation before the FBI Symposium on Familial Searching and Genetic Privacy, Arlington, VA, March 17–18, 2008. Committee recommendations presented available at *Forensic Science Communications*, F.B.I., SWGDAM Recommendations to the FBI Director on the "Interim Plan for the Release of Information in the Event of a 'Partial Match' at NDIS," vol. 11, no. 4, October 2009, http://www.fbi.gov/hq/lab/fsc/current/standard_guidlines/swgdam.html (accessed April 17, 2010).

16. Greely et al., "Family Ties," 252.

17. George Carmody, professor of biology, Carleton University, Ottawa, Ontario, statement at New York State Forensic Science Commission hearing, January 2008.

18. Greely et al., "Family Ties," 253.

19. David R. Paoletti, Travis E. Doom, Michael L. Raymer, and Dan E. Krane, "Assessing the Implications for Close Relatives in the Event of Similar but Non-matching DNA Profiles," *Jurimetrics* 46 (Winter 2006): 161–175.

20. Bruce Weir, "DNA Evidence: Inferring Identity," in *Encyclopedia of Life Sciences*, Gerry Melino, editor-in-chief, Volume on Genetics and Molecular Biology (Chichester, UK: John Wiley, 2006), 2, http://www.els.net, doi: 10.1038/npg.els.0005452 (accessed May 3, 2010).

21. For familial searches the likelihood ratio is the chance that a sibling (relative) has a matching profile divided by the chance that a randomly chosen unrelated person has a matching profile.

22. Charles Brenner, consultant in forensic mathematics, Oakland, CA, presentation before the New York State Forensic Science Commission, January 2008.

23. Frederick R. Bieber, Charles H. Brenner, and David Lazer, supporting online material for "Finding Criminals Through DNA of Their Relatives," *Science* 312 (June 2, 2006): 1315–1316 (originally published in *Science* Express on May 11, 2006), http://www.sciencemag.org/cgi/content/full/sci;1122655/DC1 (accessed April 15, 2010).

24. Frederick R. Bieber, Charles H. Brenner, and David Lazer, "Finding Criminals Through DNA of Their Relatives," *Science* 312 (June 2, 2006): 1315–1316, quotation at 1315.

25. George Carmody, professor of biology, Carleton University, Ottawa, Ontario, statement at New York State Forensic Science Commission hearing, January 2008.

26. Kristen E. Lewis, Bruce S. Weir, and Mary-Claire King, "Genomic Approaches to the Identification of Individuals Through Familial Database Searches" (paper presented at the American Academy of Forensic Sciences 60th Anniversary Scientific Meeting, Washington, DC, February 20–23, 2008).

27. Frederick Bieber and David Lazer, "Guilt by Association," *New Scientist* 184, no. 2470 (October 23, 2004), http://www.newscientist.com/article/mg18424703.000-guilt-by-association.html (accessed April 15, 2010).

28. Scott Michels, "Using a Relative's DNA to Catch Criminals," *U.S. News and World Report* (August 3, 2006), http://www.usnews.com/usnews/news/articles/060803/3data.htm (accessed April 15, 2010).

29. Federal Bureau of Investigation, CODIS Bulletin, "Interim Plan."

30. Ibid.

31. Ibid.

32. Thomas Callaghan, presentation at the 13th National CODIS Conference, CODIS Unit, Burlingame, CA, October 29–30, 2007.

33. Tom Callaghan, chief, CODIS Unit, FBI, presentation before the FBI Symposium on Familial Searching and Genetic Privacy, Arlington, VA, March 17–18, 2008.

34. DNA Identification Act of 1994 (Title XXI, Subtitle C of the Violent Crime Control and Law Enforcement Act, Public Law 103-322), 108 Stat. 1796.

35. Senator Herb Kohl (D-WI), statement in support of the DNA Analysis and Backlog Elimination Act of 2000 during the Senate floor debate, December 6, 2000.

36. National Research Council, *DNA Technology in Forensic Science* (Washing ton, DC. National Academy Press, 1992), 86–87.

37. U.S.C. § 552a(e)(4): "Each agency that maintains a system of records shall (4) . . . publish in the Federal Register upon establishment or revision a notice of the

existence and character of the system of records, which notice shall include A) the name and location of the system; B) the categories of individuals on whom records are maintained in the system; C) the categories of records maintained in the system; D) each routine use of the records contained in the system, including the categories of users and the purpose of such use; E) the policies and practices of the agency regarding storage, retrievability, access controls, retention, and disposal of the records."

38. Department of Justice, "Privacy Act of 1974: New System of Records," *Federal Register* 61, no. 139 (July 18, 1996): 37497 (emphasis added).

39. In response to a question raised by Barry Scheck, co-director of the Innocence Project, whether states had authority to do full-scale familial searching, only 3–4 hands were raised out of approximately 150 individuals in the room.

40. Tom Callaghan, quoted in Summary Report of the Virginia Scientific Advisory Committee's Subcommittee on Familial Searches, August 2007.

41. Molly McDonough, "Familial DNA Searches Are Creating Genetic Informants," *ABA Journal*, April 21, 2008, http://www.ABAJournal.com/mobile/article/familial_DNA_searches_are_creating_genetic_informants (accessed July 30, 2010).

42. Richard Willing, "DNA 'Near Matches' Spur Privacy Fight," *USA Today*, August 3, 2007, 3A.

43. Rockne Harmon, comment in response to question posed by Louisiana State Crime Lab director, FBI Symposium on Familial Searching and Genetic Privacy, Arlington, VA, March 17–18, 2008.

44. California Department of Justice, Division of Law Enforcement, "DNA Partial Match (Crime Scene DNA Profile to Offender) Policy," April 24, 2008.

45. Code of Massachusetts, Regulations, CMR Title 515 § 2.14, "Mutual Exchange, Use and Storage of DNA Records" (emphasis added).

46. Richard Pinchin, North American Operations Manager, Forensic Science Service, U.K., presentation before the New York State Forensic Science Commission hearing, January 2008. See also Robin Williams, "Making Do with Partial Matches: DNA Intelligence and Criminal Investigations in the United Kingdom" (presentation at DNA Fingerprinting and Civil Liberties: Workshop #2, American Society for Law, Medicine and Ethics, September 17–18, 2004); and Andrew Barrow, "Sex Attacker Snared by Family DNA," *Press Association Limited*, September 23, 2005.

47. Pinchin, presentation before the New York State Forensic Science Commission hearing.

48. Ibid. Mitch Morrissey, the district attorney in Denver, has also claimed that his pilot familial searching program has identified three cases in which there was a 90 percent chance of identifying a sibling link between crime-scene evidence and an individual in the Denver database. See Jeffrey Rosen, "Genetic Surveillance for All," *Slate* (March 17, 2009), http://www.slate.com/id/2213958/pagenum/all/ (accessed April 12, 2010).

49. Lewis et al., "Genomic Approaches to the Identification of Individuals."

50. Hugh Whittall, director, Nuffield Council on Bioethics, presentation before the FBI Symposium on Familial Searching and Genetic Privacy, Arlington, VA, March 17–18, 2008.

51. Robin Williams and Paul Johnson, "Inclusiveness, Effectiveness and Intrusiveness: Issues in the Developing Uses of DNA Profiling in Support of Criminal Investigations," *Journal of Law, Medicine and Ethics* 33, no. 3 (2005): 545–558, quotation at 555.

52. Nuffield Council on Bioethics, *The Forensic Use of Bioinformation: Ethical Issues* (London: Nuffield Council on Bioethics, September 2007), 80, http://www.nuffieldbioethics.org/fileLibrary/pdf/The_forensic_use_of_bioinformation_-_ethical_issues.pdf (accessed May 23, 2010).

53. Jeffrey Rosen, professor of law, George Washington University, presentation before the FBI Symposium on Familial Searching and Genetic Privacy, Arlington, VA, March 17–18, 2008.

54. U.S. Const., Art. III, § 3, Cl. 2.

55. Chamberlain, Memorandum to Brown, 5.

56. *Hill v. NCAA*, 7 Cal. 4th (1994).

57. Bieber et al., "Finding Criminals Through DNA of Their Relatives," 1316.

58. New York Civil Rights Law, Section 79-1, "Confidentiality of Records of Genetic Tests."

59. Sonia Suter, professor of law, George Washington University, presentation before the FBI Symposium on Familial Searching and Genetic Privacy, Arlington, VA, March 17–18, 2008.

60. Erica Haimes, "Social and Ethical Issues in the Use of Familial Searching in Forensic Investigations: Insight from Family and Kinship Studies," *Journal of Law, Medicine and Ethics* 34 (Summer 2006): 263–276, quotation at 271.

61. Bieber quoted in Rick Weiss, "Vast DNA Bank Pits Policing vs. Privacy," *Washington Post*, June 3, 2006, A1.

62. Daniel J. Grimm, "The Demographics of Genetic Surveillance: Familial DNA Testing and the Hispanic Community," *Columbia Law Review* 107 (June 2007): 1164–1194, quotation at 1164.

63. Bieber et al., "Finding Criminals Through DNA of Their Relatives."

64. Cook, "Near Match of DNA Could Lead Police to More Suspects."

5. Forensic DNA Phenotyping

1. Lindsy A. Elkins, "Five Foot Two with Eyes of Blue: Physical Profiling and the Prospect of a Genetics-Based Criminal Justice System," *Notre Dame Journal of Law, Ethics and Public Policy* 17 (2003): 269–305, quotation at 269.

2. Quoted in Jessica Snyder Sachs, "DNA and a New Kind of Racial Profiling," *Popular Science* (December 1, 2003), 16–20, quotation at 20.

3. Richard Willing, "DNA Tests Offer Clues to Suspect's Race," *USA Today*, August 16, 2005.

4. Elkins, "Five Foot Two with Eyes of Blue," 282.

5. "DNA analysis could serve as an antidote to racial profiling in that reliance on genetic information in crime scene samples could correct tendencies to pursue one group disproportionately." Ibid.

6. Ibid.

7. Mark D. Shriver and Rick A. Kittles, "Genetic Ancestry and the Search for Personalized Genetic Histories," *Nature Reviews: Genetics* 5 (August 2004): 611–618, quotation at 613.

8. Willing, "DNA Tests Offer Clues to Suspect's Race."

9. Susanne B. Haga, "Policy Implications of Defining Race and More by Genome Profiling," *Genomics, Society and Policy* 2, no. 1 (2006): 57–71, quotation at 59–60.

10. Sachs, "DNA and a New Kind of Racial Profiling," 16. DNAPrint Genomics was a genetics company that offered a wide range of products and services related to genetic profiling. The company suddenly ceased operations in February 2009.

11. See DNAPrint Genomics, Press Release, "DNAPrint Genomics Is Encouraging Law Enforcement Agencies to Include DNA Witness™ in Their NIJ Grant Proposals," August 16, 2004. While DNAPrint Genomics has gone out of business, currently its press releases can be found at http://www.dnaprint.com/welcome/press/press_recent/ (accessed May 23, 2010).

12. See DNAPrint Genomics, Press Release, "DNAPrint Announces the Release of RETINOME™ for the Forensic Market: Eye Color Prediction from Crime Scene DNA," August 17, 2004, http://www.dnaprint.com/welcome/press/press_recent/ (accessed May 23, 2010).

13. DNAPrint Genomics, Corporate Profile, http://www.dnaprint.com/welcome/corporate/ (accessed May 23, 2010).

14. DNAPrint Genomics, "DNAPrint Genomics Helps Boulder Police Solve 10-Year-Old Rape Murder Case Using Cutting Edge DNA Technology," press release, January 30, 2008, www.dnaprint.com/welcome/press/press_recent/2008/0130/DNAG-Colo.pdf (accessed November 3, 2008); also picked up by Reuters newswire under the same title, January 30, 2008, www.reuters.com/article/pressRelease/idUS144134+30-Jan-2008+MW20080130 (accessed November 3, 2008).

15. John Aguilar, "DNA Hit Leads to Arrest in Susannah Chase Slaying," *Daily Camera and County News*, January 28, 2008.

16. Julie Poppen, "DNA Match Nets Arrest in '97 Boulder Rape Murder," *Rocky Mountain News*, January 28, 2008, www.rockymountainnews.com/news/2008/jan/28/dna-match-made-a-decade-later (accessed November 3, 2008).

17. Aguilar, "DNA Hit Leads to Arrest."

18. Tom McGhee, "Felon's DNA Clogs System," *DenverPost.Com*, February 4, 2008, http://www.denverpost.com/news/ci_8159567 (accessed November 3, 2008).

19. As noted in note 10, DNAPrint Genomics stopped its operations suddenly as of February 2009. Its last filing with the Securities and Exchange Commission took place on February 9, 2009. Financial analysts described the company as a cash-strapped firm with good products and a poor business plan.

20. Sachs, "DNA and a New Kind of Racial Profiling," 16.

21. Mathew Graydon, François Cholette, and Lay-Keow Ng, "Inferring Ethnicity Using Autosomal STR Loci—Comparisons Among Populations of Similar and Distinctly Different Physical Traits," *Forensic Science International* 3 (September 2009): 251–254.

22. A portion of this section was based on an unpublished paper by Noam Biale and Tania Simoncelli, "Applying Behavioral Science to Policy: A Civil Liberties Perspective," November 2008.

23. See, for example, Ruth Hubbard and Elijah Wald, *Exploding the Gene Myth* (Boston: Beacon Press, 1999).

24. See Lori B. Andrews, "Predicting and Punishing Antisocial Acts: How the Criminal Justice System Might Use Behavioral Genetics," in *Behavioral Genetics: The Clash of Culture and Biology*, ed. Ronald A. Carson and Mark A. Rothstein (Baltimore: Johns Hopkins University Press, 1999), 116–155. See also Troy Duster, "Behavioral Genetics and Explanations of the Link Between Crime, Violence, and Race," in *Wrestling with Behavioral Genetics: Science, Ethics, and Public Conversation*, ed. Erik Parens, Audrey R. Chapman, and Nancy Press (Baltimore: Johns Hopkins University Press, 2006), 150–175.

25. See Troy Duster, *Backdoor to Eugenics*, 2nd ed. (New York: Routledge, 2003); see also Edwin Black, *War Against the Weak: Eugenics and America's Campaign to Create a Master Race* (New York: Four Walls Eight Windows, 2003).

26. Patricia A. Jacobs, A. M. Brunton, M. Melville, R. Brittain, and W. McClemont, "Aggressive Behavior, Mental Sub-normality and the XYY Male," *Nature* 208, no. 17 (1965): 1351–1352. This study found that 7 of 197 inmates (3.5 percent) were XYY, while the ratio of XYY in the general population was 1.3/1,000 (< 1 percent).

27. Richard Moran, "The Search for the Born Criminal and the Medical Control of Criminality," in *Deviance and Medicalization: From Badness to Sickness,* ed. P. Conrad and J. W. Schneider (New York: Temple University Press, 1992), 228. See also "Dr. Hutschenecker's Modest Proposal," editorial, *Washington Post*, April 10, 1970; and "Physician, Heal Thyself," *Time* (April 20, 1970), 8.

28. M. Rutter, H. Giller, and A. Hagell, *Antisocial Behavior by Young People* (Cambridge: Cambridge University Press, 1998).

29. A. Caspi, J. McClay, T. E. Moffitt, et al., "Role of Genotype in the Cycle of Violence of Maltreated Children," *Science* 297 (August 2, 2002): 851–854.

30. Council of State Governments, "Consensus Report 2002" (Lexington, KY: Council of State Governments, 2002), 136.

31. See Human Rights Watch, "Ill-Equipped: U.S. Prisons and Offenders with Mental Illness," 2003, http://www.hrw.org/en/reports/2003/10/21/ill-equipped, 109 (accessed April 3, 2010).

32. Wojciech et al., "Determination of Phenotype Associated SNPs in the MC1R Gene," 349.

33. National Center for Biotechnology Information (NCBI), "GeneTests: Growth of Laboratory Directory," http://www.ncbi.nlm.nih.gov/projects/GeneTests/static/whatsnew/labdirgrowth.shtml (accessed April 4, 2010).

34. Hui Huang, Eitan E. Winter, Huajun Wang, et al., "Evolutionary Conservation and Selection of Human Diseases Gene Orthologs in the Rat and Mouse Genomes," *Genome Biology* 5, no. 7 (June 28, 2004), http://www.pubmedcentral.nih.gov/articlerender.fcgiartid=463309 (accessed September 13, 2009).

35. Jennifer Brevorka, "Police Call on Psychic to Help in Bennett Case," *News and Observer* (Raleigh, NC), October 24, 2004.

36. Mark A. Rothstein and Meaghan K. Talbott, "The Expanding Use of DNA in Law Enforcement: What Role for Privacy?" *Journal of Law, Medicine and Ethics* 34, no. 2 (2006): 153–164. See also Chris Calabrese, testimony before the National Committee on Vital and Health Statistics (NCVHS), Subcommittee on Privacy and Confidentiality, February 18, 2004, http://www.aclu.org/technology-and-liberty/aclus-chris-calabreses-testimony-ncvhs (accessed November 8, 2008).

37. K. Thangaraj, A. G. Reddy, and L. Singh, "Is the Amelogenin Gene Reliable for Gender Identification in Forensic Casework and Prenatal Diagnosis?" *International Journal of Legal Medicine* 116, no. 2 (April 2002): 121–123.

38. F. Mohammed and S. M. Tayel, "Sex Identification of Normal Persons and Sex Reversal Cases from Blood Stains Using FISH and PCR," *Journal of Clinical Forensic Medicine* 12, no. 3 (June 2003): 122–127.

39. Elkins, "Five Foot Two with Eyes of Blue," 282.

40. Eileen A. Grimes, Penny J. Noake, Lindsey Dixon, and Andrew Urquhart, "Sequence Polymorphism in the Human Melanocortin 1 Receptor Gene as an Indicator of the Red Hair Phenotype," *Forensic Science International* 122 (2001): 124–129.

41. B. Wojciech, U. Brudnik, T. Kupiec, et al., "Determination of Phenotype Associated SNPs in the MC1R Gene," *Journal of Forensic Sciences* 52, no. 2 (March 2007): 349–354, quotation at 354.

42. P. Valverde et al., "Variants of the Melanocyte-Stimulating Hormone Receptor Gene Are Associated with Red Hair and Fair Skin in Humans," *Nature Genetics* 11 (1995): 328–330.

43. Grimes et al., "Sequence Polymorphism in the Human Melanocortin 1 Receptor Gene."

44. Tony Frudakis, Timothy Terravainen, and Mathew Thomas, "Multilocus OCA2 Genotypes Specify Human Iris Colors," *Human Genetics* 122, nos. 3–4 (November 2007): 311–326.

45. Boonsri Dickinson, "Eye Color Explained," *Discover Magazine* (March 13, 2007), http://discovermagazine.com/2007/mar/eye-color-explained (accessed November 7, 2008).

46. R. A. Sturm, "Human Pigmentation Genes and Their Response to Solar UV Radiation," *Mutation Research Review* 422 (1998): 69–76.

47. A. H. Robins, *Biological Perspectives on Human Pigmentation* (Cambridge: Cambridge University Press, 1991), 42–58.

48. Frudakis, *Molecular Photofitting*, 585.

49. Troy Duster, "The Implications of Behavioral Genetics Inquiry for Explanations of the Link Between Crime, Violence and Race" (unpublished manuscript, December 15, 2003).

50. I. W. Evett, J. S. Buckleton, A. Raymond, and H. Roberts, "The Evidential Value of DNA Profiles," *Journal of the Forensic Science Society* 33, no. 4 (1993): 243–244.

51. Bert-Jaap Koops and Maurice Schellekens, "Forensic DNA Phenotyping: Regulatory Issues," *Columbia Science and Technology Law Review* 9 (2008): 158–202, quotation at 191.

52. Sachs, "DNA and a New Kind of Racial Profiling," 20.

53. Koops and Schellekens, "Forensic DNA Phenotyping," 171.

54. Wyoming Statutes Ann. § 7-19-404 (2007).

55. Michelle Hibbert, "DNA Databanks: Law Enforcement's Greatest Surveillance Tool?" *Wake Forest Law Review* 34 (Fall 1999): 767–825, at 819.

56. Koops and Schellekens, "Forensic DNA Phenotyping," 166–167.

57. Ibid., 168–169.

58. Ibid., 168.

59. See N. Farahany and W. Bernet, *Behavioral Genetics in Criminal Cases: Past, Present and Future*, Vanderbilt Public Law Research Paper no. 06-15 2 (1) (Nashville: Vanderbilt Law School, 2006), 72–79.

60. Affymetrix Corp., Santa, Clara, CA, http://www.affymetrix.com/index.affx (accessed July 5, 2007).

61. Sachs, "DNA and a New Kind of Racial Profiling," 20.

62. Koops and Schellekens, "Forensic DNA Phenotyping," 201.

63. Ibid., 174.

6. Surreptitious Biological Sampling

1. Alex Kozinski, minority dissent, *United States v. Kincade*, U.S. Court of Appeals for the 9th Circuit, No. 02-50380, D.C. No. CR-93-00714-RAG-01, April 18, 2004, 11468.

2. *State of Washington v. Athan*, 160 Wn. 2d 354 (2007), ¶ 35 (Case No. 75312-1, Filed May 10, 2007), See FindLaw, http://caselaw.findlaw.com/wa-supreme-court/1495639.html (accessed May 23, 2010).

3. Associated Press, "Covert DNA Collection Prompts Questions," *Forensic Magazine* (March 19, 2007), http://www.forensicmag.com/News_Print.asp?pid=137 (accessed July 30, 2007).

4. *State of Washington v. Athan.*

5. *State of Washington v. Athan*, Brief of *Amicus Curiae*, American Civil Liberties Union of Washington, in support of Appellant, John Nicholas Athan, Supreme Court of the State of Washington (No. 75312-1), prepared by Douglas B. Klunder (2007).

6. *State of Washington v. Athan*, ¶ 35.

7. Ibid, ¶ 60.

8. Ibid., ¶ 23.

9. Ibid., ¶ ¶ 41–42.

10. *State of Washington v. Jackson*, 150 Wn.2d at 262.

11. *State of Washington v. Young*, 123 Wn.2d at 181–182.

12. *State of Washington v. Gunwall*, 720 P.2d 808.

13. *State of Washington v. Boland*, 115 Wn.2d 571, 578, 800 P.2d 1112 (1990).

14. Shankar Vedantam, "Study Links Gene Variant in Men to Marital Discord," *Washington Post*, September 2, 2008, A2.

15. *State of Washington v. Athan* (Dissenting Opinion, written by Mary E. Fairhurst).

16. Elizabeth E. Joh, "Reclaiming 'Abandoned' DNA: The Fourth Amendment and Genetic Privacy," *Northwestern University Law Review* 100 (2006): 857–884, quotation at 874.

17. Elizabeth Joh, "Reclaiming 'Abandoned' DNA: The Fourth Amendment and Genetic Privacy," UC Davis Legal Studies Research Paper Series, Research Paper no. 40 (April 2005), 1–31, quotation at 14; Social Science Research Network Electronic Paper Collection, http://ssrn.com/abstract=702571 (accessed April 18, 2010).

18. Sheldon Krimsky, "The Right to Our DNA," letter to the editor, *New York Times*, April 18, 2007, A22.

19. Robert C. Green and George J. Annas, "The Genetic Privacy of Presidential Candidates," *New England Journal of Medicine* 359 (November 20, 2008): 2192–2193.

20. See, e.g., *State of Nebraska v. Wickline,* 232 N.W. 2d 253 (Neb. 1989) (finding that the police did not need a warrant to collect cigarettes left at the police station since these items were "abandoned" and "sufficiently exposed" to the officer and the public).

21. *California v. Greenwood*, 486 U.S. 35 (1988).

22. Joh, "Reclaiming 'Abandoned' DNA," (2006), 881.

23. *Kyllo v. United States* 533 U.S.27 (2001).

24. Erin Murphy, "The New Forensics: Criminal Justice, False Certainty, and the Second Generation of Scientific Evidence," *California Law Review* 95 (June 2007): 721–797, quotation at 736.

25. Amy Harmon, "The DNA Age: Stalking Strangers' DNA to Fill In the Family Tree," *New York Times*, April 2, 2007.

26. Paul S. Applebaum, "Behavioral Genetics and the Punishment of Crime," *Psychiatric Services* 56, no. 1 (January 2005): 25–27.

27. Joh, "Reclaiming 'Abandoned' DNA" (2006), 882–883.

7. Exonerations

1. Jessica Blank and Erik Jensen, *The Exonerated* (New York: Faber and Faber, 2004), 45.

2. Samuel R. Gross, Kristen Jacoby, Daniel J. Matheson, Nicholas Montgomery, and Sujata Patil, "Exonerations in the United States, 1989 through 2003," *Journal of Criminal Law and Criminology* 95, no. 2 (Winter 2005): 523–560, quotation at 525.

3. See Susan Haack, *Defending Science—Within Reason: Between Scientism and Cynicism* (New York: Prometheus Books, 2003), 233–264 (on science and law); Sheila Jasanoff, *Science at the Bar: Law, Science, and Technology in America* (Cambridge, MA: Harvard University Press, 1995).

4. Robert Merton, "Science and Democratic Social Structure," in Merton, *Social Theory and Social Structure* (Glencoe, IL: Free Press, 1957), 550–561.

5. In the 1993 Supreme Court decision *Daubert v. Merrill Dow Pharmaceuticals, Inc.* (113 S.Ct. 2786 [1993]) the Court established a gatekeeper role for the trial judge in deciding whether to allow expert testimony into the courtroom. When the Supreme Court issued its opinion in *Daubert*, it suggested four criteria that judges can use to determine whether scientific testimony was reliable and therefore admissible: (1) the evidence should be based on a testable theory or technique; (2) the theory or technique has been peer reviewed; (3) the particular technique has a known error rate; (4) the underlying science is generally accepted.

6. *Berger v. United States*, 295 U.S. 78, 68 (1935).

7. Susan Rutberg, "Anatomy of a Miscarriage of Justice: The Wrongful Conviction of Peter J. Rose," *Golden Gate University Law Review* 37 (2006): 26.

8. Karl Popper was a leading twentieth-century philosopher who, while critical of inductivism, advanced a method of empirical falsificationism. Popper, *The Logic of Scientific Discovery* (New York: Harper & Row, 1968).

9. Thomas Kuhn, *The Structure of Scientific Revolutions* (Chicago: University of Chicago Press, 1962), 66–76.

10. Sheldon Krimsky, *Science in the Private Interest* (Lanham, MD: Rowman and Littlefield, 2003).

11. Stephen Breyer, *Breaking the Vicious Circle* (Cambridge, MA: Harvard University Press, 1993), 53–81.

12. *People v. Dotson*, 99 Ill. App. 3d 117 (1981).

13. Edward Connors, Thomas Lundregan, Neal Miller, and Tom McEwen, *Convicted by Juries, Exonerated by Science: Case Studies in the Use of DNA Evidence to Establish Innocence After Trial* (Washington, DC: National Institute of Justice, U.S. Department of Justice, June 1996), 60–61.

14. This was based on the court's reading of *Herrera v. Collins*, 506 U.S. 390, 400 (1993).

15. *Darryl E. Hunt v. Martin J. McDade*, Appeal from the U.S. District Court for the Middle District Court of North Carolina at Winston-Salem, argued January 24, 2000; decided February 25, 2000, http://pacer.ca4.uscourts.gov/opinion.pdf/986808.U.pdf (accessed April 6, 2010).

16. Mark Rabil, telephone conversation with the authors, January 15, 2008.

17. Office of the City Manager, *Sykes Administrative Review Committee Report* (Winston-Salem, NC, February 2007), 93, http://www.ci.winston-salem.nc.us/Home/CityGovernment/CityManager/Articles/SARCReportSykes-Hunt (accessed April 6, 2010).

18. Paula Zahn, interview with Marvin Lamont Anderson and Peter Neufeld, August 23, 2002, CNN.com transcripts, http://archives.cnn.com/TRANSCRIPTS/0208/23/ltm.06.html (accessed April 6, 2010).

19. Frank Green, "Eyewitness ID Fallibility Shown," *Richmond Times Dispatch*, March 16, 2003.

20. Zahn, interview with Anderson and Neufeld.

21. Green, "Eyewitness ID Fallibility Shown."

22. Zahn, interview with Anderson and Neufeld.

23. Brandon L. Garrett, "Judging Innocence," *Columbia Law Review* 108 (January 2008): 55–142, quotation at 116.

24. Personal correspondence between Rebecca Brown, the Innocence Project, and Sheldon Krimsky, February 17, 2009.

25. Personal correspondence between Rebecca Brown, the Innocence Project, and Sheldon Krimsky, November 14, 2008.

26. Garrett, "Judging Innocence," 116.

27. Nina Morrison, staff attorney, the Innocence Project, affidavit, July 24, 2006.

28. Ibid.

29. Miles Moffeit, "Prosecutors Resist Retrial in DNA Case," *Denver Post*, March 18, 2008.

30. *Arizona v. Youngblood*, Supreme Court No. 86-1904, 488 U.S. 51, decided November 29, 1988.

31. Susan Greene and Miles Moffeit, "Bad Faith Difficult to Prove," *Denver Post*, July 22, 2007, http://www.denverpost.com/evidence/ci_6429277 (accessed April 6, 2010).

32. Barry Scheck, "Innocence, Race, and the Death Penalty," *Howard University Law Journal* 50 (Winter 2007): 445–469, statement at 449.

33. Garrett, "Judging Innocence," 88.

34. Northwestern Law School, Bluhm Legal Clinic, "Christopher Ochoa: DNA Exonerated Christopher Ochoa of a Crime to Which He Had Confessed," http://www.wrongfulconvictionscenter.com/wrongfulconvictions/exonerations/txOchoa Summary.html (accessed April 7, 2010).

35. Alexandra Perina, "The False Confession," *Psychology Today Magazine* (March/April 2003), http://www.psychologytoday.com/articles/pto-20030430-000002.html (accessed November 28, 2008).

36. Innocence Project, "Understand the Causes: False Confessions," http://www.innocenceproject.org/understand/False-Confessions.php (accessed November 28, 2008).

37. Leon Harris, interview with Barry Scheck and Pater Neufeld, February 6, 2002, CNN.Com transcripts, http://transcripts.cnn.com/TRANSCRIPTS/0202/06/lt.24.html (accessed April 6, 2010).

38. Personal correspondence between Vanessa Potkin, staff attorney, the Innocence Project, and Sheldon Krimsky, January 6, 2009.

39. Garrett, "Judging Innocence," 119.

40. Ibid., 64.

41. Ibid., 120.

42. Personal correspondence between Rebecca Brown, the Innocence Project, and Sheldon Krimsky, February 17, 2009.

43. Gross et al., "Exonerations in the United States," 523.

44. Solomon Moore, "Study Calls for Oversight of Forensics in Crime Labs," *New York Times*, February 19, 2009, A12; National Research Council, Committee on Identifying the Needs of the Forensic Sciences Community, *Strengthening Forensic Science in the United States: A Path Forward* (Washington, DC: National Academy Press, 2009).

45. Personal correspondence between Rebecca Brown, the Innocence Project, and Sheldon Krimsky, February 17, 2009.

8. The Illusory Appeal of a Universal DNA Data Bank

1. "Lord Justice Stephen Sedley's Proposal for a Universal DNA Database in the UK—The BBC Radio 4 Interview" (Danny Shaw, interviewer), September 5, 2007, http://ecologics.wordpress.com/2007/09/11/lord-justice-stephen-sedley%e2%80

%99s-proposal-for-a-universal-dna-database-in-the-uk-the-bbc-interview/ (accessed April 9, 2010).

2. *State v. Olivas*, 856 P.2d 1076, 1094 (Wash. 1993), J. Utter, concurring.

3. "Let's Catch More Rapists Before They Strike Again," editorial, *Glamour* (January 2007): 102.

4. *2003 DNA Database Expansion Legislation*, prepared by Smith Alling Lane on behalf of Applied Biosystems (December 2003), http://www.dnaresource.com/2003%20DNA%20Expansion%20bills.pdf (accessed April 12, 2005); see also D. McCullagh, "What to Do with DNA Data?" *Wired News* (November 19, 1999), http://www.wired.com/politics/law/news/1999/11/32617 (accessed April 9, 2010).

5. *Michigan Commission on Genetic Privacy and Progress: Final Report and Recommendation* (Lansing: Michigan Department of Community Health, February 1999), http://www.michigan.gov/documents/GeneticsReport_11649_7.pdf (accessed April 9, 2010).

6. James Sturcke, "Code Cracking: 25 Years of DNA Detection," *Guardian UK*, May 7, 2009.

7. "Let's Catch More Rapists Before They Strike Again," 102.

8. "Of the sexual assault victims in the NSA, 74 percent reported that the assault was committed by someone they knew well. Almost one-third (32.5 percent) of sexual assault cases involved perpetrators who were friends, 21.1 percent were committed by a family member, and 23.2 percent were committed by strangers." U.S. Department of Justice, "Rape and Sexual Assault," http://www.ojp.usdoj.gov/ovc/ncvrw/2005/pg5o.html (accessed September 16, 2009).

9. "In the first rape experience of female victims, perpetrators were reported to be intimate partners (30.4%), family members (23.7%), and acquaintances (20%). In the first rape experience of male victims, perpetrators were reported to be acquaintances (32.3%), family members (17.7%), friends (17.6%), and intimate partners (15.9%)." Centers for Disease Control, "Sexual Violence," Facts at a Glance, Spring 2008, http://www.cdc.gov/ViolencePrevention/pdf/SV-DataSheet-a.pdf (accessed September 16, 2009).

10. "Incest offenders ranged between 4 and 10%; Rapists ranged between 7 and 35%; Child molesters with female victims ranged between 10 and 29%; Child molesters with male victims ranged between 13 and 40%; Exhibitionists ranged between 41 and 71%." U.S. Department of Justice, Center for Sex Offender Management, "Recidivism of Sex Offenders," May 2001, http://www.csom.org/pubs/recidsexof.html (accessed September 16, 2009).

11. Kristen M. Zgoba and Lenore M. J. Simon, "Recidivism Rates of Sexual Offenders up to 7 Years Later," *Criminal Justice Review* 30, no. 2 (September 2005): 155–173.

12. Rebecca Sasser Peterson, "DNA Databases: When Fear Goes Too Far," *American Criminal Law Review* 37, no. 3 (Summer 2000): 1219–1238, quotation at 1234.

13. D. H. Kaye and Michael E. Smith, "DNA Identification Databases: Legality, Legitimacy, and the Case for Population-Wide Coverage," *Wisconsin Law Review* (2003): 413–459, quotation at 459.

14. Ben Quarmby, "The Case for National DNA Identification Cards," *Duke Law and Technology Review* (January 31, 2003), 1–9, quotation at 7.

15. Christine Rosen, "Liberty, Privacy and DNA Databases," *New Atlantis* 1 (Spring 2003): 37–52, http://www.thenewatlantis.com/publications/liberty-privacy-and-dna-databases (accessed April 9, 2010).

16. See Al Baker, "Effort to Reinstate Death Penalty Law Is Stalled in Albany," *New York Times*, November 18, 2004, B6; see also John Paul Truskett, "The Death Penalty, International Law, and Human Rights," *Tulsa Journal of Comparative and International Law* 11 (2004): 557, at 589–593 (discussing empirical studies worldwide).

17. Raymond Bonner, "Absence of Executions: A Special Report; States with No Death Penalty Share Lower Homicide Rates," *New York Times*, September 22, 2000, A1. For 2007, the average murder rate for death penalty states is 5.83 (per 100,000) and for non-death penalty states it is 4.10 (per 100,000). Death Penalty Information Center, "Deterrence: States Without the Death Penalty Have Had Consistently Lower Murder Rates," http://www.deathpenaltyinfo.org/deterrence-states-without-death-penalty-have-had-consistently-lower-murder-rates (accessed April 7, 2010).

18. Ryan S. King, Marc Mauer, and Malcolm C. Young, *Incarceration and Crime: A Complex Relationship* (Washington, DC: The Sentencing Project, 2005), 3, http://www.sentencingproject.org/doc/publications/inc_iandc_complex.pdf (accessed April 9, 2010).

19. Scottish Office Central Research Unit, "Crime and Criminal Justice Research Findings No. 30: The Effect of Closed Circuit Television on Recorded Crime Rates and Public Concern About Crime in Glasgow," July 7, 1999, http://www.scotcrim.u-net.com/researchc2.htm (accessed April 7, 2010).

20. U.K. Parliament, Select Committee on Home Affairs, "Fifth Report," June 8, 2008, http://www.publications.parliament.uk/pa/cm200708/cmselect/cmhaff/58/5810.htm (accessed April 10, 2010).

21. Rosen, "Liberty, Privacy and DNA Databases," 41.

22. Paul M. Monteleoni, "DNA Databases, Universality, and the Fourth Amendment," *New York University Law Review* 82 (April 2007): 247–280, quotation at 279.

23. Richard A. Posner, *The Economic Analysis of Law* (Boston: Little, Brown, 1973).

24. U.K. National DNA Database, *Annual Report 2003–4*, 23.

25. Amitai Etzioni, "A Communitarian Approach: A Viewpoint on the Study of the Legal, Ethical and Policy Considerations Raised by DNA Tests and Databases,"

Journal of Law, Medicine and Ethics 34, no. 2 (Summer 2006): 214–221, quotation at 214.

26. Innocence Project, "Know the Cases: Darryl Hunt," http://www.innocenceproject.org/Content/181.php (accessed October 31, 2008).

27. D. H. Kaye and Michael E. Smith, "DNA Databases for Law Enforcement: The Coverage Question and the Case for a Population-Wide Database," in *DNA and the Criminal Justice System: The Technology of Justice*, ed. David Lazer (Cambridge, MA: MIT Press, 2004), 269.

28. Kaye and Smith, "DNA Identification Databases," 415.

29. Kaye and Smith, "DNA Databases for Law Enforcement," 272.

30. Troy Duster, *Backdoor to Eugenics*, 2nd ed. (New York: Routledge, 2003).

31. Pilar Ossorio and Troy Duster, "Race and Genetics: Controversies in Biomedical, Behavioral, and Forensic Sciences," *American Psychologist* 60, no. 1 (January 2005): 115–128; Harry Levine and Deborah Peterson Small, *Marijuana Arrest Crusade: Racial Bias and Police Policy in New York City, 1997–2007* (New York: New York Civil Liberties Union, April 2008), http://www.nyclu.org/node/1736 (accessed April 10, 2010).

32. Ossorio and Duster, "Race and Genetics," 125.

33. David Dudley, "A Cold Hit," *Cornell Magazine* 109, no. 1 (July/August 2006), http://www.cornellalumnimagazine.com/Archive/2006julaug/features/Feature2.asp (accessed April 10, 2010).

34. Nuffield Council on Bioethics, *The Forensic Use of Bioinformation: Ethical Issues* (London: Nuffield Council on Bioethics, September 2007), 56, http://www.nuffieldbioethics.org/fileLibrary/pdf/The_forensic_use_of_bioinformation_-_ethical_issues.pdf (accessed October 23, 2007).

35. "Lord Justice Stephen Sedley's Proposal for a Universal DNA Database in the UK" (interview).

36. Ossorio and Duster, "Race and Genetics," 125.

37. Peterson, "DNA Databases," 1228.

38. Meredith A. Bieber, "Meeting the Statute or Beating It: Using 'John Doe' Indictments Based on DNA to Meet the Statute of Limitations," *University of Pennsylvania Law Review* 50 (January 2002): 1079–1097.

39. See T. Simoncelli, "Retreating Justice: Proposed Expansion of Federal DNA Database Threatens Civil Liberties," *Genewatch* 17, no. 2 (March–April 2004): 3–6. See also National Association of Criminal Defense Lawyers (NACDL), "Resolution of the Board of Directors Regarding John Doe DNA Warrants/Indictments," May 1, 2004, http://www.nacdl.org/public.nsf/26cf10555dafce2b85256d97005c8fd0/3c5c2f17e8194b4785256eba006bb259?OpenDocument (accessed April 10, 2010).

40. Kaye and Smith, "DNA Databases for Law Enforcement," 267.

41. Mark A. Rothstein and Sandra Carnahan, "Legal and Policy Issues in Expanding the Scope of Law Enforcement DNA Databanks," *Brooklyn Law Review* 67 (Fall 2001): 127–170, quotation at 155.

42. John Joseph, "Compulsory DNA Database Ruled Out," *Reuters UK*, February 23, 2008, http://uk.reuters.com/article/idUKL2370029120080223 (accessed April 10, 2010).

43. BBCTV, *Daily Politics*, "A Time for a Universal DNA Database?" (Interviewer: Anita Anand; Interviewees: Greg Hands and Jill Saward), March 27, 2009, http://news.bbc.co.uk/2/hi/programmes/the_daily_politics/7968121.stm (accessed April 10, 2010).

44. Michael E. Smith, "Let's Make the DNA Identification Database as Inclusive as Possible," *Journal of Law, Medicine and Ethics* 34, no. 2 (Summer 2006): 385–389, quotation at 388. See also Dudley, "A Cold Hit."

45. Smith, "Let's Make the DNA Identification Database as Inclusive as Possible," 388.

46. Richard Stacy, "The Future of DNA," *Denver Post*, January 25, 2009.

47. Rosen, "Liberty, Privacy and DNA Databases."

48. U.S. Department of Justice, Bureau of Justice Statistics, "Census of Publicly Funded Forensic Crime Laboratories: 50 Largest Crime Labs, 2002," September 2004, http://bjs.ojp.usdoj.gov/content/pub/ascii/50lcl02.txt (accessed April 10, 2010).

49. U.S. Department of Justice, National Institutes of Justice, "Report to the Attorney General on Delays in Forensic DNA Analysis," March 2003, http://www.ncjrs.gov/pdffiles1/nij/199425.pdf (accessed April 10, 2010).

50. U.S. Department of Justice, "Audit of the Convicted Offender DNA Backlog Reduction Program," March 2009, http://www.justice.gov/oig/reports/OJP/a0923/final.pdf (accessed April 10, 2010).

51. National Bioethics Advisory Commission, *Research Involving Human Biological Materials: Ethical Issues and Policy Guidance*, vol. 1: *Report and Recommendations of the National Bioethics Advisory Commission* (Rockville, MD: National Bioethics Advisory Commission, August 1999), 13, http://bioethics.georgetown.edu/nbac/hbm.pdf (accessed April 11, 2010).

52. Nigel Morris, "A 'Chilling' Proposal for a Universal DNA Database," *Independent* (London), September 6, 2007.

53. Whatman Ltd., "Biobanking/DNA Repositories," http://www.whatman.com/CatBioBankingDNARepositories.aspx (accessed April 9, 2010).

54. National Commission on the Future of DNA Evidence, "Proceedings," September 27, 1999 (Legal Issues Working Group Report and Discussion, Professor Michael Smith, Working Group Chair), http://www.ojp.usdoj.gov/nij/topics/forensics/events/dnamtgtrans7/trans-l.html (accessed April 11, 2010).

55. BBC Radio 4 (*Today* program), interview with Richard Thomas, information commissioner, Home Office, September 5, 2007, www.bbc.co.uk/radio4/today/listenagain/listenagain_20070905.shtml (accessed May 23, 2010).

56. Eric Lipton, "U.S. Requiring Port Workers to Have I.D.s and Reviews," *New York Times*, January 4, 2007, A14.

57. *Davis v. Mississippi*, 394 U.S. 721 (1969).

58. Peterson, "DNA Databases," 1219.

59. Jeffrey Rosen, "Genetic Surveillance for All," *Slate* (March 17, 2009), http://www.slate.com/id/2213958/pagenum/all/ (accessed April 12, 2010).

9. The United Kingdom

1. Association of Chief Police Officers, United Kingdom, *ACPO DNA Good Practice Manual*, 3rd ed. (2007), 35, http://www.denverda.org/DNA_Documents/Final%20Document%20-%20DNA%20Good%20Practice.%20August%20'05.pdf (accessed May 26, 2010).

2. Thomas Ross, police liaison officer/office manager, Scottish Police DNA Database, "Police Retention of Prints and Samples: Proposals for Legislation," 2005, http://www.scotland.gov.uk/Resource/Doc/77843/0018258.pdf, 3 (accessed April 13, 2010).

3. For a comparison of key features of forensic DNA data banks in six countries, see the appendix to this volume.

4. For the per capita murder rate, see NationMaster.Com, "Crime Statistics Murders (per Capita) (Most Recent) by Country," http://www.nationmaster.com/graph/cri_mur_percap-crime-murders-per-capita (accessed July 29, 2009).

5. P. Roberts and C. Willmore, *The Role of Forensic Science Evidence in Criminal Proceedings*, Royal Commission on Criminal Justice Study 11 (London: HMSO, 1993), 9.

6. Robin Williams and Paul Johnson, "Inclusiveness, Effectiveness and Intrusiveness: Issues in Developing Uses of DNA Profiling in Support of Criminal Investigations," *Journal of Law, Medicine and Ethics* 33, no. 3 (2005): 545–558, quotation at 547, http://www.ncbi.nlm.nih.gov/pmc/articles/PMC1370918/pdf/nihms-6441.pdf (accessed April 13, 2010).

7. Paul Johnson, Paul Martin, and Robin Williams, "Genetics and Forensics: Making the National DNA Database," *Science Studies* 16, no. 2 (2003): 22–37 (2003), quotation at 30.

8. The laws that are relevant include the Police and Criminal Evidence Act 1984 and the Serious and Organized Crime Act 2005.

9. Carole McCartney, "The DNA Expansion Programme and Criminal Investigation," *British Journal of Criminology* 46, no. 2 (2006): 175–192, at 176.

10. GeneWatch UK, "Police Retention of DNA from Northern Ireland," June 2007, 1, http://www.genewatch.org/uploads/f03c6d66a9b354535738483c1c3d49e4/MLAbrief07_fin.pdf (accessed May 26, 2010).

11. GeneWatch UK, "Parliamentary Questions on DNA (Forensics), December 2005 to September 2006," October 2006, http://www.genewatch.org/uploads/f03c6d66a9b354535738483c1c3d49e4/PQs_Oct06.doc (accessed April 13, 2010).

12. Supplementary Letter from Vernon Coaker, M.P., Minister of State, Home Office to House of Lords, Constitution Committee, December 11, 2008, published in *Surveillance: Citizens and the State* (House of Lords, Constitution Committee—Second Report, January 21, 2009), http://www.publications.parliament.uk/pa/ld200809/ldselect/ldconst/18/8111907.htm (accessed May 9, 2010).

13. H. Wallace, "Prejudice, Stigma and DNA Databases" (paper for the Council for Responsible Genetics, July 2008).

14. For a chart of forensic DNA data banks in EU countries, see Michael Townsley and Gloria Laycock, eds., *Forensic Science Conference Proceedings: Beyond DNA in the UK—Integration and Harmonization*, Newport, South Wales, May 17–19, 2004 (London: Home Office Policy Unit, 2004), 36, http://police.homeoffice.gov.uk/publications/police-reform/Forensics_Part_12835.pdf?view=Binary (accessed April 13, 2010).

15. "DNA of Under-10s on Government Database," *Guardian Unlimited*, June 14, 2007, http://www.guardian.co.uk/print/0,,330031140-111274,00.html (accessed July 5, 2007).

16. John Lettice, "UK Gov Seeks 'Scientific Basis' for Nationality," *Register*, June 22, 2007.

17. GeneWatch UK, "Human Genetics Parliamentary Briefing No. 6: The Police National DNA Database: An Update," July 2006, http://www.genewatch.org/sub-539478 (accessed April 14, 2010). See also A. Barnett, "Police DNA Database 'Is Spiraling Out of Control,'" *Observer*, July 16, 2006.

18. Parliamentary Office of Science and Technology, "The National DNA Database," *Postnote*, no. 258 (February 2006): 2, http://www.parliament.uk/documents/upload/postpn258.pdf (accessed April 14, 2010).

19. Home Office Circular 1/2006: "If you should receive a request in the future for access to a DNA sample or profile taken under PACE for use in paternity testing you should therefore refuse the request, citing the statutory prohibition on such use and referring to the London Borough of Lambeth case." http://www.homeoffice.gov.uk/about-us/publications/home-office-circulars/circulars-2006/001-2006/ (accessed June 22, 2009).

20. Chris Williams, "Minister Pledges No Complete DNA Database," *Register*, March 30, 2006.

21. Richard Pinchin, North American Operations Manager, Forensic Science Service, U.K., presentation before the New York State Forensic Science Commission hearing, January 2008. See also Robin Williams, "Making Do with Partial Matches: DNA Intelligence and Criminal Investigations in the United Kingdom" (presentation for DNA Fingerprinting and Civil Liberties: Workshop #2, American Society for Law, Medicine and Ethics, September 17–18, 2004); and Andrew Barrow, "Sex Attacker Snared by Family DNA," *Press Association Limited*, September 23, 2005.

22. V. Kearney, "Approval for Low Copy DNA Testing," *BBC News*, April 11, 2008, http://news.bbc.co.uk/2/hi/uk_news/northern_ireland/7341782.stm (accessed April 14, 2010).

23. Bruce Budowle, executive director, Institute of Investigative Genetics, University of North Texas Health Science Center at Fort Worth, statement at the 19th Annual Computers, Freedom and Privacy Conference, Washington, DC, June 2–4, 2009.

24. *The Queen v. Sean Hoey*, Neutral Citation Number [2007] NICC 49, http://www.denverda.org/DNA_Documents/Hoey.pdf.

25. Carole McCartney, "Ethics Watch," *Nature Reviews Genetics* 9 (May 2008): 325, http://www.nature.com/nrg/journal/v9/n5/pdf/nrg2362.pdf (accessed April 14, 2010).

26. "Scientists Affirm DNA Test Criticized in Omagh Trial," *News Letter*, April 11, 2008, http://www.newsletter.co.uk/news/Scientists-affirm-DNA test -criticised.3974115.jp (accessed April 14, 2010). See also Bruce Budowle, Arthur J. Eisenberg, and Angela van Daal, "Validity of Low Copy Number Typing and Applications to Forensic Science," *Croatian Medical Journal* 50 (2009): 207–219, http://www.ncbi.nlm.nih.gov/pmc/articles/PMC2702736/ (accessed April 14, 2010).

27. James Randerson, "'We've Now Pushed the Technology to the Absolute Limit . . .': The Case Against the Latest DNA Evidence," *Guardian*, January 16, 2008, http://www.guardian.co.uk/uk/2008/jan/16/ukcrime.forensicscience1 (accessed April 14, 2010).

28. Budowle et al., "Validity of Low Copy Number Typing and Applications to Forensic Science," 207.

29. Budowle, statement at the 19th Annual Computers, Freedom and Privacy Conference.

30. Jason Gilder, Roger Koppl, Irving Kornfeld, et al., "Comments on the Review of Low Copy Number Testing" (letter), *International Journal of Legal Medicine* 123, no. 6 (2009): 535–536.

31. Randerson, "'We've Now Pushed the Technology to the Absolute Limit.'"

32. Carole E. McCartney, "Forensic DNA Sampling and the England and Wales National DNA Database: A Skeptical Approach," *Critical Criminology* 12 (2004): 157–178, quotation at 174.

33. See Human Genetics Commission, "Inside Information," May 2002, http://www.hgc.gov.uk/UploadDocs/DocPub/Document/insideinformation_summary.pdf (accessed April 14, 2010). See also Human Genetics Commission, "HGC Response to the Scottish Executive Consultation on Police Retention of Prints and Samples," September 30, 2005, http://www.scotland.gov.uk/Resource/Doc/77843/0018244.pdf (accessed April 14, 2010).

34. Home Office, "Supplementary Memorandum, Appendix 20," in House of Commons Science and Technology Committee, *Forensic Science on Trial*, vol. 2, HC 96-II (2005), www.publications.parliament.uk/pa/cm200405/cmselect/cmsctech/96/96ii.pdf (accessed April 14, 2005).

35. Home Office, *Keeping the Right People on the DNA Database: Science and Public Protection*, May 2009, http://www.statewatch.org/news/2009/may/uk-ho-dna-consult.pdf (accessed June 23, 2009).

36. See generally Council for Responsible Genetics, http://www.councilforresponsiblegenetics.org/. See also Helen Wallace, "The UK National DNA Database," *EMBO Reports* 7 (2006) (Special Issue): S26–S30.

37. Helen Wallace, "Prejudice, Stigma and DNA Databases," *GeneWatch* 21, no. 3–4 (November–December 2008): 14–16.

38. Nick Taylor, "Genes on Record—One Size Fits All?" *New Law Journal* 156 (September 8, 2006): 1354.

39. Helen Wallace, "GeneWatch PR: Home Office Drags Its Feet on DNA Database Removals," May 7, 2009, http://www.genewatch.org/article.shtml?als%5Bcid%5D=539478&als%5Bitemid%5D=564505 (accessed April 14, 2010).

40. Select Committee on Home Affairs, Parliament, United Kingdom, "Second Report," June 15, 2007, http://www.publications.parliament.uk/pa/cm200607/cmselect/cmhaff/181/18103.htm (accessed June 20, 2007).

41. U.K. Parliament, House of Commons, Select Committee on Home Affairs, Second Report, Part I: "Nature and Extent of Young Black People's Overrepresentation," May 22, 2007, http://www.publications.parliament.uk/pa/cm200607/cmselect/cmhaff/181/18105.htm (accessed June 29, 2007).

42. Richard Ford, "DNA of Children Under Ten Are on National Database," *Times Online*, June 15, 2007, http://www.timesonline.co.uk/tol/news/uk/crime/article1934914.ece (accessed April 15, 2010).

43. Ibid.

44. Parliamentary Office of Science and Technology, "The National DNA Database."

45. GeneWatch UK, "Human Genetics Parliamentary Briefing No. 6."

46. Dr. Helen Wallace, statement in the Grand Chamber of the European Court of Human Rights Between *"S" and Marper v. The United Kingdom*, application nos. 30562/04 and 30566/04.

47. Nuffield Council on Bioethics, *The Forensic Use of Bioinformation: Ethical Issues* (London: Nuffield Council on Bioethics, September 2007), xv, http://www.nuffieldbioethics.org/fileLibrary/pdf/The_forensic_use_of_bioinformation_-_ethical_issues.pdf (accessed April 14, 2010).

48. Ibid., xv.

49. Ibid., 54.

50. Ibid., 52.

51. BBC News, "Teacher Wins Police DNA Battle," March 23, 2006, http://news.bbc.co.uk/2/hi/uk_news/england/west_midlands/4837206.stm (accessed April 14, 2010).

52. *S. and Marper v. The United Kingdom* (*Marper*), European Court of Human Rights, Grand Chamber, nos. 30562/04 and 30566/04.

53. *S. and Marper v. The United Kingdom*, 30562/04 [2008] ECHR 1581 (December 4, 2008), ¶ 69, http://www.bailii.org/eu/cases/ECHR/2008/1581.html (accessed May 9, 2010).

54. Ibid., ¶ 71.

55. Ibid., ¶ 121.

56. Ibid., ¶ 75.

57. Ibid., ¶ 124.

58. Ibid., ¶ 47.

10. Japan's Forensic DNA Data Bank

1. Japan Federation of Bar Associations (JFBA), "Opinion on the National Police Agency DNA Database System," December 21, 2007, English summary and full text of the opinion at http://www.nichibenren.or.jp/en/activities/statements/071221.html. (accessed May 18, 2010).

2. The loci that are currently used in generating a profile are as follows: TPOX, D3S1358, FGA, D5S818, CSF1PO, D7S820, D8S1179, TH01, vWA, D13S317, D16S539, D18S51, D21S11, D2S1338, and D19S43. E. Omura, discussion with T. Simoncelli and written comments by T. Yamamoto, professor, Waseda Law School, Tokyo, July 30, 2007.

3. Japan's Supreme Court first established the "balancing test" in 1969 in the *Hakata Station TV Film Subpoena Case*. Supreme Court (Japan), grand bench, Keishu, vol. 23, no. 11, at 1490, November 26, 1969.

4. Omura, discussion with Simoncelli and written comments by Yamamoto.

5. Ibid.

6. *National Public Safety Commission Regulation No. 15, Based on Article 13, Section 1 of the Police Law Enforcement Rules, Regulations on Treatment of DNA Type Profiles*, enacted August 26, 2005, Article 1.

7. Omura, discussion with Simoncelli and written comments by Yamamoto.

8. Dr. Kazumasa Sekiguchi, National Research Institute of Police Science, personal correspondence with Sheldon Krimsky, February 16, 2007.

9. National Police Agency (NPA), "Towards Activating DNA Profile Information" (publication date unknown), translated by E. Omura, January 2008.

10. National Police Agency, "Guidelines on Operating Expert Examination of DNA Profiles," referred to in Toshikazu Shimizu, National Policy Agency, Criminal Bureau, "Commencing the Operation of the Crime Scene DNA Profiles Database System, 2005" (English summary prepared for authors by E. Omura).

11. Tania Simoncelli, interview with H. Tokunaga, professor of law, Konan University, Osaka, Japan, July 18, 2007.

12. These cases all rested on intentionality; the courts found that the police agency did not exhaust the sample intentionally. See H. Tokunaga, "Sample Preservation for Evaluation by Experts," *Konan Law Review* 45, nos. 1–2 (2004): 229–257.

13. Omura, discussion with Simoncelli and written comments by Yamamoto.

14. Sekiguchi, personal correspondence with Krimsky.

15. Japan's population is approximately 127 million. See National Institute of Population and Social Security Research, "Population Statistics of Japan: 2006," http://www.ipss.go.jp/p-info/e/PSJ2006.pdf (accessed April 15, 2010).

16. See, for example, H. Tokunaga, "DNA Database Without Proper Legislation," *Konan Law Review* 46, no. 3 (2005): 115–134; Tania Simoncelli, interview with T. Yamamoto, H. Sato, and K. Kai, Waseda Law School, Tokyo, July 30, 2007, translation by E. Omura.

17. JFBA, "Opinion on the National Police Agency DNA Database System."

18. Omura, discussion with Simoncelli and written comments by Yamamoto.

19. JFBA, "Opinion on the National Police Agency DNA Database System." See also Tokunaga, "DNA Database Without Proper Legislation," and Omura, discussion with Simoncelli and written comments by Yamamoto.

20. Article 31 of Japan's Constitution states, "No person shall be deprived of life or liberty, nor shall any other criminal penalty be imposed, except according to procedure established by law."

21. H. Tokunaga, "Quality Assurance System of Forensic DNA Typing," *DNA Polymorphism* 15 (2007): 349–353, summarized and translated by E. Omura, January 2008.

22. Tania Simoncelli, interview with H. Tokunaga, professor of law, Konan University, Osaka, Japan, July 19, 2007. In addition, in June 2004 the NPA held a forum

titled "Use of DNA Profiles: Reference to the UK System" that featured a U.K. law-enforcement official as the keynote speaker. See H. Tokunaga, "Issues on DNA Database" (presentation to the Human Rights Committee, Japan Federation of Bar Associations, August 2005), summarized and translated by E. Omura, January 2008.

23. NPA, "Towards Activating DNA Profile Information," sec. III; Simoncelli, interview with Tokunaga, July 18, 2007.

24. Toshikazu Shimnizu, NPA Criminal Bureau, "Operation of Crime Scene DNA Database System," April 2005, summarized and translated by E. Omura, January 2008.

25. Simoncelli, interview with Tokunaga, July 19, 2007.

26. David Cyranoski, "Japan's Ethnic Crime Database Sparks Fears Over Human Rights," *Nature* 427 (January 29, 2004): 383.

27. Ibid.

28. None of the legal scholars we spoke with in Japan were aware of any ethnicity testing occurring to date or of the initial proposal.

11. Australia

1. Australian Law Reform Commission (ALRC), *Essentially Yours: The Protection of Human Genetic Information in Australia* (Sydney: Australian Law Reform Commission, 2003), pt. J, chap. 39, para. 39.14 (March 14), http://www.austlii.edu.au/au/other/alrc/publications/reports/96/ (accessed April 19, 2010).

2. Office of the Federal Privacy Commissioner, Submission to the Senate Legal and Constitutional Legislation Committee, Commonwealth of Australia, *Inquiry into the Crimes Amendment (Forensic Procedures) Bill 2000,* November 2000, www.privacy.gov.au/materials/types/download/8630/6479 (accessed April 18, 2010).

3. New South Wales, "Review of the Crimes (Forensic Procedures) Act 2000," Parliamentary Paper no. 1118, February 2002, 13.

4. Deborah Crosbie, *Protection of Genetic Information: An International Comparison* (report to the Human Genetics Commission, U.K., September 2000), 83, http://www.hgc.gov.uk/UploadDocs/DocPub/Document/international_regulations.pdf (accessed May 27, 2010).

5. Mark Findlay and New South Wales Government, *Independent Review of the Crimes (Forensic Procedures) Act 2000*, Criminal Law Division, Attorney-General's Department (Sydney, 2003), 22, http://www.lawlink.nsw.gov.au/lawlink/clrd/ll_clrd.nsf/pages/CLRD_forensics_report (accessed April 18, 2010). See also Ian Frekelton, barrister, Owen Dixon Chambers, Melbourne, "DNA Profiling: Forensic DNA Under the Microscope," *Criminal Law Journal* 14 (1990): 23–41.

6. Frekelton, "DNA Profiling," 29.

7. Transcript from the *PM* radio program titled "DNA Laws Begin in Victoria," September 2, 2002, Kate Tozer, reporter. "[Mark Colvin:] In Victoria, the police force has been given the power to take forced DNA samples from people who aren't serving any sentence." http://www.abc.net.au/pm/stories/s665063.htm (accessed April 18, 2010).

8. Crimes Act 1914 (Cth), pt. 1D, div. 8.

9. Findlay and New South Wales Government, *Independent Review of the Crimes (Forensic Procedures) Act 2000*, 12, 17.

10. Richard Hindmarsh, "Australian Biocivic Concerns and Governance of Forensic DNA Technologies: Confronting Technocracy," *New Genetics and Society* (September 1, 2008), 269.

11. Smith Alling Lane, "DNA Legislation and News," September 22, 2000, http://www.dnaresource.com/documents/2000_09_22.pdf (accessed April 18, 2010).

12. Justice Action, "The Crimes (Forensic Procedures) Act 2000: Justice Action's Submission to the Standing Committee on Law and Justice Inquiry," June 6, 2001, http://www.justiceaction.org.au/index.php?option=com_content&task=blogsection&id=28&Itemid=142 (accessed April 19, 2010).

13. Parliament, New South Wales, Crimes (Forensic Procedures) Bill, June 28, 2000, full-day Hansard transcript, http://www.parliament.nsw.gov.au/prod/parlment/hansart.nsf/V3Key/LC20000628005 (accessed April 19, 2010).

14. CrimTrac, *Annual Report 2006–7* (Canberra: CrimTrac, 2007), 19.

15. Rachel Lebihan, "DNA Database Not a Perfect Match," *Australian Financial Review*, April 3, 2006.

16. Hindmarsh, "Australian Biocivic Concerns," 272.

17. Ibid.

18. Australian Law Reform Commission, *Essentially Yours*.

19. Ibid., "List of Recommendations," pt. J.

20. Charles Lawson, Griffith University Law School, personal correspondence with Sheldon Krimsky, November 12, 2007.

21. CrimTrac, *Annual Report 2006–7*, 19.

22. Hindmarsh, "Australian Biocivic Concerns."

23. Colin James, "Officers Acted Illegally, Judge Rules," *Australian Business Intelligence*, May 28, 2006.

24. Genetic Future, "Australian State Government Outsources Forensic DNA Testing," January 21, 2008, http://www.genetic-future.com/2008/01/australian-state-government-outsources.html (accessed April 19, 2010).

25. Justice Action, "Submission to the Senate Legal and Constitutional Committee Inquiry into the Commonwealth Crimes Amendment (Forensic Procedures) Bill 2000," October 30, 2000, http://www.justiceaction.org.au/index.php?option=com_content&task=view&id=179&Itemid=149 (accessed April 10, 2010).

12. Germany

1. Martin Kreickenbaum, "Germany: Expansion of DNA Testing—A Step Towards Genetic Registration," *World Socialist Website*, February 24, 2005. http://www.wsws.org/articles/2005/feb2005/dna-f24.shtml (accessed April 20, 2010).

2. Ian Mader, "Mass DNA Test Nets Murder Suspect," *San Diego Union-Tribune,* May 3, 1998.

3. Hermann Schmitter and Peter M. Schneider, "Legal Aspects of Forensic DNA Analysis in Germany," *Forensic Science International* 88 (1997): 95–98, citing *Neue Juristische Wochenschrift* (NJW, translated as *New Legal Weekly*) 1990, 2944–2945, http://rsw.beck.de/rsw/shop/default.asp?site=njw (accessed May 27, 2010).

4. Ibid., 96 (citing NJW 1996, 3071–3073).

5. Ibid.

6. Ibid.

7. See Federal Constitutional Court, 2nd Chamber of the Second Division of the Federal Constitutional Court, August, 2 1996, 2 BvR 273/06; 1511/96; NJW 1996, 3071, http://www.bundesverfassungsgericht.de/entscheidungen/rk20070212_2bvr027306.html (accessed April 20, 2010). A summary of the case results is given by Schmittter and Schneider: "In serious crime cases the judge may order the taking of blood samples for DNA analysis even if there is no strong suspicion against the persons the blood is taken from. Thus, under certain circumstances mass screening of a large number of potential suspects does not interfere with constitutional rights." Schmitter and Schneider, "Legal Aspects of Forensic DNA Analysis in Germany," 96.

8. German Code of Criminal Procedure, Section 81g [DNA Analysis] (2): "The cell tissue collected may be used only for the molecular and genetic examination referred to in subsection (1); it shall be destroyed without delay once it is no longer required for that purpose. Information other than that required to establish the DNA code may not be ascertained during the examination; tests to establish such information shall be inadmissible." http://www.iuscomp.org/gla/statutes/StPO.htm#81g (accessed April 20, 2010).

9. German Code of Criminal Procedure, Section 81a (3), http://www.iuscomp.org/gla/statutes/StPO.htm#81a (accessed April 20, 2010).

10. Ibid., 96–97.

11. Alexander Hanebeck, "DNA Analysis and the Right to Privacy: Federal Constitutional Court Clarifies Rules on the Use of Genetic Fingerprints," *German Law Journal* 2, no. 3 (February 15, 2001), http://www.germanlawjournal.com/index.php?pageID=11&artID=50 (accessed April 20, 2010).

12. Peter M. Schneider, Institut für Rechtsmedizin, Universität Mainz, Mainz, Germany, personal correspondence with Tania Simoncelli, August 28, 2008.

13. See Bundeskriminalamt, Bundeskriminalamtgesetz (BKAG) § 32 II.

14. Carole McCartney, *Forensic Identification and Criminal Justice: Forensic Science, Justice and Risk* (Uffculme, UK: Willan Publishing Co., 2006), 162. See also Hanebeck, "DNA Analysis and the Right to Privacy."

15. Schneider, personal correspondence with Simoncelli.

16. The provision of the German Code of Criminal Procedure allowing for a determination of gender was added to Section 81e by Parliament.

17. Hermann Schmitter, Bundeskriminalamt, Wiesbaden, Germany, "Promoting DNA in Germany" (First International DNA User's Conference, Lyons, France, November 24–26, 1999, hosted by Interpol), http://www.interpol.int/public/Forensic/dna/conference/Promoting03.asp (accessed April 20, 2010).

18. Schmitter and Schneider, "Legal Aspects of Forensic DNA Analysis in Germany," 95–98.

19. See Bundeskriminalamt, BKAG §8 III.

20. Christopher H. Asplen and Smith Alling Lane,, "International Perspectives on Forensic DNA Databases," ISRCL Conference, The Hague, August 24–28, 2005, http://www.isrcl.org/Papers/Asplenn.pdf (accessed April 20, 2010). See also Nuffield Council on Bioethics, *The Forensic Use of Bioinformation: Ethical Issues* (London: Nuffield Council on Bioethics, September 2007), 52.

21. Eric Topfer, "Searching for Needles in an Ever Expanding Haystack: Cross-Border DNA Data Exchange in the Wake of the Prüm Treaty," *Statewatch* 18 (July–September 2008): 14–16, information at 16.

13. Italy

1. Giuseppe Novelli, professor of medical genetics at Tor Vergata University in Rome and adjunct professor for medical sciences at the University of Arkansas, quote from a personal interview in December 2008 by Marina Semiglia, author of "DNA in the Mass Media," a thesis discussed in February 2009 at the Scuola Internazionale Superiore di Studi Avanzati (SISSA) in Trieste.

2. Nathan van Camp and Kris Dierickx, *National Forensic DNA Databases, Socio-Ethical Challenges and Current Practices in the EU*, European Ethical-Legal Papers 9 (Leuven: Katholieke Universiteit Leuven, Centre for Biomedical Ethics and Law, 2007), https://www.kuleuven.be/cbmer/page.php?LAN=E&ID=399&FILE=subject&PAGE=1 (accessed April 21, 2010).

3. The Agency for the Protection of Personal Data, also called the Data Protection Authority or the Privacy Authority, is an institution created in 2003 by law decree no. 153, the so-called Privacy Bill, available at http://www.garanteprivacy.it/garante/navig/jsp/index.jsp. The Privacy Authority is an administrative institution that is *super partes* (completely independent of others), and its management is sui generis.

4. Agenzia Giornalistica Italiana (AGI), April 28, 2004.

5. "Authorization to Treat Genetic Data," February 22, 2007, *Gazzetta Ufficiale*, no. 65 (March 19, 2007), written by the authority, http://www.garanteprivacy.it/garante/navig/jsp/index.jsp, document 1395420 (accessed April 23, 2010).

6. "Segnalazione al Parlamento e al Governo su dati genetici per fini di giustizia," Archive Document no. 1456163, September 19, 2007, http://www.garanteprivacy.it/garante/navig/jsp/index.jsp (accessed April 23, 2010).

7. Ibid.

8. See Garante Per La Protezione Dei Dati Personali, Bulletin No. 87, October 15, 2007, http://www.garanteprivacy.it/garante/doc.jsp?ID=1448799 (accessed April 21, 2010).

9. Silvio Berlusconi (Forza Italia) led the government. The law decree—a legislative instrument provided in the constitution—is immediately operative. The "Pisanu project" was supported by Alleanza Nazionale, whose leader, Gianfranco Fini, was at the time foreign minister, while Federation of the Greens (Federazione dei Verdi) and the Communist Refoundation Party (Partito della Rifondazione Comunista)—extreme leftist formations in the Italian political panorama—voted against it.

10. Bill no. 905, under examination in the Senate, http://mobile.senato.it/documenti/repository/dossier/studi/2008/Dossier_037.pdf (accessed April 21, 2010).

11. The Prüm Treaty (May 27, 2005) aims to foster ways of tackling cross-border crime by allowing for individual DNA profiles to be directly compared with those from computerized databases of other member states, for instance, for identification and prosecution purposes. In particular, it contains dispositions about the exchange of data banks and fingerprints. See Europa Press Release Rapid, "The Integration of the 'Prüm Treaty' into EU-legislation—Council Decision on the Stepping Up of Cross-Border Co-operation, Particularly in Combating Terrorism and Cross-Border Crime," http://europa.eu/rapid/pressReleasesAction.do?reference=IP/07/803 (accessed April 23, 2010). The full text of the Prüm Treaty is available at Balzacq Thierry, "Challenge, Liberty and Security: The Treaty of Prüm and the Principle of Loyalty," December 4, 2006, http://www.libertysecurity.org/IMG/pdf/Prum-ConventionEn.pdf (accessed April 22, 2010).

12. The 17 signatory nations to the Treaty of Prüm as of 2010 are Germany, Austria, the Netherlands, Belgium, Luxembourg, Spain, France, Finland, Italy, Portugal, Slovenia, Sweden, Bulgaria, Romania, Greece, Slovakia, and Hungary.

13. See Chamber of Deputies, XVI Legislature, Service Studies, Bills, Italy's Accession to the Treaty of Prüm, http://documenti.camera.it/leg16/dossier/testi/GI0135.htm#_Toc221950440 (accessed April 23, 2010).

14. Two police corps operate in Italy and intervene in the case of serious crimes: the Carabinieri, mostly present in medium and small urban centers and under the direct control of the Ministry of Defense, and the State Police, above all present in big urban centers, under the control of the home secretary.

15. The attorney general of the Court of Appeal represents the public prosecution in the second (and final) trial in Italian justice. See http://www.codicisimone.it/codici/index0.htm (accessed April 23, 2010).

16. The rules for the collection of evidence are listed in the Code of Criminal Procedure (articles 187 and following). The onus of proof of innocence does not lie with the defendant, whose innocence is always presumed. His conviction (Code of Criminal Procedure, article 533) must come after a full positive demonstration of his guilt; where no claim is made regarding the insufficiency of proof, the judge pronounces the sentence "Not guilty," and the defendant is acquitted. If the defendant is proved guilty of the crime he or she has been accused of, then it has been proved "beyond every reasonable doubt." In spite of the emphatic tones of the national media, the mere finding of somebody's (and especially of a defendant's) DNA at the scene of the crime does not represent a condition either necessary or sufficient for a conviction.

17. Attilio Bolzoni, "I bimbi fantasma di Lampedusa," *La Repubblica,* October 7, 2008.

18. Beatrice Montini, "I Carabinieri hanno un archivio DNA illegale," *L'Unità,* May 17, 2006; Beatrice Montini, "DNA, già 15 mila identità conservate fuorilegge," *L'Unità,* May 18, 2006.

19. Colonel Dr. Luciano Garofano, Carabinieri, Italy, personal correspondence with Sheldon Krimsky, January 28, 2009.

20. Declan McCullagh, "Global Police Database for Fingerprints, Airline Data?" *Zdnet Asia,* July 13, 2007, http://www.zdnetasia.com/news/security/0,39044215,62028349,00.htm (accessed April 23, 2010).

21. Interpol media release, " INTERPOL Underlines G8's Vital Role in Global Effort Against Transnational Crime in 21st Century," June 10, 2008, http://www.interpol.int/public/ICPO/PressReleases/PR2008/PR200823.asp (accessed April 23, 2010).

22. The draft law crafted by the under secretary for justice, Luigi Li Gotti, in the previous Prodi government (Act of Senate 1877) to establish national DNA data banks was in fact accepted by the XVI legislature, with the Berlusconi government, first by Act of the Senate 586 and later modified by Act of the Senate 905. From comparison of these two bills (1877 and 905) it emerges that (summary no. 1, July 30, 2008) "the legislative suggestions of parliamentary and government initiatives have an identical content, with the exception of the regulation for the financial backing," and that "the measures have the same content of a similar bill (Act of Senate 1877) of governmental initiatives, presented in the past legislature."

23. Security dispositions (also known as "Security Package"), bill no. 773, introduced on June 3, 2008, http://www.governo.It/GovernoInforma (search "Pacchetto Sicurezza") (accessed April 23, 2010).

24. Home Secretary Roberto Maroni, during his half-yearly report about people missing, announced in December 2008: "To this date the dead bodies found by police

and impossible to identify are 648. Each of them was certainly a lost person; therefore it is necessary to have a system to identify them without any possible doubt, accelerating the approval of rules to create a DNA databank." See Corriere Della Serat.it, http://www.corriere.it/cronache/08_dicembre_05/privacy_banche_dati_italia_gran_bretagna_2d0ed3e4-c2ec-11dd-8440-00144f02aabc.shtml (accessed April 23, 2010).

25. Schede di lettura, XVI Legislaturea Disegni di legge, http://mobile.senato.it/documenti/repository/dossier/studi/2008/Dossier_037.pdf. (accessed April 23, 2010).

26. In 1997 the European Union officially recognized the importance of DNA data banks as an instrument to contribute to penal investigations. The EU Council's resolution of June 9, 1997, about the exchange of DNA profiles states that "every country belonging to the Community is invited to organize the formation of national DNA databanks," and "such system should offer sufficient guarantees for security and the protection of personal data." *Gazzetta Ufficiale*, no. C193 (June 24, 1997): 2–3. See Eur-Lex Access to European Law, European Council Resolution of June 25, 2001, recalling Resolution of June 9, 1997, http://eurlex.europa.eu/smartapi/cgi/sga_doc?smartapi!celexapi!prod!CELEXnumdoc&lg=EN&numdoc=32001G0703(01)&model=guichett (accessed April 23, 2010).

27. Garofano, personal correspondence with Krimsky.

28. Maria Fronthaler was a 74-year-old woman who was raped and murdered in her home at Valle San Silvestro, a village near Dobbiaco, during the night of March 31 and April 1, 2002. The Carabinieri of the Parma RIS found a suspect after conducting a dragnet of approximately 200 village male inhabitants. They all accepted the mouth-swabbing test, and their DNA was compared with that in the seminal liquid found on the corpse. At a certain point the Carabinieri found a near match to the sample that belonged to the father of Andreas Kristler, a 20-year-old who was not living in the village at the time at which the dragnet was conducted. He was sentenced to 18 years. Pierluigi Deppentori, "Test DNA su tutti maschi del paese scopesto l'assassino di unlanziana," *La Repubblica*, June 8, 2002.

29. See We the People Will Not Be Chipped.Com, "Italy to Create DNA Databank," September 22, 2007, http://www.wethepeoplewillnotbechipped.com/main/news.php?readmore=543 (accessed April 23, 2010).

30. Colonel Dr. Luciano Garofano, Carabinieri, Italy, personal correspondence with Sheldon Krimsky, January 5, 2009. Before legislation setting up the national forensic DNA database, the Italian courts had faced the case of DNA and the "tears of a statue of the Madonna." A small statue of the Madonna belonging to a local family living in Civitavecchia, Italy was alleged to have shed tears of blood. Many local residents proclaimed this a miracle. When police had the red liquid analyzed it turned out to be human blood belonging to a male individual. A judge ordered the blood of the owner of the statue to be analyzed, in order to compare his DNA with that of the statue's tears, The owner did not give his consent. The Constitutional Court ruled that the owner of the Madonna had a legitimate right to refuse the DNA

test. "The Crying Game," *The Guardian*, December 9, 2000, http://www.guardian.co.uk/theguardian/2000/dec/09/weekend7.weekend1 (accessed April 22, 2010).

31. See note 11.

32. *Relazione 2006: Discorso del Presdiente Francesco Pizzetti* (Rome: Garante Per La Protezione Dei Dati Personali, July 12, 2007, http://www.garanteprivacy.it/garante/document?ID=1422370 (accessed April 21, 2010).

33. Reservations were also expressed by the Communist Refoundation Party, worried about the "cataloguing" of citizens, while the Green Party feared that this could lead to DNA records being collected for the entire population. But other members of Parliament supported the initiative as "a measure that in terms of security finally puts Italy at the same level as the rest of Europe," said Maurizio Fistariol, with the center-left La Margherita. See Frank M. Pizzorusso, "Will Italy Set Up a DNA Databank?" *i-Italy Magazine*, September 24, 2007, http://www.i-italy.org/305/will-italy-set-dna-databank (accessed April 22, 2010).

34. Rosaria Amato, "Rapporto Eurispes 2007: Italia—Giustizia al collasso" [Italy—Justice to collapse], Wallstreetrack, http://wallstreetrack.wordpress.com/2007/02/01/rapporto-delleurispes-2007-italia-giustizia-al-collasso/ (accessed April 22, 2010).

35. Qualitative and quantitative research on the two Italian leading dailies, *Corriere della Sera* and *La Repubblica*, by Marina Semiglia for her thesis for the master's degree in scientific communication at the Scuola Internazionale Superiori di Studi Avanzati (SISSA), "DNA in Mass Media," Trieste, February 2009. This unpublished thesis was based on the articles that appeared in 2006–2007 in which the word "DNA" is used in the forensic sphere. The reflections on bioethical themes are scanty, while the prevailing idea is that everything based on genetic data is infallible and a decisive factor.

36. As of 1992, Giuseppe Novelli was a consultant for Home Affairs, Scientific Section, and was also a member of the Bio-security Commission for the President and of the National Commission for Genetic Tests for the Ministry of Health. Telephone conversation with Gianna Milano, February 2, 2009.

37. Andrea Monti, personal correspondence with Gianna Milano, February 2, 2009.

38. European Digital Rights International (EDRI), "Italian DNA Database: The Devil Is in the Details," August 26, 2009, http://www.edri.org/edri-gram/number7.16/dna-database-italy (accessed April 22, 2010).

14. Privacy and Genetic Surveillance

1. *Osborn v. United States*, 385 U.S. 343 (December 12, 1966).

2. Statement of Senator Edward Kennedy in support of the Genetic Information and Nondiscrimination Act, *Congressional Record—Senate* 153, no. 12 (January 22, 2007): S847.

3. Anita L. Allen, "Genetic Privacy: Emerging Concepts and Values," in *Genetic Secrets: Protecting Privacy and Confidentiality in the Genetic Era*, ed. Mark A. Rothstein (New Haven, CT: Yale University Press, 1997), 39.

4. Samuel Warren and Louis D. Brandeis, "The Right to Privacy," *Harvard Law Review* 4, no. 5 (1890): 193, http://groups.csail.mit.edu/mac/classes/6.805/articles/privacy/Privacy_brand_warr2.html (accessed May 23, 2010).

5. In *Griswold v. Connecticut*, 381 U.S. 479 (1965), the Supreme Court ruled that the Bill of Rights entails a broad right to privacy that prohibits states from criminalizing an individual's decision to use contraception.

6. Human Genetics Commission, *Nothing to Hide, Nothing to Fear? Balancing Individual Rights and the Public Interest in the Governance and Use of the National DNA Database* (London: Human Genetics Commission, November 2009), 45.

7. Centers for Disease Control and Prevention, "Genetic Testing," http://www.cdc.gov/genomics/gtesting/ (accessed April 23, 2010).

8. George J. Annas, "Genetic Privacy," in *DNA and the Criminal Justice System: The Technology of Justice*, ed. David Lazer (Cambridge, MA: MIT Press, 2004), 136.

9. P. R. Billings, M. A. Kohn, M. de Cuevas, J. Beckwith, J. S. Alper, and M. R. Natowicz, "Discrimination as a Consequence of Genetic Testing," *American Journal of Human Genetics* 50 (March 1992): 476–482.

10. U.S. Equal Employment Opportunity Commission, "EEOC Settles ADA Suit Against BNSF for Genetic Bias," April 18, 2001, http://www.eeoc.gov/eeoc/newsroom/release/archive/4-18-01.html (accessed May 23, 2010).

11. Declaration of Aakash Desai in support of motion for preliminary injunction in *Haskell v. Brown*, 677 F. Supp. 2d 1187 (N.D. Cal. 2009).

12. J. L. Mnookin, "Fingerprint Evidence in an Age of DNA Profiling," *Brooklyn Law Review* 67 (2001): 13–70, http://papers.ssrn.com/abstract=292087 (accessed April 24, 2010).

13. William Moschella, assistant attorney general, letter to the Honorable Orrin Hatch, April 28, 2004, *Congressional Record—Senate* 151, no. 162 (December 16, 2005): S13749–S13766, http://www.gpo.gov/fdsys/pkg/CREC-2005-12-16/html/CREC-2005-12-16-pt1-PgS13749.htm (accessed April 24, 2010).

14. Statement of Senator Jon Kyl, *Congressional Record—Senate* 151, no. 162 (December 16, 2005): S13756–S13759, http://0-www.gpo.gov.library.colby.edu/fdsys/pkg/CREC-2005-12-16/pdf/CREC-2005-12-16-pt1-PgS13749.pdf (accessed April 24, 2010).

15. See John D. H. Stead, Jérôme Buard, John A. Todd, and Alec J. Jeffreys, "Influence of Allele Lineage on the Role of the Insulin Minisatellite in Susceptibility to Type 1 Diabetes," *Human Molecular Genetics* 9, no. 20 (2000): 2929–2935. See also D. Concar, "Fingerprint Fear," *New Scientist* (May 2, 2001), http://www.newscientist.com/article/dn694-fingerprint-fear.html (accessed April 24, 2010).

16. Jean McEwen, "DNA Sampling and Banking: Practices and Procedures in the United States," in *Human DNA: Law and Policy: International and Comparative Perspectives*, ed. Bartha Maria Knoppers (The Hague: Kluwer Law International, 1997), 410.

17. Mark Rothstein and Sandra Carnahan, "Legal and Policy Issues in Expanding the Scope of Law Enforcement DNA Databanks," *Brooklyn Law Review* 67 (2001): 127–168, quotation at 156.

18. Randall S. March and Bruce Budowle, "Are Developments in Forensic Applications of DNA Technology Consistent with Privacy Protections?" in *Genetic Secrets: Protecting Privacy and Confidentiality in the Genetic Era*, ed. Mark A. Rothstein (New Haven, CT: Yale University Press, 1997), 226.

19. D. H. Kaye, "Behavioral Genetics Research and Criminal DNA Databases," *Law and Contemporary Problems* 69 (Winter/Spring 2006): 259–299, information at 273.

20. B. Steinhardt, "Privacy and Forensic DNA Data Banks," in *DNA and the Criminal Justice System: The Technology of Justice*, ed. David Lazer (Cambridge, MA: MIT Press, 2004), 173–196. See also Kaye, "Behavioral Genetics Research and Criminal DNA Databases," 273.

21. Davina Dana Bressler, "Criminal DNA Databank Statutes and Medical Research," *Jurimetrics* 43 (Fall 2002): 51–67, quotation at 66.

22. Ibid.

23. Kaye, "Behavioral Genetics Research and Criminal DNA Databases," 282.

24. 42 U.S.C. §14132(b).

25. Bressler, "Criminal DNA Databank Statutes and Medical Research," 64.

26. Alabama Code, para. 36-18-31, 2001.

27. Bressler, "Criminal DNA Databank Statutes and Medical Research," 67.

28. Kaye, "Behavioral Genetics Research and Criminal DNA Databases," 298.

29. Antony Barnett, "Police DNA Database 'Is Spiraling Out of Control,'" *Observer*, July 16, 2006.

30. Kristina Staley, GeneWatch UK, *The Police National DNA Database: Balancing Crime Detection, Human Rights and Privacy* (London: GeneWatch UK, January 2005), 46.

31. GeneWatch UK, "Using the Police National DNA Database—Under Adequate Control?" July 2006, http://www.genewatch.org/uploads/f03c6d66a9b354535738483c1c3d49e4/research_brief_fin.doc (accessed January 10, 2010).

32. Human Genetics Commission, *Nothing to Hide, Nothing to Fear?* 70.

33. *Olmstead v. United States*, 277 U.S. 438 (1928).

34. *Katz v. United States*, 389 U.S. 347 (1967), argued October 17, 1967, decided December 18, 1967.

35. Ibid.

36. Ibid.

37. *Schmerber v. California*, 384 U.S. 757 (1966).

38. *Ferguson v. City of Charleston*, 532 U.S. 67 (2001).

39. *Skinner v. Railway Labor Executives' Ass'n*, 489 U.S. 602 (1989). The Court described a compelled urine sample as a "host of private medical facts . . . which might be revealed by the chemical analysis of the sample fluid."

40. *Cupp v. Murphy*, 412 U.S. 291 (1973).

41. For example, Judge Alex Kozinski, in his dissent in *United States v. Kincade,* stated, "It is important to recognize that the Fourth Amendment intrusion here is not primarily the taking of the blood, but the seizure of the DNA fingerprint and its inclusion in a searchable database." See *United States v. Kincade*, 379 F.3d 813 (9th Cir. 2004), at 873 (J. Kozinski, dissenting).

42. Rothstein and Carnahan, "Legal and Policy Issues," 148n132.

43. Ibid., 144.

44. *United States v. Kincade.*

45. *Rise v. Oregon,* 59 F.3d 1556 (9th Cir. 1995).

46. *United States v. Mitchell*, 2009 U.S. Dist. Lexis 103575 (W.D. Pa. November 6, 2009).

47. *Anderson v. Com. Va.*, S.E.2d, 2007 WL 2683734 (Va.) (September 14, 2007).

48. For a legal analysis of familial searching, see Lindsey Weiss, "All in the Family: A Fourth Amendment Analysis of Familial Searching," 2008, http://works.bepress.com/cgi/viewcontent.cgi?article=1001&context=lindsey_weiss (accessed April 25, 2010).

49. American Society of Human Genetics, Statement, "Professional Disclosure of Familial Genetic Information," *American Journal of Human Genetics* 62 (1998): 474–483, quotation at 474.

50. Annas, "Genetic Privacy," 136.

51. Human Genetics Commission, "Baroness Helena Kennedy Welcomes DNA Testing Ban," August 30, 2006, http://www.hgc.gov.uk/Client/news_item.asp?NewsId=63 (accessed April 24, 2010).

52. Amy Harmon, "Stalking Strangers' DNA to Fill in the Family Tree," *New York Times*, April 2, 2007.

53. Shankar Vedantam, "Study Links Gene Variant in Men to Marital Discord," *Washington Post*, September 2, 2008, A02.

54. James H. Fowler and Christopher T. Dawes, "Two Genes Predict Voter Turnout," *Journal of Politics* 70, no. 3 (July 2008): 579–594.

15. Racial Disparities in DNA Data Banking

1. Susanne B. Haga, "Policy Implications of Defining Race and More by Genetic Profiling," *Genomics, Society and Policy* 2, no. 1 (2006): 57–71, quotation at 64.

2. Troy Duster, "Behavioral Genetics and Explanations of the Link Between Crime, Violence, and Race," in *Wrestling with Behavioral Genetics: Science, Ethics, and Public Conversation*, ed. Erik Parens, Audrey R. Chapman, and Nancy Press (Baltimore: Johns Hopkins University Press, 2006), 158. See also Troy Duster, "Selective Arrests, an Ever-Expanding DNA Forensic Database, and the Specter of an Early Twenty-First Century Equivalent of Phrenology," in *DNA and the Criminal Justice System: The Technology of Justice*, ed. David Lazer (Cambridge, MA: MIT Press, 2004), 315–334.

3. Alex R. Piquero and Robert W. Brame, "Assessing the Race-Crime and Ethnicity-Crime Relationship in a Sample of Serious Adolescent Delinquents," *Crime and Delinquency*, February 29, 2008, 390–422.

4. Michael Risher, "Racial Disparities in Databanking of DNA Profiles," *GeneWatch* 22, nos. 3–4 (July–August 2009), http://www.councilforresponsiblegenetics.org/GeneWatch/GeneWatchPage.aspx?pageId=204 (accessed May 23, 2010).

5. Troy Duster, *Backdoor to Eugenics*, 2nd ed. (New York: Routledge, 2003), 150–151.

6. Barbara S. Meierhoefer, "The Role of Offense and Offender Characteristics in Federal Sentencing," *Southern California Law Review* 66 (November 1992): 367–398, at 388, http://www.lexisnexis.com/hottopics/lnacademic/ (accessed April 26, 2010).

7. Justice Policy Institute, *The Vortex: The Concentrated Racial Impact of Drug Imprisonment and the Characteristics of Punitive Counties* (Washington, DC: Justice Policy Institute, December 2007), 7, http://www.justicepolicy.org/images/upload/07-12_REP_Vortex_AC-DP.pdf (accessed April 26, 2010).

8. Harry I. Levine, Jon Gettman, Craig Reinarman, and Deborah Peterson Small, "Drug Arrests and DNA: Building Jim Crow's Database" (paper presented at the Forum on Racial Justice Impacts of Forensic DNA Databanks, New York University, June 19, 2008), citing U.S. Department of Health and Human Services, SAMHSA, Office of Applied Studies, *2005 National Survey on Drug Use & Health*, Detailed Tables, Table 1.80B, "Marijuana Use in Lifetime, Past Year, and Past Month Among Persons Aged 18 to 25, by Racial/Ethnic Subgroups: Percentages, Annual Averages Based on 2002–2003 and 2004–2005," http://www.oas.samhsa.gov/NSDUH/2k5NSDUH/tabs/Sect1peTabs67to132.htm#Tab1.80B (accessed April 26, 2010).

9. Levine et al., "Drug Arrests and DNA."

10. Vincent Schiraldi and Jason Ziedenberg, "Race and Incarceration in Maryland" (Washington, DC: Justice Policy Institute, October 23, 2003), 1–29, at 1, http://www.justicepolicy.org/images/upload/03-10_REP_MDRaceIncarceration_AC-MD-RD.pdf (accessed April 26, 2010).

11. Justice Policy Institute, *Vortex*, 8.

12. Justice Mapping Center, *NYC Analysis—October 2006*, Slide #7, "Men Admitted to Prison: New York City," http://www.justicemapping.org/expertise/ (accessed May 23, 2010). The Justice Mapping Center uses computer mapping and other graphical depictions of quantitative data to analyze and communicate social policy information.

13. Pilar Ossorio and Troy Duster, "Race and Genetics," *American Psychologist* 60, no. 1 (January 2005): 115–128, at 122.

14. Risher, "Racial Disparities in Databanking of DNA Profiles."

15. Aleksandar Tomic and Jahn K. Hakes, "Case Dismissed: Police Discretion and Racial Differences in Dismissals of Felony Charges," *American Law and Economics Review* 10, no. 1 (2008): 110–141.

16. Justice Policy Institute, *Vortex*, 3.

17. Data obtained from Stephen Saloom, policy director, the Innocence Project, May 30, 2007.

18. Rick Weiss, "Vast DNA Bank Pits Policing vs. Privacy," *Washington Post*, June 3, 2006, A1.

19. Duster, "Selective Arrests," 329.

20. Pew Center on the States, *One in 100: Behind Bars in America, 2008* (Washington, DC: Pew Charitable Trusts, 2008), 6, http://www.pewcenteronthestates.org/uploadedFiles/One%20in%20100.pdf (accessed April 26, 2010). See also N. C. Aizenman, "New High in U.S. Prison Numbers," *Washington Post*, February 29, 2008, A01.

21. Brett E. Garland, Cassia Spohn, and Eric J. Wodahl, "Racial Disproportionality in the American Prison Population: Using the Blumstein Method to Address the Critical Race and Justice Issue of the 21st Century," *Justice Policy Journal* 5 (Fall 2008): 1–42, at 4.

22. Henry T. Greely, Daniel P. Riordan, Nanibaa' A. Garrison, and Joanna L. Mountain, "Family Ties: The Use of DNA Offender Databases to Catch Offenders' Kin," *Journal of Law, Medicine and Ethics* 34, no. 2 (Summer 2006): 248–262, quotation at 258.

23. CODIS, "Measuring Success," http://www.fbi.gov/hq/lab/codis/clickmap.htm (accessed May 28, 2010). The National DNA Index (NDIS) contains over 8,201,707 offender profiles and 315,789 forensic profiles as of April 2010. Ultimately, the success of the CODIS program will be measured by the crimes it helps to solve. CODIS's primary metric, the "Investigation Aided," tracks the number of criminal investigations where CODIS has added value to the investigative process. As of April 2010 CODIS has produced over 116,200 hits assisting in more than 114,000 investigations. The CODIS Web site statistics are updated every month.

24. D. H. Kaye and Michael E. Smith, "DNA Identification Databases: Legality, Legitimacy, and the Case for Population-Wide Coverage," *Wisconsin Law Review* (2003): 413–459, quotation at 452.

25. Ibid., 454.

26. U.S. Department of Justice, Federal Bureau of Investigation, "Data on Arrests (18 Years and Older)," http://www.fbi.gov/ucr/cius2007/data/table_43.html (accessed October 12, 2008).

27. The government reports the number of arrests each year, as opposed to the number of persons arrested each year. Some individuals are arrested more than once in a given year.

28. U.S. Department of Justice, Bureau of Justice Statistics, *Compendium of Federal Justice Statistics, 2004*, http://www.ojp.usdoj.gov/bjs/pub/pdf/cfjs04.pdf (accessed April 26, 2010).

29. California Department of Justice, *Crime in California, 2002*, 29, 66, 68, http://ag.ca.gov/cjsc/publications/candd/cd02/preface.pdf (accessed April 26, 2010). See also California Department of Justice, Justice Information Services, Bureau of Criminal Information and Analysis, *Crime in California 2007 Data Tables*, Table 37, http://ag.ca.gov/cjsc/publications/candd/cd07/preface.pdf (accessed April 26, 2010).

30. Jerome G. Miller, "African American Males in the Criminal Justice System," *Phi Delta Kappan* 78 (June 1997): K1–K12.

31. Steven R. Donziger, ed., *The Real War on Crime: The Report of the National Criminal Justice Commission* (New York: HarperPerennial, 1996), 107. See also J. G. Miller, *Search and Destroy: African-American Males in the Criminal Justice System* (New York: Cambridge University Press, 1996).

32. In 2007 the Federal Bureau of Investigation's Uniform Crime Reports (UCR) estimated that there were about 1,702,537 state and local arrests for drug-abuse violations in the United States. http://www.fbi.gov/ucr/cius2008/data/table_29.html (accessed April 26, 2010).

33. Greely et al., "Family Ties," 259.

34. B. Leapman, "Three in Four Young Black Men on the DNA Database," *Sunday Telegraph*, November 5, 2006, http://www.telegraph.co.uk/news/uknews/1533295/Three-in-four-young-black-men-on-the-DNA-database.html (accessed April 26, 2010).

35. Belle Dumé, "A Portable DNA Detector," *Technology Review* (September 24, 2008). http://www.technologyreview.com/biomedicine/21415/ (accessed April 26, 2010).

36. See John D. H. Stead, Jérôme Buard, John A. Todd, and Alec J. Jeffreys, "Influence of Allele Lineage on the Role of the Insulin Minisatellite in Susceptibility to Type 1 Diabetes," *Human Molecular Genetics* 9, no. 20 (2000): 2929–2935. See also D. Concar, "Fingerprint Fear," *New Scientist Space* (May 2, 2001).

37. National Institutes of Health, Office of Human Subjects Research, *The Belmont Report: Ethical Principles and Guidelines for the Protection of Human Subjects of Research* (Washington, DC: National Commission for the Protection of Human

Subjects of Biomedical and Behavioral Research, April 18, 1979), 7–9, http://ohsr.od.nih.gov/guidelines/belmont.html (accessed May 23, 2010).

38. ACLU Policy on Medical Experimentation, Policy no. 266 (1981). "It is the policy of the ACLU to oppose all nontherapeutic medical experimentation on persons held involuntarily in public institutions."

39. Sheldon Krimsky and Tania Simoncelli, "Testing Pesticides in Humans: Of Mice and Men Divided by Ten," *JAMA* 297, no. 21 (June 5, 2007): 2405–2407.

40. Ibid.

41. Avshalom Caspi, Joseph McClay, Terrie E. Moffitt, Jonathan Mill, Judy Martin, Ian W. Craig, Alan Taylor, and Richie Poulton, "Role of the Genotype in the Cycle of Violence in Maltreated Children," *Science* 297 (2002): 851–854.

42. B. J. Culliton, "XYY: Harvard Researcher Under Fire Stops Newborn Screening," *Science* 188 (June 17, 1975): 1284–1285.

43. J. Kim-Cohen, A. Caspi, A. Taylor, B. Williams, R. Newcombe, I. W. Craig, and T. E. Moffitt, "MAOA, Maltreatment, and Gene-Environment Interaction Predicting Children's Mental Health: New Evidence and a Meta-analysis," *Molecular Psychiatry* 11, no. 10 (October 2006): 903–913.

44. Ewen Callaway, "'Gangsta Gene' Identified in US Teens," *New Scientist* (June 19, 2009), http://www.newscientist.com/article/dn17337-gangsta-gene-identified-in-us-teens.html (accessed April 26, 2010).

45. Elisa Pieri and Mairi Levitt, "Criminality in Our Genes?" *Genomics Network*, no. 5 (March 2007): 4–5, quotation at 5.

46. Duster, "Selective Arrests," 331.

47. R. C. Lewontin, "The Apportionment of Human Diversity," *Evolutionary Biology* 6 (1972): 381–398.

48. H. Tang, T. Quertermous, B. Rodriguez, et al., "Genetic Structure, Self-Identified Race/Ethnicity, and Confounding in Case-Control Association Studies," *American Journal of Human Genetics* 76 (2005): 268–275, quotation at 268.

49. Ibid., 274.

50. DNAPrint Genomics ceased operations in February 2009. See "DNAPrint Genomics Goes Bust," *GenomeWeb*, March 3, 2009, http://www.genomeweb.com/node/912684?emc=el&m=325264&l=1&v=e993a10706 (accessed May 23, 2010).

51. Ted Kessis, "Racial Identification and Future Application of SNPs," DNA-Print Genomics, http://www.dnaprint.com/welcome/productsandservices/forensics/ (accessed December 27, 2007).

52. Richard S. Cooper, Jay S. Kaufman, and Ryk Ward, "Race and Genetics," *New England Journal of Medicine* 348 (March 20, 2003): 1166–1170.

53. Tony N. Frudakis, *Molecular Photofitting: Predicting Ancestry and Phenotype Using DNA* (Amsterdam: Elsevier, 2008), 440–441.

54. Ibid., 16.

55. Michael Lynch, Simon A. Cole, Ruth McNally, and Kathleen Jordan, *Truth Machine: The Contentious History of DNA Fingerprinting* (Chicago: University of Chicago Press, 2008), 37.

56. Frudakis, *Molecular Photofitting*, 477.

57. Ibid., 481.

58. Ibid., 485.

59. Duana Fullwiley, "Can DNA 'Witness' Race? Forensic Uses of an Imperfect Ancestry Testing Technology," in *Race and the Genetic Revolution: Science, Myth and Culture*, ed. S. Krimsky and K. Sloan (New York: Columbia University Press, forthcoming), 3.

60. Frudakis, *Molecular Photofitting*, 489.

61. Fullwiley, "Can DNA 'Witness' Race?" 11.

62. Ibid.

63. Frudakis, *Molecular Photofitting*, 489.

64. Ibid., 483.

16. Fallibility in DNA Identification

1. William A. Tobin and William C. Thompson, "Evaluating and Challenging Forensic Identification Evidence," *The Champion* (July 2006): 12–21, quotation at 12.

2. Innocence Project, "Mission Statement," http://www.innocenceproject.org/about/Mission-Statement.php (accessed April 28, 2010).

3. William C. Thompson, "Tarnish on the 'Gold Standard': Recent Problems in Forensic DNA Testing," *The Champion* (January/February 2006): 10–16, at 10.

4. William Thompson, "Actual Innocence: Lessons Learned from Incorrect Declarations of Matches" (paper presented at the Third Annual Conference of Forensic Bioinformatics, "DNA from Crime Scene to Court Room: An Expert Forum," University of Dayton School of Law, Dayton, OH, August 20–22, 2004), http://www.bioforensics.com/conference04/Actual_Innocence/index.html (accessed April 28, 2010).

5. In the early days of DNA typing it was harder for DNA tests to be contaminated because more of the sample was needed in order for it to be detected. Today a lab needs only 40 cells to produce a DNA profile. Ruth Teichroeb, "Rare Look Inside State Crime Labs Reveals Recurring DNA Test Problems," *Seattle Post-Intelligencer*, July 22, 2004.

6. D. R. Paoletti, C. M. Krane, M. L. Raymer, and D. Krane, "Empirical Analysis of the STR Profiles Resulting from Conceptual Mixtures," *Journal of Forensic Analysis* 50 (2005): 1–6.

7. Thompson, "Tarnish on the 'Gold Standard,'" 10–11.

8. Ibid., 12.

9. Seattle Post-Intellegencer Staff, "DNA Testing Mistakes at the State Patrol Crime Labs," *Seattle Post-Intelligencer*, July 22, 2004, http://seattlepi.nwsource.com/local/183018_crimelabboxesweb22.html (accessed April 28, 2010).

10. Thompson, "Tarnish on the 'Gold Standard,'" 14.

11. Erin Murphy, "The Art in the Science of DNA: A Layperson's Guide to the Subjectivity Inherent in Forensic DNA Typing," *Emory Law Journal* 58 (2008): 489–512, at 503.

12. William C. Thompson, Simon Ford, Travis E. Doom, Michael L. Raymer, and Dan E. Krane, "Evaluating Forensic DNA Evidence, Part 2," *The Champion* (April 2003): 16–25, quotation at 21.

13. This assertion of twin genetic identity has recently been questioned because new evidence reveals that so-called identical adult twins may not have identical DNA sequences. Charles Q. Choi, "Identical Twins Are Not Genetically Identical," *Scientific American* 298 (May 2008): 24–26.

14. See Dan Krane, "Random Match Probability," presentation at the Forensic Bioinformatics 2nd Annual Conference, "Statistics and DNA Profiling," August 29–30, 2003, Wright State University, Dayton, OH, http://www.bioforensics.com/conference/RMP/ (accessed April 28, 2010).

15. Sheri Fink, "Reasonable Doubt," *Discover Magazine* 27, no. 8 (July 2006): 54–58, quotation at 57.

16. This example comes from William C. Thompson, "The Potential for Error in Forensic DNA Testing (and How That Complicates the Use of DNA Databases for Criminal Investigation)" (paper produced for the Council for Responsible Genetics [CRG] and its national conference, "Forensic DNA Databanks and Race: Issues, Abuses and Action," New York University, June 19–20, 2008).

17. William Thompson, personal correspondence, with Sheldon Krimsky, August 2008.

18. Dan Krane, personal correspondence with Sheldon Krimsky, December 6, 2007.

19. As of 2008 the FBI reported that 248,943 forensic profiles and 6,539,919 known profiles had been accumulated on CODIS, making it the largest DNA data bank in the world, surpassing the United Kingdom's National DNA Database, which consisted of an estimated 5,617,604 subject profiles as of March 2009. http://www.fbi.gov/hq/lab/html/codisbrochure_text.htm (accessed April 28, 2010); http://www.npia.police.uk/en/docs/NDNAD07-09-LR.pdf (accessed April 28, 2010).

20. Keith Devlin, "Devlin's Angle," Mathematical Association of America, http://www.maa.org/devlin/devlin_10_06.html (accessed April 28, 2010).

21. David H. Kaye. "Trawling DNA Databases for Partial Matches: What Is the F.B.I. Afraid of?" *Cornell Journal of Law and Public Policy* 19 (2009): 145–171, at 155–158.

22. National Policing Improvement Agency, Association of Chief Police Officers, *DNA Good Practice Manual*, 3rd ed. (2007), 12: "Although the technology

used for profiling DNA is extremely refined, it does not enable scientists to say with complete certainty that a DNA sample taken from an individual person is unique. As a result, a match is a matter of very high probability but not absolute certainty. Therefore, evidence of a match between a sample recovered from a scene of crime and a DNA sample taken from a suspect can be compelling but not conclusive evidence on its own, of presence at the crime scene. A corroborative piece of evidence is needed to remove all doubt." http://www.npia.police.uk/en/docs/Stage_One_EIA__ACPO_DNA.pdf (accessed April 28, 2010).

23. Committee on DNA Forensic Science: An Update, National Research Council, *The Evaluation of Forensic DNA Evidence* (Washington, DC: National Academy Press, 1996), 32.

24. Devlin, "Devlin's Angle."

25. Ibid.

26. Thompson, "Potential for Error in Forensic DNA Testing."

27. David J. Balding, *Weight-of-Evidence for Forensic DNA Profiles* (Chichester, UK: John Wiley and Sons, 2005), 32, 93.

28. Thompson, "Potential for Error in Forensic DNA Testing," 8.

29. Erica Haimes, "Social and Ethical Issues in the Use of Familial Searching in Forensic Investigations: Insights from Family and Kinship Studies," *Journal of Law, Medicine and Ethics* 34 (Summer 2006): 263–276.

30. Richard Willing, "Criminals Try to Outwit DNA," *USA Today*, August 28, 2000.

31. Joseph Hixson, *The Patchwork Mouse* (Garden City, NY: Anchor Press/Doubleday, 1976).

32. Gina Kolata, "Clone Scandal: 'A Tragic Turn' for Science," *New York Times*, December 16, 2005, A6.

33. Arthur Koestler, *The Case of the Midwife Toad* (New York: Random House, 1972).

34. Amnesty International, "USA: A Life in the Balance: The Case of Mumia Abu-Jamal," 1–33, at 6, http://www.mumia.de/special/a%20life%20in%20the%20balance.eng.pdf (accessed April 28, 2010).

35. David P. Leonard, "Different Worlds, Different Realities," *Loyola of Los Angeles Law Review* 34 (January 2001): 863–894, quotation at 885.

36. Merrick Bobb, "Symposium: New Approaches to Ensuring the Legitimacy of Police Conduct: Civilian Oversight of the Police in the United States," *Saint Louis University Public Law Review* 22 (2003): 151–166, quotation at 151.

37. William C. Thompson, "DNA Evidence in the O. J. Simpson Trial," *University of Colorado Law Review* 67 (Fall 1996): 827–857.

38. C. Rosen, "Liberty, Privacy and DNA Databases," *New Atlantis* 1 (2003): 37–52, at 52.

39. Willing, "Criminals Try to Outwit DNA."

40. Dan Frumkin, Adam Wasserstrom, Ariane Davidson, and Arnon Grafit, "Authentication of Forensic DNA Samples," *Forensic Science International: Genetics* 4, no. 2 (June 2009): 95–103.

41. Andrew Pollack, "DNA Evidence Can Be Fabricated, Scientists Show," *New York Times,* August 18, 2009.

42. Eleona Mayne and Sophie Borland, "The Mother and Three Children Who Didn't Share Her DNA," *Mail on Sunday* (London), March 5, 2006.

43. Extraordinary People, "Lydia Kay Fairchild: The Twin Inside Me," http://wwwmymultiplesclerosis.co.uk/misc/chimera.html (accessed April 28, 2010).

44. O. L. Strain, J. C. S. Dean, M. P. R. Hamilton, and D. T. Bonthron, "Brief Report: A True Hermaphrodite Chimera Resulting from Embryo Amalgamation After In Vitro Fertilization," *New England Journal of Medicine* 338 (January 15, 1998): 166–169.

45. Howard Wolinsky, "A Mythical Beast: Increased Attention Highlights the Hidden Wonders of Chimeras," *European Molecular Biology Organization (EMBO) Reports* 8, no. 3 (2007): 212–214, quotation at 212.

46. Catherine Arcabascio, "Chimeras: Double the DNA—Double the Fun for Crime Scene Investigators, Prosecutors and Defense Attorneys?" *Akron Law Review* 40 (2007): 435–464, statistics at 444.

47. Gina Kolata, "Cheating, or an Early Mingling of the Blood," *New York Times,* May 10, 2005.

48. Wolinsky, "Mythical Beast," 214.

49. Arcabascio, "Chimeras," 454.

17. The Efficacy of DNA Data Banks

1. Carole McCartney, "The DNA Expansion Programme and Criminal Investigation," *British Journal of Criminology* 20 (2005): 175–192, quotation at 178.

2. Senator Joseph L. Bruno, majority leader, New York State Senate, news release, July 16, 2007.

3. A lifelong estrogen supplement for women was the recommendation made in Robert A. Wilson's best-selling book *Feminine Forever* (New York: Lippincott, 1966).

4. As of March 2008 the FBI reported that the database contained 98,748 arrestee profiles. Tom Callaghan, director, FBI CODIS Unit, presentation at the National Symposium on Familial Searching, Arlington, VA, March 17–18, 2008.

5. The FBI CODIS Web site provides the following statistics for the database. As of March 2010, NDIS contained 8,080,941 "offender" profiles and 311,560 crime scene profiles. http://www.fbi.gov/hq/lab/codis/clickmap.htm (accessed April 30, 2010).

6. Ibid.

7. William C. Thompson, "Tarnish on the 'Gold Standard': Recent Problems in Forensic DNA Testing," *The Champion* (January/February 2006): 10–16, at 13.

8. Helen Wallace, GeneWatch UK, "The DNA Expansion Programme: Reporting Real Achievement?" February 2006, 4, http://www.genewatch.org/uploads/f03c6d66a9b354535738483c1c3d49e4/DNAexpansion_brief_final.pdf (accessed April 30, 2010).

9. McCartney, "DNA Expansion Programme and Criminal Investigation," 182.

10. Ibid.

11. U.K. National DNA Database, *Annual Report 2003–4* (Birmingham, UK: Forensic Science Service, 2004), 23, http://www.forensic.gov.uk/pdf/company/publications/annual-reports/annual-report-NDNAD.pdf (accessed May 29, 2010).

12. McCartney, "DNA Expansion Programme and Criminal Investigation," 183.

13. U.K. National DNA Database, *Annual Report 2005–6* (London: Home Office, 2006), 37.

14. Ibid., 36.

15. Ibid., 37.

16. The Home Office claims that the decline in the match rate was largely the result of replicate sampling following the introduction of new sampling kits; however, a major increase in subject testing was due to implementation of the Criminal Justice Act, which requires DNA profiling of anyone arrested for any recordable offense.

17. Crown Prosecution Service (CPS), London, "Guidance for Cases Involving DNA," pt. 1, 1, http://www.cps.gov.uk/legal/s_to_u/scientific_evidence/guidance_for_cases_involving_dna/ (accessed April 29, 2010). "Charges cannot be based upon a DNA profile match alone; there must always be appropriate supporting evidence."

18. U.K. National DNA Database, *Annual Report 2003–4*, 23.

19. Ibid.

20. U.K. National DNA Database, *Annual Report, 2005–6*, 14.

21. Bernard Rix, "The Contribution of Shoemark Data to Police Intelligence, Crime Detection and Prosecution," *Findings* 236 (2004): 2, published by the Home Office, United Kingdom, http://rds.homeoffice.gov.uk/rds/pdfs04/r236.pdf (accessed May 1, 2010).

22. "Violent crime" includes offenses of murder, forcible rape, robbery, and aggravated assault. "Property crime" includes offenses of burglary, larceny-theft, and motor-vehicle theft. FBI, *Crime in the United States 2002*, "Index of Crime," table 2 (Washington, DC: FBI, 2002), http://www.fbi.gov/ucr/02cius.htm (accessed May 1, 2010).

23. Wallace, "DNA Expansion Programme," 7.

24. Ibid.

25. Thomas Ross, police liaison officer/office manager, Scottish Police DNA Database, "Police Retention of Prints and Samples: Proposals for Legislation," 2005,

http://www.scotland.gov.uk/Resource/Doc/77843/0018258.pdf (accessed May 1, 2010).

26. Sarah V. Hart, Director, National Institute of Justice, *Report to the Attorney General on Delays in Forensic DNA Analysis* (Washington, DC: National Institute of Justice, March 2003), 2.

27. P. Solomon Banda, "Backlong Hinders DNA Tracking System," Associated Press, July 11, 2008, http://www.highbeam.com/doc/1G1-181197546.html (accessed April 30, 2010).

28. Ruth Teichroeb, "Rare Look Inside State Crime Labs Reveals Recurring DNA Test Problems," *Seattle-Post Intelligencer*, July 22, 2004, http://www.seattlepi.com/local/183007_crimelab22.html (accessed April 30, 2010).

29. William C. Thompson, "The Potential for Error in Forensic DNA Testing (and How That Complicates the Use of DNA Databases for Criminal Identification)" (paper produced for the Council for Responsible Genetics [CRG] and its national conference, "Forensic DNA Databanks and Race: Issues, Abuses and Action," New York University, June 19–20, 2008), 14.

30. Jenny Rushlow, "Rapid DNA Database Expansion and Disparate Minority Impact," *GeneWatch* 20, no. 4 (August 2007): 7.

31. Rockne Harmon (Alameda County District Attorney's Office), comment during afternoon breakout session, "DNA Fingerprinting and Civil Liberties: Workshop #4," hosted by the American Society of Law, Medicine and Ethics, JFK School of Government, Cambridge, MA, September 16–17, 2005.

18. Toward a Vision of Justice

1. Fred Bieber, statement in a video promoting DNA collection from arrestees, "Why Should Your State Pass DNA Arrestee Testing Laws?" available through DNA Saves, http://www.dnasaves.org/video/ (accessed May 6, 2010).

2. Sheila Jasanoff, "Just Evidence: The Limits of Science in the Legal Process," *Journal of Law, Medicine and Ethics* 34, no. 2 (Summer 2006): 328–341, quotation at 339.

3. Simon Cole, "How Much Justice Can Technology Afford? The Impact of DNA Technology on Equal Criminal Justice," *Science and Public Policy* 34, no. 2 (March 2007): 95–107, quotation at 105.

4. W. C. Thompson, F. Taroni, and C. G. Aitken, "How the Probability of a False Positive Affects the Value of DNA Evidence," *Journal of Forensic Sciences* 48, no. 1 (January 2003): 1–8, at 1, http://projects.nfstc.org/workshops/resources/articles/How%20the%20Probability%20of%20a%20False%20Positive%20Affects%20the.pdf (accessed May 6, 2010).

5. David Baugh, defense attorney, Virginia, quoted by Laura LaFay, "Reasonable Doubt," *Style Weekly* (July 6, 2005), http://www.styleweekly.com/ME2/Default.asp (accessed May 6, 2010).

6. In *Kyllo v. United States*, 533 U.S. 27 (2001), the Supreme Court reversed Kyllo's conviction for growing marijuana, finding that the use of a thermal-imaging device from a public vantage point to monitor the radiation of heat from a person's home was a "search" within the meaning of the Fourth Amendment and thus required a warrant. (See chapter 6, box 6.2, on *Kyllo v. United States*.)

7. Bruce Budowle, "Declaration in Support of Motion for Preliminary Injunction," *Haskell v. Brown*, 677 F. Supp. 2d 1187 (N.D. Cal. 2009), 11.

8. Susan Haack, "Inquiry and Advocacy, Fallibilism and Finality: Culture and Inference in Science and the Law," *Law, Probability and Risk* 2 (2003): 205–214.

9. *Hiibel v. Sixth Judicial District Court of Nevada, Humboldt County et al.*, U.S. Supreme Court, no. 03-5554, decided June 21, 2004, http://www.law.cornell.edu/supct/pdf/03-5554P.ZO (accessed May 6, 2010). The majority of the court upheld a Nevada law requiring a person to identify himself or herself to a police officer during an investigative stop. "A state law requiring a suspect to disclose his name in the course of a valid *Terry* stop is consistent with Fourth Amendment prohibitions against unreasonable searches and seizures."

10. *United States v. Mitchell*, 2009 U.S. Dist. LEXIS 103575 (W.D. Pa., November 6, 2009).

11. New York State introduced a rule permitting partial DNA matches in December 2009. Jeremy W. Peters, "New Rule Allows Use of Partial DNA Matches," *New York Times*, January 25, 2010.

12. Elizabeth E. Joh, "Reclaiming 'Abandoned' DNA: The Fourth Amendment and Genetic Privacy," *Northwestern University Law Review* 100 (2006): 857–884, quotation at 882.

13. Ibid., 860.

SELECTED READINGS

Aronson, Jay D. *Genetic Witness: Science, Law, and Controversy in the Making of DNA Profiling.* New Brunswick, NJ: Rutgers University Press, 2007.

Clark, George (Woody). *Justice and Science: Trials and Triumphs of DNA Evidence.* New Brunswick, NJ: Rutgers University Press, 2007.

Committee on DNA Forensic Science, National Academy of Sciences. *The Evaluation of Forensic DNA Evidence.* Washington, DC: National Academy Press, 1996.

Dwyer, Jim, Peter Neufeld, and Barry Scheck. *Actual Innocence: Five Days to Execution and Other Dispatches from the Wrongly Convicted.* New York: Doubleday, 2000.

Hindmarsh, Richard, and Barbara Prainsack, eds. *Genetic Suspects: Global Governance of Forensic DNA Profiling and Databasing.* Cambridge: Cambridge University Press, 2010.

Kobilinsky, Lawrence, Thomas F. Liotti, and Jamel Oeser-Sweat. *DNA: Forensic and Legal Applications.* Hoboken, NJ: John Wiley, 2005.

Lazer, David, ed. *DNA and the Criminal Justice System: The Technology of Justice.* Cambridge, MA: MIT Press, 2004.

Lynch, Michael, Simon A. Cole, Ruth McNally, and Kathleen Jordan. *Truth Machine: The Contentious History of DNA Fingerprinting.* Chicago: University of Chicago Press, 2008.

Pyrek, Kelly, M. *Forensic Science Under Siege: The Challenges of Forensic Laboratories and the Medico-Legal Investigation System.* Burlington, MA: Elsevier, 2007.

Rothstein, Mark A., ed. *Genetic Secrets: Protecting Privacy and Confidentiality in the Genetic Era.* New Haven, CT: Yale University Press, 1997.

Wambaugh, Joseph. *The Blooding.* New York: Morrow, 1989.

INDEX